ATLANTIC SALMON: PLANNING FOR THE FUTURE

Edited by Derek Mills, Department of Forestry and Natural Resources, University of Edinburgh and David Piggins, The Salmon Research Trust of Ireland, on behalf of the Atlantic Salmon Trust

The Atlantic salmon is a fish of major interest to fish biologists and the fisheries industries of countries bordering the north Atlantic. Particularly these include eastern Canada, the north-eastern maritime states of the USA, the UK, Ireland, Scandinavia, France and Spain.

This book presents a major international review of the Atlantic salmon, its biology, management and conservation. Papers are based on the Third International Atlantic Salmon Symposium held in Biarritz in October 1986. The book should represent a current state-of-the art reference for all fish biologists and fisheries' scientists.

ATLANTIC SALMON:
Planning for the Future

Edited by DEREK MILLS AND DAVID PIGGINS

The Proceedings of the Third International Atlantic
Salmon Symposium — held in Biarritz, France,
21 — 23 October, 1986

Sponsored jointly by
The Atlantic Salmon Trust
L'Association Internationale de Défense du Sauman
Atlantique

CROOM HELM
London & Sydney

TIMBER PRESS
Portland, Oregon

©1988 The Atlantic Salmon Trust
Croom Helm Ltd, Provident House, Burrell Row,
Beckenham, Kent BR3 1AT
Croom Helm Australia, 44-50 Waterloo Road,
North Ryde, 2113, New South Wales

British Library Cataloguing in Publication Data

International Atlantic Salmon Symposium
    (3rd: 1986: Biarritz)

    Atlantic salmon: planning for the future: the
    proceedings of the Third International
    Atlantic Salmon Symposium, held in Biarritz,
    France, 21-23 October, 1986
    1. Salmon-fisheries—Atlantic Ocean
    2. Atlantic salmon 3. Fishery conservation—
    Atlantic Ocean
    I. Title II. Mills, Derek III. Piggins, David
    IV. Atlantic Salmon Trust V. Association
    internationale de defense du saumon Atlantique
    639.9'7755   SH346

    ISBN 0-7099-5108-6

First published in the USA 1988 by
Timber Press,
9999 S.W. Wilshire,
Portland, OR 97225
USA

Printed and bound in Great Britain by
Biddles Ltd, Guildford and King's Lynn

CONTENTS

Steering Committee
Acknowledgements
Welcoming Remarks - Dr. Richard Vibert
Opening Address   - Mme. Michele Alliot-Marie
Royal Message   - HRH The Prince Charles, Prince of Wales
                     delivered by His Grace The Duke of
                     Wellington

Session One - Salmon in an International Context

Chairman: Dr. W.M. Carter, Atlantic Salmon Federation, Canada

1.   INTERNATIONAL CO-OPERATION THROUGH
      NASCO ....

      M. Windsor and P. Hutchinson

2.   THE COMMUNITY'S APPROACH TO INTER-
      NATIONAL SALMON MANAGEMENT ....

      J. Spencer

3.   THE NORTH AMERICAN VIEWPOINT ....

      J.T.H. Fenety

Contents

Session Two - Status of Exploitation

Chairman: Mr K. Hoydal, Føroya Landsstyri, Faroes

4. EXPLOITATION OF THE RESOURCE IN FRANCE

    J. Arrignon, J.P. Tane and M. Latreille

5. THE STATUS OF EXPLOITATION OF SALMON IN
    ENGLAND AND WALES ....

    G.S. Harris

6. STATUS OF EXPLOITATION OF ATLANTIC
    SALMON IN SCOTLAND ....

    R.B. Williamson

Chairman: Rear-Admiral D.J. Mackenzie, Atlantic Salmon
           Trust U.K.

7. HARVEST AND RECENT MANAGEMENT OF
    ATLANTIC SALMON IN CANADA ....

    T.L. Marshall

8. STATUS OF EXPLOITATION OF ATLANTIC
    SALMON IN NORWAY ....

    Lars P. Hansen

9. EXPLOITATION OF ATLANTIC SALMON IN
    ICELAND ....

    Thór Gudjónsson

Chairman: Monsieur J. Delarue, L'Association Internationale de
           Défense du Saumon Atlantique

10. THE ATLANTIC SALMON IN THE RIVERS OF
    SPAIN, WITH PARTICULAR REFERENCE TO
    CANTABRIA ....

    Carlos Garcia de Leaniz and Juan José Martinez

Contents

11. THE ATLANTIC SALMON IN ASTURIAS, SPAIN:
ANALYSIS OF CATCHES, 1985-86. INVENTORY
OF JUVENILE DENSITIES ....

Juan Antonio Martin Ventura

12. EXPLOITATION OF SALMON IN IRELAND ....

T.K. Whitaker

Session Three - Science and Management

Chairman: Dr. B. Chevassus, INRA, France

13. CATCH RECORDS - FACTS OR MYTHS ? ....

Alex T. Bielak and G. Power

14. RELATING CATCH RECORDS TO STOCKS 256

W.M. Shearer

15. THE USE OF LESLIE MATRICES TO ASSESS THE
SALMON POPULATION OF THE RIVER CORRIB   275

John Browne

16. RELATIONSHIP BETWEEN ATLANTIC SALMON
SMOLTS AND ADULTS IN CANADIAN RIVERS   301

E. Michael P. Chadwick

17. MEASUREMENT OF ATLANTIC SALMON
SPAWNING ESCAPEMENT 325

P. Prouzet and J. Dumas

18. STOCK ENHANCEMENT OF ATLANTIC SALMON   345

G.J.A. Kennedy

Chairman: Dr. D.J. Piggins, Salmon Research Trust of
Ireland

19. SALMON ENHANCEMENT: STOCK DISCRETENESS
AND CHOICE OF MATERIAL FOR STOCKING   373

J.E. Thorpe

Contents

20. ATLANTIC SALMON IN AN EXTENSIVE FRENCH
RIVER-SYSTEM; THE LOIRE-ALLIER 389

Robin Cuinat

21. THE RESTORATION OF THE JACQUES-CARTIER:
A MAJOR CHALLENGE AND A COLLECTIVE
PRIDE 400

Marcel Frenette, Pierre Dulude and Michel
Beaurivage

Chairman: Mr G.H. Bielby, South-west Water Authority, U.K.

22. ATLANTIC SALMON RESTORATION IN THE
CONNECTICUT RIVER 415

R.A. Jones

23. THE ANGLERS' POINT OF VIEW 427

G.O. Edwards

Session Four - Ocean Life of Salmon

Chairman: Mr B.B. Parrish, International Council for the
Exploration of the Sea, Denmark

24. EXPLOITATION AND MIGRATION OF SALMON
ON THE HIGH SEAS, IN RELATION TO
GREENLAND 438

Jens Møller Jensen

25. EXPLOITATION AND MIGRATION OF SALMON
IN FAROESE WATERS 458

Stein Hjalti i Jákupsstovu

26. OCEAN LIFE OF ATLANTIC SALMON IN THE
NORTHWEST ATLANTIC 483

D.G. Reddin

27. FUTURE INVESTIGATIONS ON THE OCEAN LIFE
OF SALMON 512

Svend Aage Horsted

Contents

Session Five - Illegal Exploitation

Chairman: D. Solomon, Atlantic Salmon Trust, U.K.

28.   THE IMPACT OF ILLEGAL FISHING ON SALMON
       STOCKS IN THE FOYLE AREA 524

       W. Gerald Crawford

29.   THE INDIAN ATLANTIC SALMON FISHERY ON
       THE RESTIGOUCHE RIVER: ILLEGAL FISHING
       OR ABORIGINAL RIGHT? 535

       Stephen D. Hazell

30.   ILLEGAL NET FISHING FOR SALMON IN
       NORWAY 557

       Sven Mehli

Closing Address: Stewards of the Salmon 563

       Richard Buck, Restoration of Atlantic Salmon in
       America

Summary and Recommendations: 569

       Derek Mills, Department of Forestry and Natural
       Resources, University of Edinburgh, U.K.

Index 578

STEERING COMMITTEE

Dr D.H. Mills	Chairman. University of Edinburgh
Mr A. Prichard	Secretary. Atlantic Salmon Trust
Mrs J. Botsford	Publicity Officer. Botsford Public Relations Ltd.
Ambassador C. Batault	L'Association Internationale de Défense du Saumon Atlantique
Mr G. Hadoke	Atlantic Salmon Trust
Dr G. Harris	Welsh Water Authority
Dr D.J. Piggins	Salmon Research Trust of Ireland

ACKNOWLEDGEMENTS

The Atlantic Salmon Trust and L'Association Internationale de Défense du Saumon Atlantique wish to express their great appreciation of His Royal Highness the Prince Charles, Prince of Wales, for his message to the Symposium.

The joint sponsors also wish to offer their grateful thanks to those organisations and individuals who generously provided financial support and to all who participated.

Finally the sponsors would like to express their gratitude to Madame Martine Dorado and her staff from Wagons-lits Tourisme, Biarritz, for their efficient service and organisation.

OPENING REMARKS MADE BY MONSIEUR RICHARD VIBERT AT THE OPENING OF THE THIRD INTERNATIONAL ATLANTIC SALMON SYMPOSIUM

Mr Mayor, I must thank you for having agreed to make the inaugural speech in the place of your daughter, the Secretary of State Madame Michele Alliot-Marie, since she was unable to get away from Paris today because of the transport strike.

Your Grace, Messrs Chairmen, distinguished guests, ladies and gentlemen. It is a great honour for me to welcome you to Biarritz for the Third International Atlantic Salmon Symposium. I welcome you in the name of France and of the town of Biarritz by whom we are received. I welcome you in the name of its two sponsors, the Atlantic Salmon Trust and the Association Internationale de Défense du Saumon Atlantique.

For all the younger members of the audience, and, indeed, even for some of the older ones who, like myself, took part in the first Symposium at St Andrews in 1972 and in the second one at Edinburgh in 1978, I shall try to explain briefly the ways in which these two earlier Symposia differed from one another and how they differ from the one which is about to take place in the light of the major events which have taken place between these three gatherings.

In 1972, the St Andrews Symposium, which brought together over 400 persons from twelve countries, dealt in essence with the scientific and technical aspects of the exploitation of the Atlantic salmon. It was in 1971, the previous year, that the catches on the feeding grounds of West Greenland which had only been 60 tonnes in 1960, reached their maximum: 2,689 tonnes, that is to say, 35 per cent of the total catch of the countries of the North Atlantic (7,631 tonnes). This posed a problem.

The catches on the feeding grounds around the Faroe Islands were still insignificant: zero in 1971 and 9 tonnes in 1972. In 1978, the Edinburgh Symposium, which brought together more than 250 participants from eleven countries, dealt more especially with the role which the salmon could and should play in the future. This Symposium concluded that to ensure the survival of the salmon it was necessary to set up an International Convention between salmon-producing and - harvesting nations. It is this Convention for the Conservation of the North Atlantic Salmon which was adopted at a Diplomatic Conference held in Reykjavik on 18-22 January 1982 which created the North Atlantic Salmon Conservation Organisation (NASCO). This organisation, which is established in Edinburgh and which has three Regional Commissions, has as its objective 'to contribute through consultation and co-operation to the conservation, restoration, enhancement and rational management of the salmon stocks which are the subject of the present Convention whilst taking into account the best scientific information available'.

This organisation, which held its first meeting on 17-21 January 1984, already has achievements to its credit: the encouraging decision taken on 18 July 1984 to reduce the quota for the West Greenland catch from 1,190 to 870 tonnes, and the decision to reduce the quota for the Faroe catch from 625 to 550 tonnes.

In 1986, this Third Symposium is being held in Biarritz within a regulatory framework which recognises that on the high seas feeding grounds there is a 'grazing right' of those States within whose fishing jurisdiction these feeding grounds lie. Fishing for salmon is forbidden beyond the fishing limits of all States (200 nautical miles). Within these zones fishing for salmon is forbidden beyond 12 nautical miles from the coastline except in the case of West Greenland where this distance is increased to 40 nautical miles, and for the Faroe Isles where the distance is 200 nautical miles. It is in this context we shall be debating more particularly, firstly the problems arising from the essential international co-operation over all matters concerned with the exploitation and the future of the salmon beyond territorial waters: the feeding grounds off the West coast of Greenland and the Faroe Isles.

Secondly, the problems of the biology of salmon and the movements in the status of their stocks as well as the social problems which have to be considered by interested States in order to achieve a rational policy for the exploitation of their salmon stocks in their territorial waters.

We must strive by the close of this Symposium, to identify some tangible ideas (or even a single idea) which the North Atlantic Salmon Conservation Organisation might adopt which would persuade all our Governments to perceive and to resolve the essential problems which may affect the future survival of the species.

OPENING ADDRESS SPEECH –
MADAME MICHELE ALLIOT-MARIE, SECRETARY OF STATE FOR EDUCATION

The speech was read for her by her father, Monsieur Marie, the Mayor of Biarritz

I am particularly happy to have the honour of opening the Third International Symposium on the Atlantic Salmon which is taking place in a town with which I have close ties. Monsieur Alain Carignon, Minister of the Environment, would have wished to participate in this important event, but other compelling obligations have prevented him from doing so and he asks that you will forgive him.

It is not by chance, nor because of the unique charm of its situation and its range of hotel accommodation, that the town of Biarritz was chosen by the two prestigious organisers of this meeting: the Atlantic Salmon Trust from the United Kingdom and the Association Internationale de Défense du Saumon Atlantique, to whose joint action I pay tribute. Biarritz is more than a simple seaside town, a town open to the world of the ocean and in the forefront of ecological defence. Thus it was that in 1928 a small hydrobiological centre was set up in Biarritz sponsored by organisations and administrations which were interested in the salmon and in salmon fishing. In 1945, on the initiative of Monsieur Vibert, a bigger scientific centre was created in Biarritz, and today this centre, whose premises are at St Pée sur Nivelle, is an important multidisciplinary research laboratory on the salmon. It is therefore particularly appropriate that Biarritz should today be hosting this Symposium which will study both the problems of the exploitation of this fish in home waters as well as its exploitation and life at sea. The Atlantic salmon - Salmo salar - is, for scientists, the only 'real' salmon: Pacific varieties being only close cousins. The Atlantic salmon has more taste, its flesh is paler and firmer and 'crumbles' much less. It is also the species which would appear to be most threatened. There is still time to react and to realise, as has happened in countries such as

Canada, the value of the patrimony which this beautiful animal constitutes, and to take adequate steps in its defence.

In the month of April the Kwakiutl Indians of Vancouver Island celebrate the rituals presaging 'the month of the salmon'. With the return of this king of fish, which they considered a gift from the Gods, the populations of the region rejoiced in abundance and prosperity. Everywhere else in the world where this creature is known for the succulence of its flesh and its eggs, it is surely considered to be a most prestigious prey, a sort of exemplar of strength and beauty, and a wonderful illustration of tenacity and power in its instinct to survive. In addition to its aesthetic qualities which have inspired many works of art, it is the courage of the creature, its perseverance in fulfilling its destiny and its faithfulness to its place of birth which have always captured the imagination of those who have observed it. It is a curious destiny that this animal, by definition an orphan from birth, after slowly maturing in the waters of the river in which it came to life (from one to two years whilst it evolves from alevin to fry, parr and smolt), the fish makes its way to the vast expanses of the ocean where it has to adapt itself to salt water and to face unknown predators. The distance covered by this creature during its marine existence is astounding: for the most part it completes its growth in the waters off the West coast of Greenland, rich in shrimp and sprats. In some cases maturity is achieved after about a year when the fish is called a grilse, or in two or three years when it becomes a true salmon.

Once maturity has been reached the salmon makes the journey in the opposite direction, which generally takes the fish back to the very river in which it first saw the light. It is driven by a powerful instinct in order to make this migration, and it is also guided by an extremely faithful olfactory memory as recent scientific studies have demonstrated. The salmon thus move once again from salt water to fresh, provided the latter presents the same qualities of purity and oxygenation. They ascend the water courses, often torrents, thanks to prodigious energy, but, in the course of these runs, many of them lose their life. In autumn, the hen fish lays her eggs in a redd which she has scooped out in the gravel, and the male fertilises them with its milt. They die immediately afterwards, and the cycle would start again if the industrial societies which we have created did not imperil this equilibrium. For many years the considerable fall off in salmon populations in the river systems flowing into the Atlantic has been very frequently emphasised. Since the salmon provides a good indication of the state of health of our estuaries and our rivers, this regrettable situation highlights the effort necessary in the struggle against water pollution and the rehabilitation of rivers.

The reduction in salmon populations is due to many factors which have been made the subject of numerous studies. Industrial or domestic pollution, the many dams and abstractors of water placed in the rivers, the destruction of habitat by the extraction of sand and gravel and, in some cases, over-fishing both in the rivers and at sea are important factors which have contributed to the shrinking of stocks in Europe. In France, the situation is not very good, but great efforts have been made over the past 30 years to rehabilitate the salmon rivers. At the present time the various Ministerial Departments concerned (environment, agriculture, the sea, equipment, tourism) are working together to evaluate the natural resources of our rivers and streams including their tourist and fishery values. The Ministry of the Environment, with the participation of the Conseil Supérieur de la Pêche and various organisations on the spot, especially fishing and fish farming associations, have taken many initiatives toward restoring the populations of Atlantic salmon. I wish to emphasise that, in the area of the Adour-Garonne, programmes have been started up by the two regions of Aquitaine and Midi-Pyrénées within the framework of the plan for the development of aquaculture and natural fishery resources. There is hope for improvement in the future, also in improvements under way of rivers such as the Aveyron with the support of the Agence de Bassin Adour-Garonne.

In the field of research, co-ordination is carried out particularly within the amphihaline species by the setting up of a joint group from the Institut National de la Recherche Agronomique, the Institut Francais de Recherches pour l'Exploitation de la Mer, the Centre National du Machinisme Agricole, the Génie Rural, the Eaux et des Fôrets, the Conseil Supérieur de la Pêche and the French Universities most closely concerned.

We must now sustain and increase our efforts so that Salmo salar, which appeared on our planet about 135 million years ago, and which has always figured in our feasts, should be able to multiply for our greater pleasure, for the joy of the fishermen and for the considerable economic and tourist impact which it may have on our region and our country.

MESSAGE FROM HRH THE PRINCE CHARLES, PRINCE OF WALES - DELIVERED BY - HIS GRACE, THE DUKE OF WELLINGTON

When I made the Opening Address at the Second Salmon Symposium held in Edinburgh eight years ago I said that I was there because I desperately wanted to see the salmon survive as a species, and that their disappearance would matter to us all. I was therefore delighted to hear that a resolution from that Symposium had resulted in the setting up of the North Atlantic Salmon Conservation Organisation.

The meeting of nations to discuss mutual problems in a friendly and constructive way is something to which we should always aim, and I am particularly pleased to see from your programme that you have an even wider representation than last time, including Greenland and the Faroe Islands. This is an important step forward, which can only help us to understand and respect each others' interest in the salmon.

We all wish to see the salmon survive, and must be prepared to yield a little to each others demands in order to reach a mutual agreement on what needs to be done. The firm action taken recently by Norway on drift netting with monofilament nets, the legislation to control illegal fishing for salmon which is before our Parliament at present, and discussions over means for controlling the sale of salmon (either by licensing dealers or by tagging) are all encouraging developments on which we can build.

I wish you all a most successful and productive conference.

Chapter One

INTERNATIONAL COOPERATION THROUGH NASCO

M. L. Windsor and P. Hutchinson
NASCO, 11 Rutland Square, Edinburgh EH1 2AS

1.1 INTRODUCTON

On 1 October 1983, following ratification by six Parties, the Convention for the Conservation of Salmon in the North Atlantic Ocean (hereafter referred to as 'the Convention') entered into force. The Convention established a new international organisation, the North Atlantic Salmon Conservation Organisation (NASCO), with the objective of contributing through consultation and cooperation to the conservation, restoration, enhancement and rational management of salmon stocks subject to the Convention, taking into account the best scientific evidence available to it.

The need for an international treaty had previously been stressed by a number of interested bodies including the assembly of the Second International Atlantic Salmon Symposium which adopted the following resolution:

> The symposium resolves that for effective protection of North Atlantic salmon and in order to encourage the rehabilitation and enhancement of Atlantic salmon wherever they are found or once occurred and for the rational management of salmon fisheries, an International Convention for Atlantic salmon be established by those countries bordering the North Atlantic and its connected seas that produce and/or fish for Atlantic salmon. (Went, 1978).

The Symposium further resolved that the Convention should include, inter alia, provisions to:

(1) ban fishing for Atlantic salmon beyond 12 miles,
(2) provide for cooperation among all countries in conservation, regulation and enforcement measures and
(3) provide .a forum for international cooperation in research and the exchange of data on Atlantic salmon (Went, 1978).

1

A first draft of a treaty was produced in January 1979 for comment by the North Atlantic nations and following five formal Working Group Meetings the Convention was adopted at a Diplomatic Conference in Reykjavik in 1982.

The adoption of the Convention opened a new chapter in the long and distinguished legal history of the Atlantic salmon, which is now one of the few species with an international treaty devoted solely to its conservation, restoration, enhancement and rational management. This paper describes the Convention and the conservation measures which have been adopted to date through three rounds of inter-governmental negotiation within NASCO.

1.2 THE CONVENTION

In the preamble to the Convention it is stated that the Parties:

(1)   recognise that salmon originating in the rivers of different States intermingle in certain parts of the North Atlantic Ocean,

(2)   take into account international law, the provisions on anadromous stocks of fish in the Draft Convention of the Third United Nations Conference on the Law of the Sea and other developments in international fora relating to anadromous stocks,

(3)   desire to promote the acquisition, analysis and dissemination of scientific information pertaining to salmon stocks in the North Atlantic Ocean,

(4)   desire to promote the conservation, restoration, enhancement and rational management of salmon stocks in the North Atlantic Ocean through international cooperation.

A detailed review of the 21 Articles that comprise the Convention is beyond the scope of this paper but, to summarise, the Convention applies to the salmon stocks which migrate beyond areas of fisheries jurisdiction of coastal States of the Atlantic Ocean north of 36° N latitude throughout their migratory range (Figure 1.1). The Convention does not therefore apply, for example, to Baltic salmon, but it does apply to ranched salmon and fish farm escapees if they migrate beyond areas of fisheries jurisdiction. The Convention prohibits fishing of salmon beyond areas of fisheries jurisdiction of coastal States. Within areas of fisheries jurisdiction of coastal States fishing of salmon is prohibited beyond 12 nautical miles from the baselines from which the breadth of the territorial sea is measured, except in the following areas:

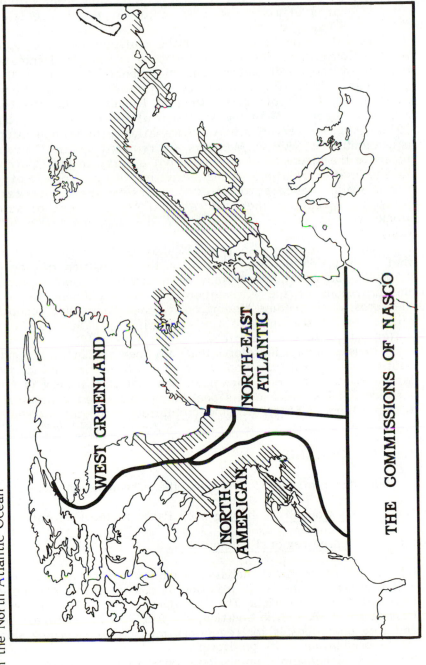

Figure 1.1: The Commissions of NASCO, the Convention area and the distribution of Atlantic salmon in the North Atlantic Ocean

(1)   in the West Greenland Commission area, up to 40 nautical miles from the baselines, and

(2)   in the North-East Atlantic Commission area, within the area of fisheries jurisdiction of the Faroe Islands.

To date, nine Parties have either ratified, or have acceded to, the Convention. The list of members is Canada, Denmark (in respect of the Faroe Islands and in respect of Greenland), the European Economic Community, Finland, Iceland, Norway, Sweden, the Union of Soviet Socialist Republics and the United States of America. With the accession of the USSR in September 1986, virtually every North Atlantic nation with salmon interests is a member of NASCO. Members are required by the Convention to invite the attention of any State not a Party to the Convention to any matter relating to the activities of the vessels of that State which appears to affect adversely the conservation, restoration, enhancement or rational management of salmon stocks subject to the Convention or the implementation of the Convention.

The Convention also requires that each Party shall ensure that such action is taken, including the imposition of adequate penalties for violations, as may be necessary to make effective the provisions of the Convention and to implement regulatory measures which become binding on it. Each Party also accepts an obligation to transmit to the Council of NASCO an annual statement of those actions taken.

It is unlikely therefore that any new fisheries beyond 12 nautical miles will develop while the Convention is in force. Moreover, the northern Norwegian Sea long-line fishery (north of latitude 67° N), which at its peak amounted to an estimated 946 tonnes (Anon, 1985 b), became prohibited under the Convention.

1.3 THE ORGANISATION

As laid down in the Convention, the Organisation consists of:

(1)   <u>a Council</u> - having each Party to the Convention as a member

(2)   <u>three regional Commissions</u>

(a)   North American Commission - consisting of maritime waters within areas of fisheries jurisdiction of coastal States off the east coast of North America. The members of this Commission are Canada and USA and, in addition, the EEC has the right to submit and vote on proposals for regulatory measures concerning salmon stocks originating in its territories.

(b)   West Greenland Commission - consisting of maritime waters within the area of fisheries jurisdiction off the coast of West Greenland west of a line drawn along 44°W longitude south to 59°N latitude thence due east to 42°W longitude and thence due south. The members of this Commission are Canada, Denmark (in respect of Greenland), the EEC and USA. Iceland, Norway and Sweden have expressed an interest in membership of this Commission and the Council will consider these applications at its Fourth Annual Meeting in 1987.

(c) North-East Atlantic Commission - consisting of maritime waters east of the line referred to in (b) above. The members of this Commission are Denmark (in respect of the Faroe Islands), the EEC, Finland, Iceland, Norway and Sweden. In addition, Canada and the USA each have the right to vote on proposals for regulatory measures concerning salmon stocks originating in the rivers of Canada or the USA, respectively, and occurring off East Greenland.

The areas of the Commission are shown in Figure 1.1.

(3)   a Secretary

The functions of the Council, the Commissions and the Secretary are shown in Figure 1.2.

Figure 1.2: The functions of NASCO

COMMISSIONS

-    to provide a forum for consultation *1
     and cooperation among the members concerning the conservation, restoration, enhancement and rational management of salmon stocks subject to the Convention.

-    to propose regulatory measures for *1
     fishing in the area of fisheries jurisdiction of a member of salmon originating in the rivers of other Parties.

-    to provide a forum for consultation *2
     and cooperation between members on matters related to minimising the catches in the area of fisheries jurisdiction of one member of salmon originating in the rivers of another Party.

-    to provide a forum for consultation *2
     and cooperation between members in cases where activities undertaken or proposed by one member affect salmon originating in the rivers of the other member because, for example, of biological interactions.

Figure 1.2 (Cont'd)

- to propose regulatory measures for salmon *2
  fisheries under the jurisdiction of a member which harvests amounts of salmon significant to the other member in whose rivers that salmon originates, in order to minimise such harvests.

- to propose regulatory measures for salmon *2
  fisheries under the jursidiction of a member which harvests amounts of salmon significant to another Party in whose rivers that salmon originates.

- to make recommendations to the Council concerning the undertaking of scientific research.

*1 West Greenland and North-East Atlantic Commission only.
*2 North American Commission only.

COUNCIL

- to provide a forum for the study, analysis and exchange of information among the Parties on matters concerning the salmon stocks subject to the Convention and on the achievement of the objective of the Convention.

- to provide a forum for consultation and cooperation on matters concerning the salmon stocks in the North Atlantic Ocean beyond Commission areas.

- to coordinate the activities of the Commission.

- to establish working arrangements with ICES and other appropriate fisheries and scientific organisations.

- to make recommendations to the Parties, to ICES or other appropriate fisheries and scientific organisations concerning the undertaking of scientific research.

- to supervise and coordinate the administrative, financial and other internal affairs of the Organisation, including the relations among its constituent bodies.

- coordinate the external relations of the Organisation.

- coordinate the initiatives of the Parties concerning the activities of the vessels of States not party to the Convention which appear to affect adversely the conservation,

Figure 1.2 (Cont'd)

restoration, enhancement or rational management of salmon stocks subject to the Convention.

SECRETARY

- to provide administrative services to the Organisation.

- to compile and disseminate statistics and reports concerning the salmon stocks subject to the Convention.

- to perform such other functions as follow from the provisions of the Convention.

Among other things, the Commissions serve as the fora for consultation and cooperation between members, they propose regulatory measures and they make recommendations to the Council concerning the undertaking of scientific research. In exercising these functions a Commission must take into account:

(1) the best available information including advice from the International Council for the Exploration of the Sea and other appropriate organisations,

(2) measures taken and other factors, both inside and outside the Commission area, that affect the salmon stocks concerned,

(3) the efforts of States of origin to implement and enforce measures for the conservaton, restoration, enhancement and rational management of salmon stocks in their rivers and areas of fisheries jurisdiction,

(4) the extent to which the salmon stocks concerned feed in the areas of fisheries jurisdiction of the respective Parties,

(5) the relative effects of harvesting salmon at different stages of their migration routes,

(6) the contribution of Parties other than States of origin to the conservation of salmon stocks which migrate into their areas of fisheries jurisdiction by limiting their catches of such stocks or by other measures,

(7) the interests of communities which are particularly dependent on salmon fisheries.

In addition to negotiation of regulatory measures, the Commissions have already provided a forum to discuss a wide range of subjects of relevance to the conservation of salmon stocks. For example, within the North American Commission of NASCO there have been discussions regarding the possible effects of surface water acidification and of the introduction of Pacific salmonids into waters containing Atlantic salmon.

The Council and regional Commissions of NASCO must take into account a complex range of subjects and interests including scientific advice, international diplomacy and procedures, statistics and commercial, sporting, recreational and community interests. Because of the particular nature of the salmon resource and its many facets, there will always be political, economic, scientific and other obstacles to agreement on how it should be managed. Nevertheless, the fact that the authorities of the nine Parties approved the Convention in such a relatively short time is testimony to the concern in the North Atlantic nations regarding this resource and augurs well for international cooperation.

1.4 SALMON MANAGEMENT THROUGH NASCO - PROGRESS TO DATE

The salmon and its management elicits strong views from countries and individuals alike and it is to be expected that the first period of operation of NASCO will consist partly of the development of an understanding by each Party of the views and positions of each of the other Parties and this process is now taking place. Progress in the first three years of the Organisation has been substantial. First, the fact that nine Parties are members of NASCO and are now meeting on a regular basis to consider their interests in and their differences on the resource is, in itself, a major step forward. This international cooperation between governments simply did not exist on this scale before.

Secondly, the birth of NASCO has stimulated scientific attention to the resource. The Council and regional Commissions have, on an annual basis, defined scientific questions concerning the resource which they consider are important in order that they may accomplish their tasks. This is the first time that such comprehensive lists of scientific questions, approved by all the nations concerned, have been defined. The questions have stimulated detailed scientific responses from ICES in the form of the Reports of the Working Group on North Atlantic Salmon (Anon, 1984; 1985 a, b, c; 1986) and this information has been used in the negotiations concerning regulatory measures. The scientists concerned have identified further areas of scientific ignorance and in this way the establishment of NASCO has stimulated the acquisition, analysis and dissemination of the scientific data required for the rational management of the stocks.

Thirdly, Parties to the Convention agree to provide the Council with the following information concerning salmon stocks subject to the Convention in its rivers and areas of fisheries jurisdiction:

(1)   available catch statistics for salmon stocks at such intervals as the Council determines,
(2)   such other statistics for salmon stocks as required by the Council,
(3)   any other available scientific and statistical information which the Council requires for the purposes of the Convention,
(4)   copies of laws, regulations and programmes in force relating to the conservation, restoration, enhancement and rational management of salmon stocks.

In addition, the Parties shall notify Council, on an annual basis, of:

(5)   the adoption or repeal since its last notification of the laws, regulations and programmes described above,
(6)   any commitments by the responsible authorities concerning the adoption or maintenance in force of measures relating to the conservation, restoration, enhancement and rational management of salmon stocks,
(7)   any factors within its territory and area of fisheries jurisdiction which may significantly affect the abundance of salmon stocks.

There is, therefore, a new exchange of information between all the nations concerned regarding measures taken to ensure the conservation, restoration, enhancement and rational management of salmon stocks subject to the Convention. Furthermore, the Parties have agreed to an analysis of laws, regulations and programmes and of catch statistics. It is unlikely that this exchange and analysis of information would have occurred without NASCO.

Fourthly, measures regulating the fisheries have been agreed. In 1984, the West Greenland Commission adopted a proposal for the establishment of a total allowable catch (TAC) for the West Greenland fishery of 870 tonnes for the 1984 season. This represented a 27 per cent reduction of the quota of 1,190 tonnes which was established in 1976 through the International Commission for the North-West Atlantic Fisheries (ICNAF). Neither the 1984 quota nor the 1983 quota of 1,190 tonnes were taken, however, as a result of lower abundance of salmon off West Greenland in both years. This lower abundance can be partly explained by low temperatures during the winters of 1982/83 and 1983/84, (Anon, 1986). Also in 1984, the North-East Atlantic Commission adopted a proposal for a regulatory measure that a TAC of 625 tonnes be established for the Faroese long-line fishery, that the fishing season should be from 1 October 1984 to

31 May 1985, that the minimum size of salmon retained on board the fishing vessels should be 60 cm and that the Faroese authorities should ensure that efforts were made to increase the survival rate of discards.

In 1985 the Greenland authorities, following the lack of agreement within NASCO, unilaterally imposed a quota of 852 tonnes and in 1986, following detailed negotiations within NASCO, a proposal limiting the TAC for the West Greenland fishery to 850 tonnes, subject to 1st August starting date, in both the 1986 and 1987 fishing seasons was adopted by NASCO. The provisional catch of 851 tonnes in 1985 compares with a peak catch of 2,689 tonnes in 1971 when vessels other than Greenlandic vessels were involved in the fishery. In 1985, the Faroese authorities bilaterally with the EEC reduced the TAC in the Faroese zone to 550 tonnes for the 1985/86 season, but no regulatory measures were announced in the North-East Atlantic Commission of NASCO in either 1985 or 1986. In 1986, the North American Commission adopted a proposal to close the Newfoundland and Labrador commercial salmon fisheries in 1986 from 15 October to 31 December.

Fifthly, the advent of NASCO provided an opportunity for the Parties to present to the Commissions and the Council, as part of the negotiating process, reviews of their existing and planned conservation measures.

1.5   CONCLUSION

International cooperation on fisheries matters is never easy since vital interests and livelihoods are usually at stake. The unique characteristics of the North Atlantic salmon, its prized status, its value to recreation and tourism, its role in providing employment and, not least, the high levels of public concern about the species, mean that negotiations are doubly difficult. Nevertheless, progress is being made. It is through the international cooperation set up by the NASCO Convention that the first regulatory measures have been adopted, scientific research has been stimulated and exchanges of information on laws, regulations, programmes and statistics have been initiated. Over previous decades, wide international cooperation on management of this resource has not existed. NASCO's new task of regulating the Atlantic salmon resource is complex, but as the Organisation continues its developmental and learning progress, its aim will be to help to ensure that the North Atlantic nations can conserve, restore, enhance and rationally manage the salmon stocks through international cooperation.

REFERENCES

Anon (1984) Report of Meeting of the Working Group on North Atlantic Salmon. Aberdeen, 28 April - 4 May 1984, ICES CM 1984/Assess: 16, 58pp.

Anon (1985a) Report of Meeting of the Working Group on North Atlantic Salmon. St Andrews, New Brunswick 18-20 September 1984, ICES CM 1985/Assess:5, 55pp.

Anon (1985b) Report of Meeting of the Working Group on North Atlantic Salmon. Copenhagen, 18-26 March 1985, ICES CM 1985/Assess:11, 67pp.

Anon (1985c) Report of Meeting of the Working Group on North Atlantic Salmon. Bangor, Maine, 6-8 May 1985, ICES CM 1985/Assess: 19, 37pp.

Anon (1986) Report of the Working Group on North Atlantic Salmon. Copenhagen, 17-26 March 1986, ICES CM 1986/Assess: 17, 88pp.

MacCrimmon, M.R. and Gots, B.L. (1978) World Distribution of Atlantic Salmon, Salmo salar. Journal of the Fisheries Research Board of Canada, 36, 422-457

Went, A.E.J. (ed) (1978) Atlantic Salmon: Its future. Proceedings of 2nd International Atlantic Salmon Symposium. Edinburgh 26-28 September 1978, Fishing News Books, Farnham, Surrey, 249pp.

Chapter Two

THE COMMUNITY'S APPROACH TO INTERNATIONAL SALMON
MANAGEMENT

John Spencer
Directorate General for Fisheries, EEC Commission

COMMISSION'S APPROACH TO SALMON MANAGEMENT
WITHIN COMMUNITY WATERS

Fisheries is a 'Community competence' and as such the European
Commission is responsible for the management of the European
Community's Common Fisheries Policy. Salmon as an anadromous
species is of course different from other fish stocks such as cod,
haddock or herring stocks for which the Community fixes
annually total allowable catches (TAC) and allocates quotas for
Member States. Salmon is not a 'unitary resource' nor even a
'joint stock' as are these other stocks.

The Commission, mindful of scientific advice on salmon
management in the home fisheries, has not sought to harmonise
the different national and local legislation in force in its Member
States. It considers rather that effective management of
individual salmon stocks is best left to the local administration of
rivers or river systems where the stocks may be constantly
monitored and appropriate measures adopted locally, always
within the framework of Community and national legislation in
force.

The Community legislation referred to is the ban on salmon
fishing outside 12 miles from the baselines as well as the ban on
salmon fishing in international waters outside 200 miles. Salmon
fishing therefore can only be conducted legally by Community
fishermen within the 12 mile area and even here the Member
States have enacted legislation, further limiting the fisheries.

Member States are required under Community legislation to
submit proposed national legislation to the Commission for
vetting in order to ensure that it conforms with the Common
Fisheries Policy.

The pattern and nature of salmon fisheries within the
Member States of the Community have evolved substantially over
the years. Regulations differ between and within Member States

regarding mesh sizes, net sizes, opening and closing dates for the fisheries, licensing methods, equipment authorised, catch methods and marketing methods. These differences reflect different salmon fishing traditions and methods, as well as different socio-economic groups.

The Commission does not seek to pronounce on the different means of harvesting the salmon resources within the Community. There are 'pros' and 'cons' related to each form of fishery. Suffice to state that it is not, of course, always the best fishing method from the cost-benefit point of view that should determine necessarily how fisheries are conducted. Socio-economic factors must also be taken into account.

Other speakers will outline different facets of the 'home water' fisheries within the Community so that the objective here is to outline the role the Community has exercised in the last ten years in seeking agreement on measures relating to the conservation and rational management of the North Atlantic salmon stocks.

The essential Community objective in this period has been, through international cooperation, to protect the Community's salmon stocks during their migration pattern in a manner consistent with the LOSC and NASCO Conventions.

Why has the Community a major role in salmon management in the international context and what has been, and is currently, its approach?

In addressing the first part of the question 'why a major role', the rivers in the Community, notably, in France, Ireland and the United Kingdom contribute the largest input to the salmon stocks occurring in the North Atlantic Ocean. An agreement on international salmon management is of vital importance to the Community in order for it to achieve its objective of protecting its stocks during their period of migration. The return of these salmon stocks to the home-waters is of critical importance both for the fisheries and the replenishment of the stocks through spawning.

SALMON MEASURES WITHIN THE COMMUNITY

To illustrate the 'input' made within the Community one can take 1986 as an example. Salmon measures envisaged in the Member States of the Community in 1986 include:

(1) Stocking - Stocking programmes to ensure the renewal and increase in the level of the salmon stocks will involve a financial cost in 1986 of £2,500,000 to the Member States.

(2) Research - Continuous programmes of research on the state of the salmon stocks are conducted by the Member States of the Community on the basis of which management measures are adopted. In 1986 the financial cost of this research is estimated at £3,000,000.

(3) Development - Improvements in the river systems (e.g. fish passes, environmental control) which will facilitate salmon returning to spawn will cost £2,000,000 in 1986.

(4) Control - In the context of protecting our salmon stocks, considerable financial resources are devoted to the campaign to reduce and eradicate illegal salmon fishing. In 1986, the cost of control and enforcement in relation to salmon is estimated at £5,000,000.

(5) Management - In the overall management of the salmon resources, including the constant review of the state of the stocks to assess the effectiveness of the measures in force, an expenditure of £6,500,000 will be incurred in 1986.

The above measures, costing a minimum of some £20,000,000 (= 200 million FF) in one year reflect the Community's continuing commitment as a State of Origin in conformity with the Convention of the Law of the Sea to ensuring the conservation, rational management, and enhancement of its salmon stocks.

SOCIO-ECONOMIC IMPORTANCE

The importance of salmon to the Community can also be expressed in terms of the catch levels in the Community's home water fisheries compared to the total home water fisheries in the North Atlantic in the period 1960-1985 (Table 2.1).

DEVELOPMENT OF THE 'INTERCEPTORY' FISHERIES

The development of what are referred to as 'interceptory fisheries' in the 1960s and 1970s, firstly, in the waters off West Greenland, and secondly, in the North-East Atlantic area (essentially within the Faroes Island 200 mile zone) had grave implications for the level of returns of salmon to Community rivers. In addressing this new threat to its salmon stocks, the Community sought, through international cooperation, to have limits agreed for the level of catches in these fisheries.

The Community's Approach to International Salmon Management

Table 2.1: Average annual Community catches 1960-1985

Period	Total average annual catch in the Community's home water fisheries	Community catch as % of total average annual home water catches in the North Atlantic Area
1960 - 69	3700 t	44
1970 - 79	3676 t	45
1980 - 85	3303 t	41

Source: ICES

BILATERAL EC/CANADA AGREEMENTS ON WEST GREENLAND FISHERY

Within the context of overall bilateral fisheries agreements with Canada since 1976 and up to the time of the establishment of NASCO in late 1983, the Community agreed to fix the fishing limit at 1,190 t for the West Greenland fishery. This agreement should be considered in the context that 50 per cent of the salmon taken at West Greenland are of Community origin and the other 50 per cent are of North American origin, essentially Canadian.

BILATERAL EC/FAROES AGREEMENTS

The long-line salmon fishery at Faroes had a catch of 5 t in 1968 but the major expansion in the fishery came from 1979 onwards (119 t) and peaked in 1981 with a catch of 1,125 t (see Figure 2.1). Under annual bilateral arrangements concluded by the Community with the Faroes Islands, the catch at Faroes was fixed at 750 t for the 1981/82 season and 625 t for the 1982/83 to 1984/85 seasons.

These bilateral arrangements have had to continue unfortunately even after the creation of NASCO (in contrast to the situation relating to the West Greenland fishery) due to the Faroese refusal, at three successive NASCO Annual Meetings, to agree to NASCO regulatory measures limiting their fisheries.

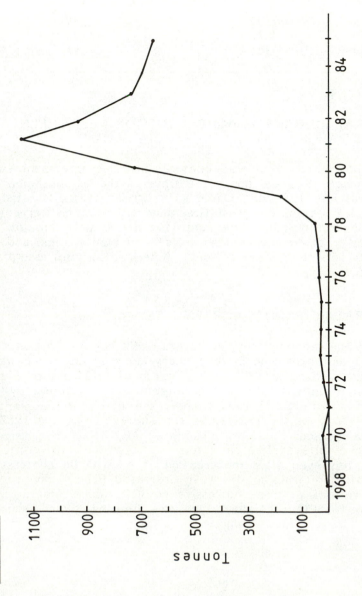

Figure 2.1: Growth of long line fishery in Faroese area 1969-1985

Tonnes

Source: ICES Working Group on North Atlantic salmon Copenhagen 18/26 March 1985
          ICES Working Group on North Atlantic salmon Copenhagen 17/26 March 1986

16

Figure 2.2: Faroese salmon catches and consequent losses to home waters stocks

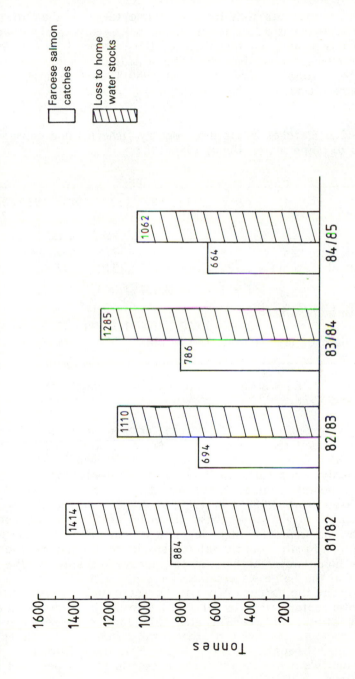

Source: a, ICES, Report of Working Group on North Atlantic Salmon, Copenhagen 17/26 March 1986.

IMPACT OF THE 'INTERCEPTORY FISHERIES' ON THE HOME WATER STOCKS

Scientists estimate that for each tonne taken in the interceptory fisheries there is a loss to the home water stocks of between 1.3 and 2 t depending on the origin of the salmon. If one takes 1.6 t as the average loss, then the total loss to the home water stocks may be estimated for the period 1980-1985 (see Figure 2.2 for losses resulting from Faroese fishery) (Table 2.2).

Table 2.2: Catches in the interceptory fisheries and consequent losses to home water stocks 1980-1985

	1980	1981	1982	1983	1984	1985	Total 80-85
Faroese catches	718	1,125	960	783	697	672	4,925
Greenland catches	1,194	1,264	1,077	310	297	851	4,993
Total	1,912	2,389	2,037	1,063	994	1,523	9,918
Loss to home water stocks	3,060	3,822	3,260	1,700	1,590	2,436	15,868
EEC share of loss	1,640	2,090	1,768	970	910	1,330	8,708

Source: ICES Working Group reports on North Atlantic Salmon 1985 and 1986.

By means of annual bilateral arrangements the Community has concluded that through NASCO regulatory measures in 1984 and 1986 for the West Greenland fishery which were based on Community proposals, the level of the interceptory fisheries has been reduced. Annual losses to the home water stocks have decreased from 3,000 t in 1980 to 2,400 t in 1985. However, in the Community's view, these fisheries are still being conducted at an unacceptably high level and constitute a considerable burden on North Atlantic salmon stocks by dissipating the efforts of the home water authorities to improve the state of the salmon stocks in the home waters.

From the foregoing it can be seen that current catches in the interceptory fisheries of 1,523 t result in a loss to the home water stocks of 2,436 t and this loss represents a highly significant 42 per cent of the total catch of the home water fisheries throughout the North Atlantic. The loss to the Community's home water stocks accounts for 55 per cent of that total loss.

COMMUNITY'S APPROACH TO SALMON MANAGEMENT

The Community's approach to salmon management in the international context is determined by the provisions of the Convention of the Law of the Sea relating to anadromous species.

Article 66(1) clearly stipulates that 'States in whose rivers anadromous stocks originate shall have the primary interest in and responsibility for such stocks'.

The Community does not contest the right of the 'interceptory' States to fish salmon of Community origin nor does it seek to eliminate the interceptory fisheries. What it does contend is that level of such fisheries must be consistent with those State's obligations under the Law of the Sea Convention. Reasonable limits should be placed on them with the view to ensuring that the efforts of the States of Origin to conserve and enhance their salmon stocks are reflected in greater returns to the home rivers.

NASCO

The Community welcomed the establishment of NASCO in 1983 and indeed worked very hard with other interested parties to ensure its creation. An effective NASCO embodies multilateral cooperation on salmon conservation. This multilateral aspect reflects perfectly the international nature of the species' migration pattern in the North Atlantic. Bilateral agreements limiting the interceptory fisheries (EC/Canada, EC/Faroes) were a necessary conservation instrument prior to NASCO but now the rationale for their existence can only be justified in a situation where NASCO fails to adopt meaningful management measures.

The Community is committed to working through NASCO for agreement on meaningful international salmon management measures and it has exercised a prominent role to date within NASCO to that end. The Community, whilst recognising that the organisation is at an early stage, views with mixed feelings the results achieved within NASCO to date.

In 1984 the first management measure was adopted by NASCO to regulate interceptory fisheries. The Community, being responsible for the management of the West Greenland fishery and taking account of the available scientific advice and the management measures adopted by the home fisheries to improve returns to their rivers, agreed to a substantial reduction in the West Greenland TAC from 1,253 t to 870 t.

In 1985 when that fishery was no longer the responsibility of the Community following Greenland's withdrawal from the European Community, Greenland refused proposals on NASCO regulatory measures and unilaterally fixed its own limit.

In 1986 it was on a Community proposal that NASCO adopted a regulatory measure fixing the level of the West Greenland fishery at 850 t for both the 1986 and 1987 seasons.

FAROESE FISHERY

In respect of this fishery, the Community must express both its disappointment and concern that the effective cooperation evident on a bilateral basis between the Community and the Faroes prior to the establishment of NASCO, has not yet manifested itself within NASCO.

This year the Community, in order to satisfy what it perceived as Faroese concern that NASCO would continually adopt regulatory measures reducing the level of their fisheries, proposed a TAC for the Faroese fishery at 550 t for two successive seasons, 1986/87 and 1987/88. The level proposed was the same as that under the bilateral arrangement between EC and Faroes for the 1985/86 season. This proposal, by stabilising the fishing level for two seasons was fully in accordance with the proposal which Greenland accepted within NASCO. Regrettably the proposal was rejected by the Faroe Islands.

For the third successive year therefore, the Faroe Island authorities have refused to agree to a NASCO regulatory measure which limits their fisheries to the level fixed under the annual bilateral arrangements with the Community. In view of the Faroese attitude and the loss to the Community's home fisheries that Faroese salmon fishery represent, the Community has been obliged to seek a bilateral agreement with the Faroes. The Community does not consider such bilateral arrangements to be a satisfactory substitute for regulatory measures adopted within NASCO.

Chapter Three

THE NORTH AMERICAN VIEWPOINT

Jack T.H. Fenety

The Miramichi Salmon Association,
Central N.B. Woodmen's Museum Building,
Boiestown, New Brunswick,
Canada EOG 1AO

INTRODUCTION

Just slightly more than 8 years ago I had the pleasure of presenting a paper on illegal fishing during the second International Atlantic Symposium in Edinburgh. The paper was devoted entirely to the illegal catch of Atlantic salmon or as I termed it 'the Atlantic Salmon Rip-off'. During the course of this presentation I will express the view that very little has changed since then despite a number of positive approaches and actions by concerned governments and conservationists. My assignment is to cover the current North American situation and although it is difficult to isolate the problems of this great world traveller, I am pleased to be able to indicate some positive actions from the North American sector of the Atlantic salmon world.

THE ADVENT OF NASCO

The coming into being of an international accord on Atlantic salmon had a profound affect on the governments of the United States and Canada. Although the government of Canada and especially those within its Department of Fisheries and Oceans, still smarted from the US/Denmark negotiations of the 1970s which led to the eventual establishment of a quota system for the West Greenland Fishery, there were strong indications by 1979 that Canada could be persuaded to be a signatory to any international accord or treaty dealing exclusively with Atlantic Salmon. Canada did become a co-operative, founding member of NASCO. It is said that old wounds are sometimes slow to heal. Canada's experiences and mine personally within ICNAF were not conducive to an open arms approach to yet another association with countries which had consistently failed to recognise the high price Canada had been forced to accept: especially by those who neither ploughed nor planted but who continued to harvest Canadian salmon stocks. In turn the United States had a long-standing quarrel with Canada in the matter of Canadian interception of certain US salmon stocks in waters off Newfoundland. This is a problem which only very recently has been addressed by Canada, though possibly not yet to the complete satisfaction of the United States. Then, of course, there was and still remains the major question of interception of North American Atlantic salmon stocks on the high seas, specifically, off West Greenland and the Faroe Islands. Given the large Canadian composition of the salmon stocks in the waters of these countries, the coming into being of an international salmon accord in 1984 stands as a tribute to the overall desire of the member countries to work toward a lasting solution to a vexing problem of long standing.

CANADA'S REALISATION

The lack of progress within ICNAF, combined with a series of salmon seasons with extremely low runs eventually forced Ottawa to take domestic action in a serious attempt to halt the steadily increasing decline of its salmon numbers, while at the same time devise long-range plans for the eventual rebuilding of home water stocks.

Among the first plans on the drawing board was the creation of a 'salmon parliament' to be known as the Atlantic Salmon Advisory Board. Many of us had urged that such a body be created. Several user groups would be included - commercial fishermen, anglers, conservationists and native peoples. This board is still operative and although it has not provided direct access to the Minister as originally intended, it does continue as a forum for all interested parties. Unfortunately, the Atlantic Salmon Advisory Board has not always (in fact, has seldom) followed the wishes of its members! In 1983 the Federal Minister was Pierre Debane whose early recognition and realisation of the serious Canadian salmon problem brought about a series of radical departures from old line policies. He breathed new life into a federal department whose personnel had grown somewhat lethargic and to a degree disenchanted.

The Debane era, though short-lived, brought about a reduction of the Newfoundland commercial fishery with the traditional mid-May starting date being postponed for two weeks. Commercial salmon fishermen in Nova Scotia and New Brunswick had their seasons completely closed, with financial compensation being awarded for the years, 1983-84 and 1985. Following the general election of 1984, John Fraser, a knowledgeable and forceful personality, became Canada's Minister of Fisheries and

Oceans. Fraser introduced even stronger salmon conservation measures. Newfoundland part-time salmon fishermen were removed from the fishery altogether for the 1985 season. Anglers in some provinces (New Brunswick and Nova Scotia) were restricted to the killing of grilse only, at 10 grilse per season and two per day. An unlimited hook and release programme was put into effect for anglers. Tagging, of rod-caught fish pioneered by the Province of New Brunswick was eventually to become mandatory for anglers in all provinces by 1986. In this year too, tagging became mandatory in the Newfoundland commercial fishery. All fish sold either locally or for export required a foolproof, locked tag. This year, still further policy changes were introduced including one which has already set the stage for what could be a lengthy battle in the courts.

The governments of the Maritime Provinces and Ottawa decreed that financial compensation would no longer be paid to commercial fishermen for not fishing. This immediately brought to the fore the voluntary buy-back policy of salmon fishermen's licences, originally proposed by Mr Debane, retained by Mr Fraser and confirmed by the present Minister, Tom Siddon. Maritime commercial fishermen have until 31 December 1986 to accept or reject the buy-back scheme jointly funded by the Provinces and Ottawa. There are approximately 425 licence holders involved, and the original fund proposed by the Government amounted to some 4 million dollars. Through negotiations with its commercial fishermen, New Brunswick this year offered an additional 2 million dollars which brought the fund up to $4 million (not including Ottawa's $2 million).

Salmon conservationists, through the Atlantic Salmon Federation (ASF) have been urging the governments to proceed with the buy-back programme as expeditiously as possible. The ASF has indicated that it is prepared to add a further $2 million to the buy-back fund in the belief that an extra few thousand dollars per licence may be the incentive required to encourage the majority of commercial licence holders to accept this offer. The New Brunswick government's final offer has already gone out to licence holders, minus the $2 million contribution suggested by the ASF. This non-government money plus bank interest would be returned to the Federation over a period of years through a surcharge placed on all angling licences to be charged until such time as the money loaned to the governments is repaid.

A final determination of this most interesting adventure cannot be expected before the end of 1986.

Although the buy-back programme is contentious, it is nothing in comparison with the fury generated by Ottawa's decision to close the New Brunswick and Nova Scotia commercial fisheries, while at the same time permitting the Newfoundland fishery to remain in being. Newfoundland commercial fishermen account for almost 85 per cent of the total salmon catch on Canada's east coast, so that it is easy to see why the mainland commercial fishermen, who play a much smaller role in Atlantic Salmon exploitation would be so incensed by Ottawa's decision.

Almost immediately following this announcement from Ottawa the commercial fishermen's union expressed its frustration and anger and later stated that they would take Ottawa to court over the matter! Shortly thereafter, the fishermen's union filed an action in the Federal Court of Canada based on Ottawa's decision to allow Newfoundland commercial fishermen to fish for salmon while at the same time denying members of the mainland fishermen's union equal rights to fish.

The fishermen's entire case is based upon constitutional grounds claiming discrimination under the Canadian charter of rights. The outcome will be awaited anxiously by all parties concerned. It is possible that the Atlantic Salmon Federation, together with other groups and individuals, may intervene on the side of the Federal Government to support further the conservationist view of the Federal Government being the sole authority and administrator in all matters concerning Atlantic salmon. If the courts decide in favour of the mainland fishermen's Unions, it would mean in effect that Ottawa's authority, derived from the Canada Fisheries Act of 1920, would be eroded to the point where an entirely new Fisheries Act would be an urgent necessity. Given the fact that the provinces would have to be consulted on the point of their being totally satisfied with any new Fisheries Act, Canada could be in very serious trouble in its bid to restore its Atlantic salmon runs. At the very least a lengthy delay would result. Should the courts decide against the Unions, it is possible the Unions could find themselves as being highly privileged as opposed to other resource users, (namely the anglers) in the eyes of the court. This situation could bring about a host of possibilities for change which the fishermen might find unexpected and unpleasant. As of now, fewer than 500 mainland commercial fishermen are pitted against some 4,000 Newfoundland commercial fishermen and the Government of Canada. The stakes are unquestionably high.

Apart from this interesting court case, much remains to be done. Additional recent government recognition combined with new management programmes will be helpful. However, if Canada is to achieve its goal of a several-fold increase in its Atlantic salmon stocks still further restrictive and productive management programmes must be implemented such as unpleasant but necessary implementation of severe restrictions against its nationals. Canada must not only hope for greater co-operation from other countries, it must insist on greater protection of its stocks in the waters of other countries.

THE UNITED STATES - A MAJOR PLAYER

The growing interest in the fortunes of the Atlantic Salmon as a United States resource of some consequence, has been and will continue to be one of the major recent components of world-wide salmon management. From its post-war beginnings in the State of Maine, through its undoubted international strength in bringing about a resolution of the high seas fishery of the 1970s, the United States has heeded the pleas of its conservation-minded, outdoor-oriented people. Combined with the undoubted strong support of its east coast states, the USA now wields considerable influence wherever Atlantic Salmon matters are discussed today. Despite the small percentage of world stocks claimed by the Americans today I feel that with any kind of luck and reasonable co-operation on the part of other countries, the United States may eventually become a major salmon-producing country. I feel safe in saying that in no other country in the recent past has so much been done to restore river habitat through programmes of eliminating dams and other river constructions. No nation can match their continuously building, multi-million dollar hatchery system. There is an almost unbelievable concerted construction programme of expensive fishways and other remedial measures in the hoped for return of Atlantic Salmon to rivers where salmon have not been seen since George Washington was President! How heartening it is to witness this faith and follow the action being taken by the USA in the hope that Atlantic salmon will once again swim back to and spawn in the Connecticut and other large US river systems.

In describing the costly, and still unproven, Atlantic salmon programmes of the United States I would not want to leave the impression that there are no problems associated with their bid to restore runs to their former salmon rivers. I mentioned earlier that irritation existed between Canada and the United States regarding the interception of salmon stocks of known US origin, especially in the northeastern part of Newfoundland. Known locally as the Twillingate fishery, Newfoundland commercial fishermen in that area had over the years harvested what were to them small amounts of salmon in the late autumn. However, their catch was not considered small by the United States. Going back to 1976 the US State Department began exerting increasing pressure on Canada either to greatly reduce this fishery or eliminate it altogether. The 31 December closing date for the Newfoundland commercial fishery became a major point of conflict. In the USA during this past summer, there was widespread disenchantment with NASCO over its failure, in American eyes, to manage these migratory Atlantic Salmon stocks, with particular emphasis on implementation of appropriate harvesting regulations. Canada, for it's part, had steadfastly refused to consider the US request until some major Canadian problems were addressed in another segment of NASCO and a proper solution arrived at with expected assistance and co-operation from the United States. Apart from increasingly strongly worded diplomatic exchanges, there were suggestions made by the American sporting press, among others, that the United States should impose economic sanctions against Canada! The Canadian decision earlier this year to reduce the Newfoundland fishery from the traditional 31 December closing to 15 October has greatly helped restore a more friendly atmosphere between our two countries. There was ample evidence of the restored friendship during NASCO's 3rd Annual Meeting last June.

THE PRESENT OUTLOOK

1986 has so far produced a series of paradoxes we may never be able to resolve. In Canada, we in the conservation movement continue to believe that present policies will go a long way towards restoration of our salmon runs to somewhere near their former levels.

However, we are puzzled by the numbers of large salmon and grilse found in our rivers this season. By any standard 1981 was not a good spawning year, based on the numbers of spawners in most of our eastern North American rivers. We could not fortell the bountiful return we have witnessed this year. I believe it is not enough to explain the large numbers of returning fish as being the result of 'extremely good sea survival'. There must be and no doubt are, many contributing factors, but how many of them can we actually classify as responsible for these unexpected increases?

Given the widespread belief that most salmon originating in United States rivers follow the same general migratory patterns of their more northern cousins, how can we account for salmon returns to US rivers this year being considerably reduced (with few exceptions) when compared with Canadian salmon returns? Must this unaccountable increase in salmon numbers of Canadian rivers be classified as a mystery? How can our scientific people account for the almost opposite situation in United States salmon returns? If these returns were close to those generally predicted, how were the Canadian salmon runs far greater than had been predicted? The years 1966 and 1967 were near record years for salmon runs in Canada, both commercially and for angling. Due to vastly different regulations and controls it is not possible to measure 1986 against 1966, but if 1987 should turn out to parallel 1967, we may find a new key to salmon reproductive patterns.

1986 was a season of paradoxes. A season of shattered

scientific and 'best guess' predictions. Do we really know as much about Salmo salar as we have come to believe so confidently in recent years? Is Mother Nature playing tricks on us? Have we overlooked periodicity; forgotten the natural cycle of the Atlantic salmon which often results in periods of high returns and conversely low returns? Whatever the final answers may be, scientific or supernatural, the indications are that we still have some way to go in the learning process for Atlantic salmon.

In summing up I hope the points I have made, relative to the Atlantic Salmon's North American base have been informative. I want to stress in the strongest possible terms that both Canada and the United States have instituted extremely harsh conservation measures against those who fish for salmon within their home waters. In Canada, the sacrifices are enormous for nearly all user groups. Side by side with those restrictive harvesting measures is the multi-million dollar effort devoted to infrastructure. Given the severity of the restrictions imposed and the money spent and committed by both Canada and the United States, it seems only fair these countries should demand that all other salmon-producing or salmon-harvesting nations should demonstrate a similar concern for salmon conservation. In closing let me repeat a few of my words from the Edinburgh Symposium.

> I recommend that all countries having significant Atlantic salmon populations should meet. Each could in its turn openly discuss its policies, express its hopes and its aspirations for a more stable and profitable fishery. Each could learn from the others of the short-comings and frustrations which now exist and all could then work toward common goals. Only through joint action, only through the adoption of rules and regulations of similar intent and meaning can the existing chaotic turmoil be ameliorated. After all, the number of countries is not great. Surely, as the only providers of this great anadromous species a target of common objectives should be reasonably easy to obtain. Such a common front, with an openly expressed and fully united opinion, could not help being heard and listened to in all fora dedicated to species preservation, species protection and species harvesting. Recognition of the prior right of the producer state is not that far off as applied to anadromous fish. How much easier it would be in future if the producer states were able, among other things, to be fully in charge of and empowered to make and enforce the rules relative to the legal and illegal killing of salmon.

We have now taken a significant step toward an established form of collective responsibility with the advent of NASCO. This will either be successful because we collectively want it to be or it will fail because of greed and indifference on the part of some of its members. Only through determination and by maintaining our will to win can we collectively ensure the future of the Atlantic salmon in North America and elsewhere.

Chapter Four

EXPLOITATION OF THE RESOURCE IN FRANCE[1]

J. Arrignon,[2] J.P. Tane[3] and M. Latreille[4]

4.1 INTRODUCTION

The exploitation of a resource assumes both quantitative and qualitative knowledge. That of the salmon resource, whether exploitable or not, implies the control of its evolution in its different areas.

In space, investigations must then concern spawning grounds, nursery areas in the river, catadromous migration routes, gathering zones in estuaries, marine migration routes towards feeding grounds, feeding areas, then homing routes in the sea and in the rivers up to the upper reaches, and finally the spawning areas.

In time, a sustained effort must be made in the study of the salmon life cycle: emergence from redds, river life of parr, control of smolts at mandatory passage sites, detection of adults in the sea, of spawners during their anadromous migration, of grilse and large salmon, and of spawners present on redds.

[1]   The French version of this paper may be obtained from: Dr. J. Arrignon, Consei Supérieur de la Pêche, 75007 Paris, France.

[2]   Conseil Supérieur de la Pêche

[3,4]   Conseil Supérieur de la Pêche : Engineers and Biologists of Regional Delegations, [3] Ministry of the Environment, [4] State Secretariat in charge of the Sea.

Figure 4.1: Representation of investigation carried on in relation with the environment and salmon life cycle

A, Oceanic feeding grounds; B, territorial waters; C, estuary; D, river area; E, spawning grounds. 1, Deep-sea fisheries; 2, coastal commercial fisheries; 3, estuary commercial fisheries; 4, amateur fishing; 5, river commercial fisheries; 6, river amateur fishing. a, International scientific campaign; b, Administration of Marine Affairs; c, control trap device for ascending and descending migratory fish marking and morphometric measurements; d, fish passes; e, catch declarations; f, data processing centre of CSP; g, authorities with power of decision concerning resource management.

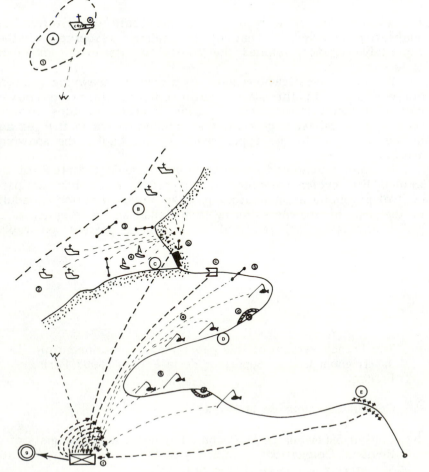

The control of resource evolution requires several investigations: census, measurement of spawning grounds visited, control of emergence, estimation of parr number, control and marking of descending smolts by means of electrical fishing or catching devices, sampling along marine migration routes and on feeding grounds within the framework of international actions, statistics of commercial fisheries in the high seas, along the coasts, in estuaries and rivers, analysis of tag returns and of data obtained by radio-tracking, recording of adults on spawning grounds.

Such are the methods and means implemented by French Services and Organisations to manage the 'Atlantic Salmon' resource (Figure 4.1).

4.2 RESOURCE EVALUATION

People in charge
Criteria summoned in Section 4.1 require an organisation for getting hydro-ecological, biological and statistical information, in the sea and in rivers. This information involves the intervention of: (a) various authorities with power of decision, regulation and policing; (b) research institutes; (c) management organisations, which, in France, differ according to the nature of the environment considered (marine or freshwater).

River environment.

Authority with power of decision, regulation and police. Ministry of the Environment, Direction of Nature Protection, Fisheries and Hydrobiology Service; the Central Administration is mainly represented in provinces by Regional Services for Water Management and Departmental Directions for Agriculture and Forestry.

Research institutes. Institut National de la Recherche Agronomique (INRA). Centre d'Etudes du Machinisme Agricole, du Génie Rural, des Eaux et des Forêts (CEMAGREF), and within this Centre, more specially the Division 'Water Quality, Fisheries and Fish Culture' and the Division 'Littoral Management and Aquaculture'.

Management organisations. Conseil Supérieur de la Pêche (CSP). Its technical direction, in collaboration with regional delegations, plays a scientific and technical part; the CSP has also a management role carried out in collaboration with the Departmental Federation of Approved Association for Fisheries and Fish Culture, Departmental Associations of Professional Fishermen, Departmental Association of Sport Fishermen (recently created) and also in collaboration with specialised associations such as the Association Internationale de Défense du Saumon Atlantique (AIDSA), Truite-Ombre-Saumon (Trout, Grayling and Salmon) (TOS), the Associations pour la Protection du Saumon (APS), etc.

Marine environment.

Authority with power of decision, regulation and police. Ministry of Transport, State Secretariat in charge of the Sea, Direction of Marine Fisheries and Marine Culture, represented for littoral zones by Regional and Departmental Directions, Marine Districts and Stations.

Research institutes. Institut Français de Recherche pour l'Exploitation de la Mer (IFREMER), which has taken over the missions of both the Institut Scientifique et Technique des Pêches Maritimes (ISTPM) and the Centre National d'Exploitation des Océans (CNEXO). These two organisations were merged in 1982.

Management organisations.

(1) Comité Central des Pêches Maritimes (CCPM)
(2) Comité Interprofessionel des Poissons d'Estuaires (CIPE), concerning more specially salmon.
(3) Local Committees for marine fisheries.

All these Organisations, at their proper level, have to take decisions, such as the establishment of fishing licences, and they have to discuss with the administration and scientific services about the management measures to take for the exploitation of these resources.

Coordination.
The number and diversity of these people in charge have induced the creation of two coordinating structures: first an organic structure (Migratory Fish Commission, at the CSP) and secondly a non-organic structure. The more recent of which, the Permanent Constitutive Group for Amphihaline Fish, is constituted by CEMAGREF, CSP, IFREMER and INRA; it is divided into vertical thematic sub-groups, each of them working on one particular migratory species - salmon included - and into horizontal sub-groups one dealing with obstacle passage facilities, the other with catch statistics. There are also interactions and co-actions between specific sub-groups.

Exploitation of the resource in France

Figure 4.2: Salmon distribution in France (1986)

Restoration programme in salmon streams: A, Bresle; B, Arques, Béthune, Valmont; C, Lower Seine - Andelle, Epte, Eure (under study); D, Meuse/Semois (under study); E, Rhine (under study); F, Loire/Vienne/Creuse/Gartempe; G, Garonne Basin; H, Dordogne Basin. Streams populated by salmon: a, coastal streams of Artois, Picardy and Upper Normandy; b, coastal streams of Lower Normandy; c, coastal streams of Brittany; d, Loire/Allier and tributaries; e, Adour/Nivelle Basin.

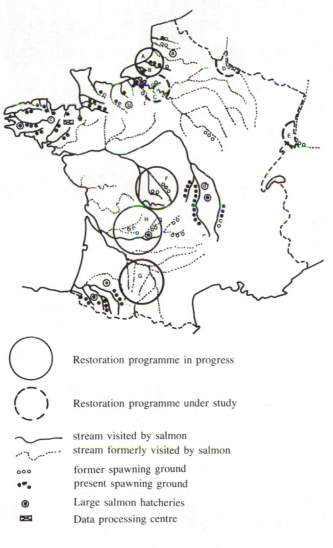

Restoration programme in progress

Restoration programme under study

stream visited by salmon
stream formerly visited by salmon

ooo   former spawning ground
•°•   present spawning ground

◉   Large salmon hatcheries

▨   Data processing centre

Table 4.1: Control trap devices installed on streams populated by migratory salmonids

Basin	Stream	Site	Number	Observations
Coastal streams North Seine	Bresle	Eu	1	Fully efficient (Table 4.2)
	Arques	Arques	1	Partly efficient
Coastal streams Lower Normandy	Calonne		1	Fully efficient
	Orne		1	Fully efficient
	Selune/Oir	Cerisel	1	Fully efficient
Brittany	Léguer	Lannion	1	Not utilised (opposition from local fishermen)
	Elorn	Kerhamon	1	Fully efficient
	Ellé	Moulin de la Mothe	1	Partly efficient
	Scorff	Moulin des Princes	1	Operational in 1987
Loire	Loire moyenne	Belleville	1	For anadromous species, efficient
		St Laurent des Eaux	1	For catadromous species, intermittent
	Allier	Vichy	1	Electrical counter, put into operation in 1985, (301 fish recorded between 1 April 1985 and 25 May 1985
Garonne	Dordogne	Bergerac	1	Operational in 1985, video, efficient
	Garonne	Golfech	1	Put into operation in 1986
Nivelle	Nivelle	St Pée	1	Efficient

Table 4.2: Migration monitoring in the Bresle stream, successfully repopulated by salmon

| Year | Marking | | Control | | Observations |
	Number	Method	Anadromous fish	Catadromous fish	
1982	–	–	59	1,115	Controls carried out in fresh and marine waters between the Bays of Seine and of Somme (inquiries and fishing books). Anglers frequently confused between sea trout and salmon. Atlantic salmon represent on average 4.3% of the catch of migratory salmonids controlled at sea (400 to 700 catches per year) and 15.7% of the catch in streams (60-100 catches per year)
1983	–	–	44	1,130	
1984	480	Freeze branding	65 (110 estimated)	750	
1985	1,410	Feeeze branding	97 (138 estimated)	1,530	
1986	732	Freeze branding	–	750	
Total	2,622		265	5,275	
7 years	5.6% (returning to stream, 2+ fish not included)		9.2%	22%	

of the total migratory salmonids (salmon and sea trout)

Means (Figure 4.2)

River environment.
Resource evaluation requires the utilisation of control devices for anadromous and catadromous migratory fish, built in suitable sites of the river.

At the same time as fish trapping operations, marking operations are carried on and utilise: (a) Carlin tags; (b) cold branding; (c) tattooing by Alcian Blue dermojet (Panjet); (d) fin clipping and (e) MK3 magnetic marking. Morphometric and scalimetric studies are also made.

From the North to the South, the control traps built between 1978 and 1986 are shown in Table 4.1. This kind of control must be utilised each time a rehabilitation effort requires the monitoring of migration evolution and each time the equipment of the more appropriate site is not impeded by technical and financial constraints.

This control may be improved by inquiries made among fishermen (Table 4.2) which, at the present time, offer a local characteristic, sometimes important and repetitive (Table 4.3).

People who have a part to play in this field are therefore engineers and biologists of the CSP, water bailiff teams of Departmental Federations and of mobile squads of intervention, associations of fishermen and particularly of salmon fishermen.

Marine environment.
The resource evaluation is jointly worked out by biologists from IFREMER and CIPE, on the basic analysis of fishing cards and inquiries.

Statistics given by the Administration of Marine Affairs are also used but they present a large margin of error. Except in the Adour river basin, the exploitation of salmon is occasional, even accidental, and geographically widely spread; it escapes the action of any organisation even well aware of these captures.

4.3 RESOURCE EXPLOITATION

River environment
In large rivers such as the Loire, Garonne, the resource is exploited by commercial and sport fisheries. Commercial fisheries differ according to the environment. In estuaries and in the 'mixed' zone, determined by legislation, it is practised by means of drag nets, in river by stop nets. Sport fishermen also take into consideration this hydrological and statutory differentiation: in estuaries and in the 'mixed' zone, they utilise nets, in rivers fishing lines.

Professional fishermen must report their catch, but the other ones not. Statistics presented in tables and histograms are thus only based on spontaneous declarations of anglers and on inquiries; they are usually approximate and under estimated.

Table 4.3: Catch control by the Association Protectrice (APS) of Massif Central, in Allier river

Year	Catch	Per catch period		catch per fisherman			
		2/3 - 15/4	15/4 - 15/6	+20	+10	+5	+3
1980	1280	825 64.5%	455 35.5%	2	9	32	502
1985	807	547 67.8%	260 32.2%[1]	greatest number per fisherman - 7[3]			

Notes: Three interesting observations:

1 Two-thirds of catches are made within the first 45 days of the authorised period (105 days).

2 93 fishermen caught at least 400 salmon, that is more than one-third of total catch for the year.

3 The general mean of catch for 1985 equals 0.37 salmon per fisherman.

Exploitation of the Resource in France

Coastal streams of Normandy.
A certain confusion exists between captures of salmon and sea trout, sometimes kept by fishermen in order to escape the constraints imposed by Fisheries Police.
    Table 4.3 and Figure 4.3 show the temporal distribution of catches. These are under-estimated insofar as it is practically impossible to know the fishing effort of amateur fishermen in estuaries who lay down stop nets on both banks and sometimes in the river mouth itself.

Coastal streams of Brittany.
In these streams, sea trout are not so numerous as in Normandy; the action of associations also induce more regular and systematic reports of catches. Data given in Figure 4.4 are therefore more representative of the real situation.

Loire-Allier river basin.
Figure 4.5 shows catch distribution over about 60 years. The exploitation by drag nets in estuary and in the 'mixed' zone is no longer recorded. The action of amateur fishermen, more important than in the past, adds to the misappreciation.

Dordogne-Garonne river basin.
The restoration on this basin is in progress; the salmon resource is not exploited, except occasionally, in the estuary; indeed the fact that, in 1985, 26 salmon were recorded at the Bergerac trap and four at the Golfech site on the Garonne suggests there are a greater number of salmon in the estuary, occasionally detected by CEMAGREF (Albiges, Rochard, Elie and Boigonties (1985)).

Rhine river basin.
Some salmon are occasionally caught in the Rhine, on the French side. Figure 4.7 shows a very progressive decline of stocks, on a level with Basle (Switzerland).

Marine environment
Sea commercial fisheries do not concern salmon, except in the marine District of Bayonne (see Figure 4.6). A survey has shown that, in 1982, a little more than 100 fishermen had directed their fishing to salmon, out of a thousand who caught salmonids.
    For amateur fishermen (including those using a boat) angling is free: this is not the case in freshwaters. There is no obligation to become a member of an Association and to pay a special tax:

Figure 4.3: Catches over a period of 20 years in four streams of Normandy (Courtesy of 'Connaissance de la Pêche')

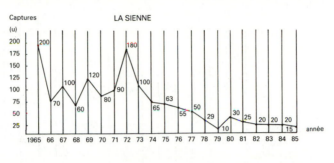

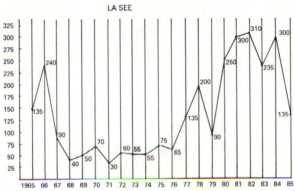

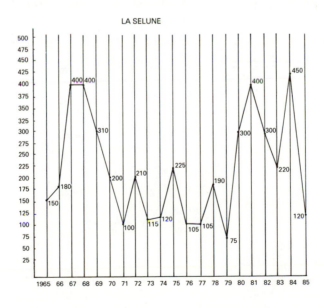

Figure 4.3 (Cont'd)

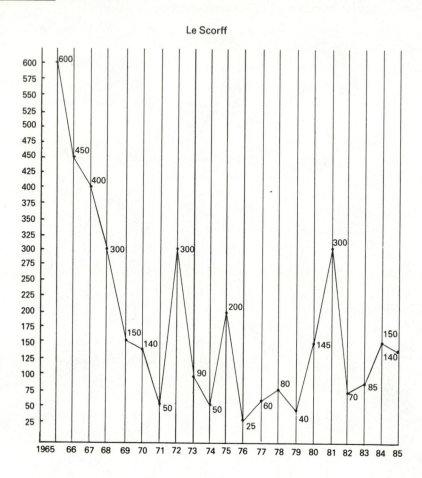

Le Scorff

Figure 4.4: Catches over a period of 20 years in 7 streams of Brittany (Courtesy 'Connaissance de la Pêche')

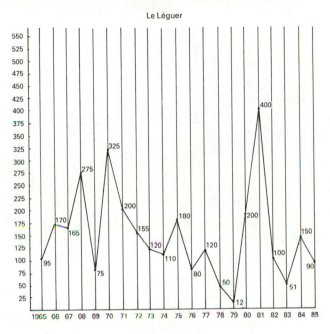

Le Léguer

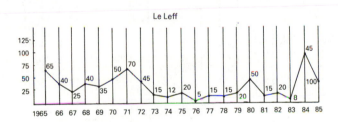

Le Leff

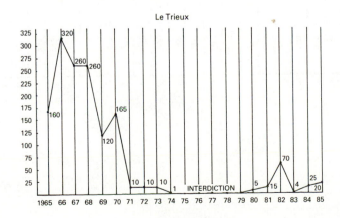

Le Trieux

Figure 4.4 (Cont'd)

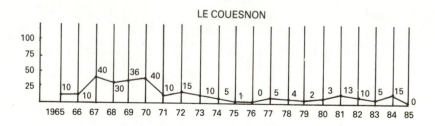

LE COUESNON

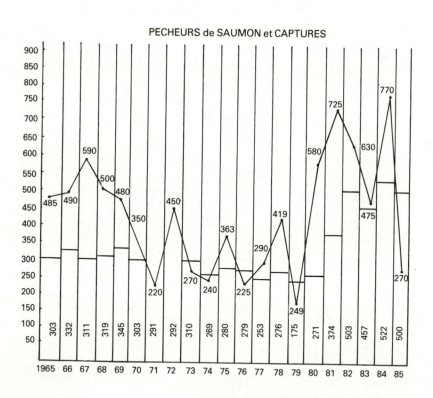

PECHEURS de SAUMON et CAPTURES

Exploitation of the Resource in France

Figure 4.4 (Cont'd)

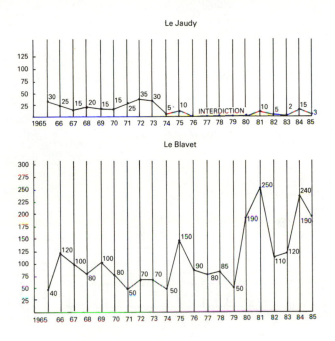

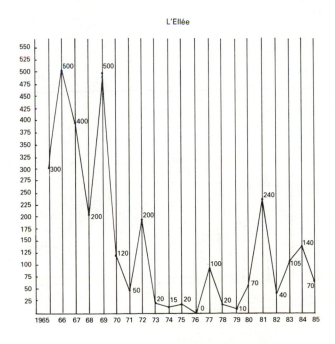

Figure 4.5: Catches in the Loire basin since 1923 (after document of the Ministry of the Environment)

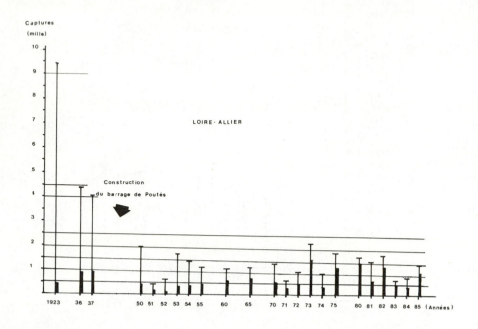

Figure 4.6: Catches in Adour basin and coastal streams of Pays Basque, from 1978 to 1985

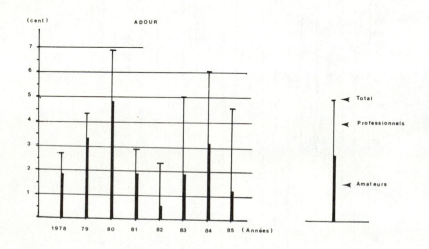

Figure 4.7: Catches in the Rhine basin since 1890 in the neighbourhood of Basle (Switzerland) (Courtesy 'Fischerei')

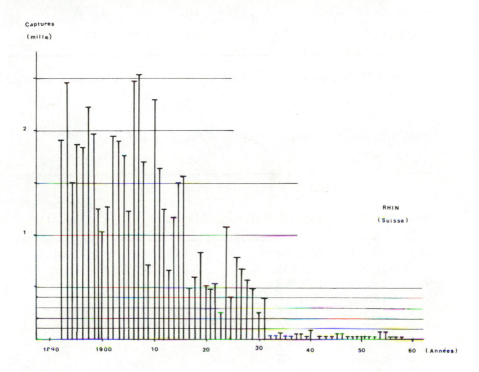

the estimation of resource exploitation is thus only possible through seldom and difficult investigations, worked out on the occasion of controls, as fishing practice is subjected to various restrictions: close and open fishing periods, minimal sizes, respect of areas considered as fishing reserves.

Official statistics delivered by the Administration of Marine Affairs are very far from reality and offer only a relative value, owing to the fact that, in 1985, under the heading 'Salmon' catches appeared which concerned mostly sea trout, a species which has shown a great increase over the last 3 years (Table 4.4).

Value of the resource

Direct commercial value.
Salmon caught in France (Tables 4.5 and 4.6) are locally consumed by most fishermen and their families, by their friends and 'relationships'. In the past, fish were eaten fresh immediately,

Exploitation of the Resource in France

Table 4.4: Salmon at sea

Salmon catch	1970	1972	1974	1976	1977	1979	1985
(tonnes)	8	10	1	2	2	29	4

Source: AIDSA and Document of the State Secretariat - Charge of the Sea.

Table 4.5: Atlantic salmon catch in France, 1955 to 1985

Year	Catch		Units of fishing effort Number[a]	Catch per unit effort[b]
	Units	Weight (kg)		
1955	36,000	226,000	5,911	6.1
1960	16,260	102,438	3,015	5.4
1965	14,388	103,593	3,126	4.6
1970	12,470	84,796	3,481	3.6
1975	7,382	47,245	3,359	2.2
1976	5,266	34,756	2,865	1.8
1977	1,988	12,723	2,572	0.77
1978	2,395	15,568	2,354	1.01
1979	1,990	12,537	2,025	0.99
1980	5,300	32,860	2,639	2.01
1981	5,200	32,344	3,287	1.58
1982	3,900	25,350	4,200	0.92
1983	3,520	22,528	3,905	0.90
1984	5,209	32,770	3,953	1.32
1985	4,443	23,000	4,483	0.99

Notes: a, The unit of fishing effort is the salmon stamp sold with fishing card.
b, Catch per unit effort corresponds to the ratio between number of salmon caught and number of salmon stamps sold.

nowadays they are deep-frozen and consumed later. The term 'relationship' refers to local restaurants which benefit from confidential sales representing an appreciable additional income for fishermen.

In 1985, the direct commercial value of salmon was estimated to be 2,300,000 francs.

Table 4.6: Salmon catch in France for 1984 and 1985, shown by salmon areas (Document CSP)

Year	1984				1985			
Type of fishing	Sea	Commercial Mixed	Commercial River	Sport	Sea	Commercial Mixed	Commercial River	Sport
Zones / Areas								
Hte Normandie-Picardie				84		100		133
Basse Normandie				785				272
Bretagne – C.d.N.				285				175
Fin.	330			1650		180		1075
Morb.				530				400
Bassin Loire – Atl.	16	66			40	160		
M. & L.		99					127	
I. & L.			2	10			1	
L. & Ch.			69	5			80	
Loiret			3				5	
Cher			16	90			25	
Loire-Allier Nièvre				102				21
Allier Allier				94				340
P. de D.				72				352
Hte L.				46				94
Bassin Garonne				4				
Gaves				294				110
Nives				14				3
Adour	231	312			745	5		
Total by zone	577	477		4965	785	445		3213
Total by category		1144		4065		1468		2975
Total number	5209 fish				4443 fish			
Total weight	32.770 kg				23.000 kg			

47

Additional value.
Statistics report the export of fresh and smoked salmon. Table
4.7 shows a rather sharp increase in 1985 in the tonnage of these
products and in the tonnage of more confidential products such as
'prepared salmon'.

Several hypotheses may be formulated: hidden increase in
salmon captures in estuaries, grouped under the heading
'Salmon', of various salmon species and of sea trout; increase in
the fresh salmon stocks imported, then smoked, prepared, even
reconditioned in France and exported again.

An inquiry carried on at the Marché d'Intérêt National
(MIN) at Rungis and among fishermen has shown that the more
appreciated and expensive smoked salmon is the fresh salmon
smoked in France (200 to 500 new francs per kg).

Indirect and accessory resources.
Salmon fishing generates an important economic activity (Table
4.8). For 1985 this activity amounted to 13,180,000 francs.

4.4   RESOURCE MANAGEMENT

Present conditions of management

River resources.
The two five-year programmes successively developed since 1976
by the Ministry of the Environment deal with four main points:

(1)   knowledge of stocks
(2)   maintenance of populations
(3)   improvement of passage facilities
(4)   improvement of environmental conditions

The knowledge of stocks has involved biological studies still
in progress and the setting of control devices (see Section 4.2 and
Tables 4.2 and 4.3). The recent law on fisheries and texts for its
application enforce and organise the control of catches, necessary
for statistical studies.

The maintenance of populations based on (1) is ensured by:
(a) egg importations from abroad; (b) breeding of fingerlings
under natural conditions; (c) construction of hatcheries; and (d)
study of egg production from local strains (Table 4.9).

Table 4.7: Exports of fresh, smoked, deep-frozen and cooked salmon from 1977 to 1985 (Document Centre Francais du Commerce Exterieur (CFCE)

Salmon Products	1977		1978		1979		1980		1981		1982		1983		1984		1985	
	T[a]	MF[b]	T	MF	T	MF	T	MF	T	MF	T	MF	T	MF	T	MF	T	MF
Fresh & Chilled	29	0.804	33	1.110	42.7	1.872	45.8	1.743	39.5	1.860	46.1	2.48	43	2.34	64.4	3.42	109.6	5.01
Smoked	279	18.203	396	20.929	367	24.863	404.4	32.071	508.5	43.020	436.6	40.17	455	42.97	552.7	58.4	701.6	70.49
Deep-Frozen	309	8.080	345	8.332	381	12.393	199	5.359	218.1	6.860	271.3	10.43	292	10.47	433.3	17.32	330.3	13.96
Cooked	1	0.112	1.7			0.125	2.6	0.184	4.3	0.290	2.3	0.15	23	1.17	8.6	0.91	124.6	3.51

Notes: a, T = tonne.
       b, MF = million francs.

Table 4.8: Derived indirect resources estimated from the number of salmon stamps sold in 1985

Item	per unit	Financial incidence in 1985	
		number of users	total
Equipment			
new fishermen	3,000	100	300,000
established fishermen	1,000	2,200	2,200,000
foreign fishermen	1,500	10	15,000
Lodging/Food			
hotel/restaurant (20% tourists)	2,500	1,800	4,500,000
picnic or inn (80% local people)	500	3,250	1,625,000
Transport			
tourists (20%)	1,000	900	900,000
local fishermen (80%)	500	3,600	1,800,000
Miscellaneous	100	4,500	450,000
Income generated by tax			
salmon stamp	400	4,483	1,793,000
Additional Value Tax	PM		13,178,200
			(13.18 million francs)

Table 4.8 (Cont'd)

Comments

The Equipment Budget is established on the following basis:

- clothes, waders, head-gear, jackets, bags, etc. .... 400 to 1,000 F
- bait casting: rod, reel, gaff, line, bait and accessories .... 800 to 1,200 F
- fly fishing: rod, reel, silk, line, flies, accessories .... 2 to 4,000 F

Let us suppose that eight out of 10 fishermen practice only bait casting but that very keen fishermen and in particular fly fishing people may have double or even treble quantities of equipment. It would then be no exaggeration to value the equipment of a new fisherman at 3,000 F on average. As for established fishermen, each season they buy lines, lures and may either change an expensive part of their equipment (rod, reel) or keep it in good repair. An average expense of 1,000 F over two years may be estimated.

Hotel/Restaurant

Tourists go to hotels or they camp, and are often accompanied. Let us take a basis of a 20-day stay twice a year. These accommodation costs, for two people, concern 20% of anglers. An inn in Bearn or Auvergne, even Brittany, generally offers accommodation at 120 F per person per day, totalling 2,500 F per person overall. Local anglers, although they often picnic, may also buy a snack or a meal at the inn. For them a budget of 500 F per year is quite reasonable.

Table 4.8 (Cont'd)

Transport

Two trips per year for tourists may represent between 2,000 and 4,000 km. An average of 60 km/day during 20 days gives us about 3,000 km, and with petrol consumption at 10 litres per 100 km, 20% of anglers spend 1,000 F per year on petrol. Local anglers travel on average 20/40 km per day when going fishing (information given by friends in Bearn and Auvergne). For about 50 days of fishing, we can divide the preceding estimate by two (1,500 km = 50 x 30 km), giving 500 F spent per year by local anglers (i.e. 80% of total anglers).

Miscellaneous

Various expenses related to the sport: specialised publications, literature, postcards, souvenirs, drinks ... 100 F per fisherman is an underestimate.

Income Generated by Taxes

- direct: sale of salmon stamps, entirely remitted to CSP.
- indirect: Additional Value Tax imposed on all derived products. In the Table, this amount is included in the revenue from salmon stamps.

Source: Lamy (1980)

Table 4.9: Production of fingerlings, parr and smolts since 1978

| Basin | Foreign contribution | | | Native contributions | | | Salmon hatcheries Annual capacity | | |
	Eggs[a]	S.O.+	S.1+	Eggs	S.O.+	S.1+	hatchery	finger-lings	Site
Bresle	-	225,000	3,860	-	560	-	-	-	Le Torpt
Eaulne	-	13,000	2,400	-	900	-	-	-	Ste Gertrude
Durdent	-	38,000	-	-	-	-	-	-	Eu
Streams of Lower Normandy	521,500	170,000	300,000	290,000	-	-	200,000	5,000	Cerisel
Brittany	3,202,000			811,000	-	-	50,000	25,000	Coquainvilliers
							200,000	15,000	La Mothe
							400,000	200,000	Le Favot
							50,000	10,000	Cardroc
LOIRE Gartempe	-	174,000	20,100	-	-	-			
Allier (1978-85)	-	252,600	93,392	-	158,000	151,241	200,000	65,000	Augerolles

Table 4.9 (Cont'd)

Basin	Foreign contribution			Native contributions			Salmon hatcheries		
							Annual capacity		
	Eggs[a]	S.0+	S.1+	Eggs	S.O+	S.1+	hatchery	finger-lings	Site
Dordogne	1,700,000	228,000	114,300	15,000	-	-	200,000	100,000	Castels
Garonne	-	-	-	-	-	-	-	15,000	Etang Ferrier
								15,000	Carderies
								-	Monna
Adour	-	1,036,000	66,700	-	-	43,000	-	-	-
Total	5,013,500	2,284,000	507,180	1,116,000	179,460	197,000			

Note: a, Eggs utilised for the production of eggs for restocking (age 0+ and 1+, also including some 2+ fish).

The improvement of passage facilities has concerned:

(1)   radio-tracking studies of homing behaviour in the large river system Loire-Allier;
(2)   destruction of useless dams;
(3)   incorporation into other dams of fish passes and lifts (Table 4.10);
(4)   working out of regulations protecting streams containing migratory species against prejudicial hydraulic structures (Figure 4.8).

In order to improve environmental conditions, pollution control has been improved in salmon streams and river cleaning operations, evaluation and creation of spawning grounds are efficiently carried out.

If the present management conditions do not differ from those existing already before 1975, it must be underlined that the promulgation on 29th June 1984 of a law on fish management in the aquatic environment represents now the essential basis for all interventions which, from 1987 on, will pursue more efficient efforts accomplished through the two successive quinquennial programmes.

Marine resources.
During the past 5 years, the reorganisation of marine legislation has been considered for a possible rehabilitation of salmon resource.

The present status of salmonid fisheries is relatively different from that existing 10 years ago. If the number of professional fishermen is not so great, they have nevertheless made efforts, in particular in organising themselves on a national scale and in introducing progressively a limitation to the profession through licences. Furthermore, the cooperation of scientists is now much appreciated.

Organisation of management according to the new legislation

River resources.
Rehabilitation of new streams: Dordogne, Garonne, Gartempe, presently in progress, should be easier with the Bresle restoration, the first stream in which salmon have been durably reacclimatised (Arrignon, 1973), thanks to new regulations which did not exist 10 years ago: prefectorial decrees concerning 'biotopes' and other texts have declared salmon streams 'reserved streams' (decrees dated 15 April 1981, 8 June 1984 and 12 March 1986). Main fluvial axes are thus protected. There new hydraulic

Table 4.10: Passage facilities existing since 1976 (by construction of new passes, destruction of useless ones, construction of fish lifts)

Basin	Site	Kind of passage facilities	Year
Canche/ Créquoise		successive pools installation of 9 dams	1980/84 1986, 1987
Bresle	Gamaches	Installation	1986, 1987
	Le Tréport	Installation	1986, 1987
Valmont	Fécamps	successive pools	1980/84
Streams of lower Normandy	See list at end of table	16 passes built and put into operation	1978/1983
Streams of Brittany	See list at end of table	54 passes built and put into operation	1978/1983
Loire	Tours Pont Wilson St Laurent	passes in dam	scheduled from 1987 to 1990
	Dampierre	passes in dam	After 1990
Loire/Vienne	Maisons rouges	pass	1983
Loire/Creuse	La Haye Descartes	pass	1984
Loire Gartempe	Nalliers	pass equipped with pools	1984
	La Brasserie	-	1984
	Mur Quéroux	pass equipped with pools	1984
	Le Verger	-	scheduled from 1986
	Chôme	-	1984
	La Chaise	pass equipped with pools	1985
	La Roche Etrangle Loup	lift	scheduled from 1986

Table 4.10 (Cont'd)

Basin	Site	Kind of passage facilites	Year
Loire/Allier	Pont du Guétin	fore dam	1985
	Pont Regemorte	fore dam	1980
	Vichy	fore dam	under study
	Pont du chateau	2 ladders	1983
	Bajasse	semi-rustic pass	1980
	Vieille Brioude	temporary pass	1981
	Chambon de Cerzat	rustic pass	1981
	Langeac	ladder	1985
	Poutès-Monistrol	lift	1986
Dordogne	Bergerac	pass with a double vertical slope, trap, video control	1985
	Tuilières	lift on right bank	scheduled from 1987
	Mauzac	pass with pools on a vertical slope	1986
Garonne	Golfech	lift	1986
	Bazacles	pass with pools	in course of study
	Ramier	pass with pools trap, video control	1986

Regional Delegation of Rennes

	River	Site	Kind of passage facilities	Year
Lower Normandy	Basse-Normandie Calvados			
	Calonne	Pont l'Eveque	Ladder R[a]	1980
	Calonne	Grimboscq	Ladder R	1980
	Orne	Hom	Ladder B	1981

Table 4.10 (Cont'd)

	River	Site	Kind of passage facilities	Year
	Orne	St-Rémy and Cossesseville	Ladder B	1982
	Orne	7 passes Haut Bassin	Ladder	1983
	Manche			
	Sienne	Cérences	Ladder R	1979
	Sienne	Gavray-Percy	Ladder R	1979
	Vire	Candol	Ladder E + R	1979
	Vire	Mancellière	Ladder E	1979
Brittany	Bretagne			
	Côtes du Nord			
	Léguer	Keryell	Ladder R	1978
	Leff	St-Jacques	Ladder B	1979
	Gouët	Plérin	Ladder e	1980
	Trieux	Pont-Caffin	Ladder B	1980
	Leff	Marie-Jeanne	Ladder B	1980
	Leguer	Kernansquillec	Lift (repairing)	1981
	Leff	Cojou	Ladder R	1981
	Leff	Moulin Poulard	Ladder R	1982
	Leff	Moulin St-Sauveur	Ladder	1982
	Trieux	Goas-Vilinic	Ladder R	1982
	Léguer	Trégrom	Ladder	1983
	Léguer	Traou	Ladder	1983
	Leff	Tonquedec	Ladder	1983
	Finistere			
	Aven	Moulin du Haut-Bois	Ladder B	1979
	Steir	Troheir	Ladder B	1979
	Pont-l'Abbé	Pont-l'Abbé	Ladder B	1979
	Hyères	Moulin Vert	Ladder B	1979
	Hyères	4 passes	Ladder B	1979
	Penzé	Borgnis	Ladder R	1979
	Odet	Moulin du Duc	Ladder R	1979
	Elle	Gorrêts	Ladder B	1980
	Aven	Piscicultre Pont-Aven	Ladder	1980
	Aven	Moulin Neuf	Ladder R	1980

Table 4.10 (Cont'd)

River	Site	Kind of passage facilities	Year
Belon	3 barrages installed	Ladder B	1980
Jet	barrage Gouiffes	Levelling	1980
Goyen	Kerlaouenan	Ladder R	1981
Aven	2 ladders	Ladder R	1982
Aven	Moulin Barbary	Ladder B	1982
Elle	Moulin de la Mothe	Ladder D	1982
Ille & Vilaine			
Couyère	Barrage Couyère	Ladder C	1979
Couesnon	Barrage d'Antrain	Ladder	1979
Couesnon	Moulin Guémain	Ladder B	1979
Couesnon	Moulin Baudry	Ladder B	1979
Couesnon	Moulin St-Jean	Ladder C	1979
Morbihan			
Scorff	Moulin Neuf	Ladder B	1979
Sebrevet	Botconan	Ladder B	1979
Blavet	Minazen	Ladder D	1979
Sarre	Bourdousse	Ladder D	1979
Blavet	Rudet	Ladder B	1979
Loc'h	Tréauray	Ladder B	1979
Ty-Mad	Inzinzac	Ladder R	1981
Scorff	Poulhibet-Coat Crenn	Ladder R	1981
Evel	Quinipily	Ladder R	1982
Blavet	Languidic	Ladder	1983
Blavet	Quellenec	Ladder B	1985

Notes: a, E = Jib; R = Braking devices; C = Value; D = Denil; e = Channel.

structures are not authorised. This measure concerns all fishermen, in the limit of their rights which are legally recognised and controlled.

The 1984 law imposes a legal flow rate in the stream, at

Figure 4.8: Reserved streams in which no hydraulic structure wil
be authorised or granted

Source: Ministry of the Environment.

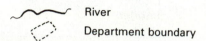 River

Department boundary

Principal water course classes (zones concerned)
Only departments concerned by this classification are
represented.

Notes: Considering the map scale, it is not possible to
         represent the 10,000 km occupied by reserved streams.

every intake; its module is determined according to hydraulic characteristics. All owners of dams built in a migratory fish stream must build, at their own expense, a fish pass, and its efficiency must be demonstrated. Passage facilities should also be improved: (a) by a greater carrying flow rate; (b) by fishways systematically arranged and more efficient than in the past; and (c) by destruction of useless dams.

Special dispositions are also taken to protect fish themselves. Independently from local reserves, full reserves will be maintained from one year to the other, to contribute to the reconstitution of populations. When salmon fisheries are authorised, a catch quota will be fixed, which will not necessarily be limited to the four fish per fisherman arbitrarily determined, as soon as a good knowledge of stocks, through compulsory catch records, allows stock management. This measure will come into effect on 1 January 1987.

The interdiction of salmon sales by sport fishermen completes the protective arrangements which make salmon tagging compulsory, according to the origin of captures: sport fishing, commercial fishing, control fishing operations worked out by the administration and research organisations. This interdiction will be extended to marine waters and to imported salmon as soon as the European Economic Community (EEC) agree.

Lastly, a modulation of fishing seasons must allow a better exploitation of certain stocks (grilse), according to salmon areas, and thus prevent genetic drift.

However, it should not be overlooked that all these new measures interfere with somewhat lawless habits of fishing. Their success will depend on a change in the behaviour of salmon fishermen rather than on penal sanctions.

Marine resources.

The adoption by the State Secretariat in charge of the Sea in 1985 of a new law will improve present texts on two precise points: (a) prohibition of the sale of fish products by anglers and (b) large increase in the fine likely to be imposed by a law Court.

Application texts of this law should also deeply re-examine fishing regulations in France, including those concerning estuarine fisheries.

The value of a text depends on its correct and persevering application. Law and regulations are not an end in themselves but a way of ensuring better means of existence to marine professional fishermen and of limiting all other kinds of fishing to sport activities.

Responsibilities of the various people in charge

Administration.
The Administration which applies decisions taken by the political power must work out, in association with users, all regulations likely to involve a good restoration of salmon resource, and to ensure their efficiency at medium term.

This efficiency is rendered precarious because of decentralisation and deconcentration of powers and local organisations more or less influenced by pressure groups. Censures may for instance follow, which are pernicious when they are repeated and do not correspond to the right adaptation of regulations to local imperatives: this is the case with the extension of the salmon fishing season in Adour estuary for the past two years, after the example of Brittany streams in which salmon characteristics are very different.

It is therefore quite necessary to be firm in the application of regional and departmental decisions, and very watchful at a national level.

Research organisations.
Research organisations, which support the Administration, are dependent upon river users and fishermen as far as stock evaluation for instance, where there is participation of users whose records supply basic information. It may be expressed at two levels: (1) Directly when the Administration wants cooperation for taking decisions or elaborating regulation texts essential to immediate or short-term management. (2) Long-term investigation: where amphihaline fish are concerned, a perfect agreement is needed in the determination of research axes as well as a constant collaboration and cooperation between laboratories of the four organisations which, in France, are involved in salmon studies. The recent creation of the Groupe Permanent de Concertation (GPC) is a promise of efficiency.

The same agreement should also prevail in the direct Research support, the Administration concerned respectively with marine and river environments consulting the organisations under its control: CSP for the Environment, INRA and CEMAGREF for Agriculture, IFREMER for the Sea. This kind of agreement should avoid the adoption of wrong criteria (see above), interfering with restoration effects carried on in one or the other environment.

Professional fishermen and river users.
A change in people's attitude with regard to lawless fishing habits implies a great campaign of information and credibility among fishermen (amateur or professional, in marine or fresh

waters) and users of salmon streams.

This effort requires competence, organisation and constancy as it must be sustained. In freshwater, the organisation of fishermen, well-defined in application texts of the law of 29 June 1984, remains the backbone of the action. It is the same for marine commercial fishing. Conversely, the lack of organised structures in amateur fishing and angling is deeply felt in the general effort and this negligent attitude sometimes reduces the efforts of other people.

To the creditable efforts made by CSP for rivers and by its related Associations (in particular the Departmental Federations of Approved Associations for Fisheries and Fish Culture) and also, for marine and estuarine waters, by the Comité Central des Pêches Maritimes (CCPM) and by CIPE, should be added those made by anglers and amateur sea fishermen whose organisation, presently based on voluntary membership of a few fishermen, is full of good but sometimes ineffectual intentions.

Evenly structured efforts should be sustained for many years in order that all French salmon fishermen (about 4,500 to 5,000 river fishermen) become respectful of a precious and rare resource, given that - according to the Gauss curve - passionate protectors of salmon will always exist, who will put their fish back into water after having had the pleasure of catching them and looking at them, while other people will only think of catching as many fish as possible.

Costs of resource rehabilitation

Table 4.11 shows the amount of investment paid out of Government stocks during these two quinquennial programmes. During this period, the cost of salmon rehabilitation has amounted to 108.6 million francs.

4.5 CONCLUSION

Considerations on the possible evolution of salmon status in France

Streams considered as reserves represent 10,000 km, 6,000 km of which are salmon streams. Out of these 6,000 km, 5,000 are now populated by Atlantic salmon (including streams under restoration).

For the coming period 1987-1991, the Ministry of the Environment considers the implementation of a full operation for Atlantic salmon management, along with that for other migratory species in estuaries. This is the SESAME project (Système d'Evaluation du Saumon Atlantique et Autres Migrateurs d'Estuaire). This project could be carried on with other partners,

Table 4.11: Salmon restoration projects given Government funding since 1976[a] (in million francs)

	Knowledge of stocks and studies	Maintenance and restoration of spawning grounds; maintenance of population stocks[b]	Equipment and passage facilities (passes)
Salaries (full- and part-time)[c]	32		
Normandy's rivers			2.6
Brittany's rivers		8.4	8.6
Loire/Allier	0.3	2.3	18
Loire/Gartempe			1.9
Dordogne	1.5	4	15
Garonne	1.2	0.8	12
Total	35	15.5	58.1
General Total		108.6	

Table 4.11 (Cont'd)

Notes:

a. Government funds: general Government budget; Fonds d'Investissement pour l'Amenagement de la Nature et de l'Environnement (FIANE) (Investment Funds for the management of nature and the environment); Fonds d'Investissement pour la Qualite de la Vie (FIQV) (Investment Funds for the quality of life); fishing tax (CSP); Basin Agencies budgets; Regional participation in programmes; Departmental contributions.

b. Part of work contributes to the general improvement of biotopes.

c. People work part of the time on tasks of general interest; on the other hand, people employed by CSP spend some of their time working on salmon rehabilitation (part-time not taken into account here).

d. Improvement in water quality is not taken into account here (investments in pollution control).

in particular the Ministry of Leisure, Hunting and Fisheries of Quebec.

For a better coordination of efforts made in France by all intervening partners in favour of Atlantic salmon, management programmes will include special dispositions for the sake of Atlantic salmon and sea trout. Indeed, it seems essential to link the future of sea trout to that of salmon, as to methods utilised for its management; the sea trout resource nowadays is fully developed in north west France. Arrangements used for recording salmon catches will then be applied to sea trout. Morphometric records and scale studies will allow a better quantification of these respective resources.

Two main trends are clearly defined for coming years:

(1)   the carrying on of a coherent effort of management which was lacking up to now, management of migratory fish streams and of their environment, management of salmon resource exploitation
(2)   the recovery of former salmon streams:

    (a)   Waterways - Dordogne/Garonne; Loire/Vienne/Gartempe; Seine/Eure/Andelle; Meuse/Semois; Rhine
    (b)   Coastal streams - Normandy: Orne/Vire/Risle/Arques; Picardy : Somme/Authie; Artois : Canche.

Thanks to the Bresle restoration started in 1967, it is obvious that a sustained effort carried on over 15 years is necessary for the steady acclimatisation of a population.

Interaction and co-action

The financial effort devoted to freshwater areas and constraints imposed to salmon fishermen in these zones may be taken into consideration only if a counterpart is accepted by national fishermen in estuaries and territorial waters, as well as by nations exploiting salmon in oceanic feeding grounds frequented by the species.

Salmon rehabilitation requires an interaction and a co-action both from national and international points of view. Repeated interventions of NASCO (North Atlantic Salmon Conservation Organisation) which gathers countries located along the Atlantic side of Europe and North America and those of ICES (International Council for the Exploration of the Sea) have led to quota limitations and appreciable agreements which should be improved and supported.

On the Atlantic side of Europe, EEC has a determining role insofar as, if the Commission may not pass a law, it may recommend the more favourable multi-national solutions to species conservation, more specially as salmon is an international resource, even if the various national structures have to manage their own stocks, in relation to their particular geopolitical conditions.

It should be advisable for the Commission to work in two directions:

(1) A standardised control of commercial exchanges of Atlantic salmon, according to a marking standard system, enabling European nations to differentiate their national production from their importations.
(2) The implementation of a European Fund for Atlantic Salmon restoration, likely to improve the salmon status as an ecological symbol and also to improve the resource and its socio-economic value, taking into consideration the fact that the salmon is a fish of international importance and therefore that the restoration effort must not be supported only by countries having the longest waterways.

4.6 SUMMARY

Two quinquennial programmes have been developed in France since 1976 to restore the salmon resource. They deal with the following points: (a) knowledge of stocks; (b) maintenance of population stocks; (c) improvement of passage facilities; and (d) improvement of environmental conditions.

In 1984 a law was passed concerning fish populations in rivers; application texts, taken in 1985 and 1986, tend to protect and increase the resource. In marine waters, legislation is under re-examination.

After the successful reacclimatisation of salmon in a stream in Normandy, some large rivers are being restored.

Management improvement is concerned with the establishment of a catch quota: four salmon per year per fisherman, from 1987, with compulsory declaration of catches and tagging of salmon.

In 1985, 4,500 amateur fishermen caught 23 tonnes of salmon, that is 0.99 fish per fisherman. The global direct value of salmon is estimated at 2.3 million francs and derived indirect resources at 13.18 million francs.

France intends to carry on the present effort which amounts already to 110 million francs. As an international fish, salmon requires international management: the implementation of a control system of catches and of trade for this species is suggested in the framework of EEC, as well as the creation of a European Fund for its protection and rehabilitation.

REFERENCES AND FURTHER READING

Albiges, C. Rochard, E. Elie, P. and Boigontier, B. (1985) Etude halieutique de l'estuaire de la Gironde - 1984. CEMAGREF - DACA, 178, 17

Arrignon, J. (1973) Tentative de réacclimatation de Salmo salar dans le bassin de la Bresle, Bulletin français Pêche Pisciculture, 248, 91-190

Arrignon, J. (1976) Aménagement écologique et piscicole des eaux douces (Ed) (Gauthier Villars), Paris, 3rd Edn., 1985, p. 276

Bagliniere, J.L. (1985) La détermination de l'âge par scalimétrie chez le saumon atlantique (Salmo salar) dans son aire de répartition méridionale : utilisation pratique et difficultés de la méthode, Bulletin français Pêche Pisciculture, 298, 69-105

Carrier, A. (1985) Comptage des saumons au pont-barrage de Vichy. FDAAPP de l'Allier, Rapport, 7 pp

Lamy, B. (1980) Essai sur les conséquences économiques d'une restauration de la pêche sportive du saumon en France, Saumons, 34, 67-71

Migaud, G. (1986) 20 années à la recherche du saumon en Bretagne et en Normandie. Connaissance de la Pêche, 91, 91, 93-95

Montignaut, J. (1980) La pêche du saumon dans la zone maritime des estuaires. Saumons, 34, 52-3

Staub, E. (1983) Wiederansiedlung des Lachses in der Schweiz. Fischerei, 91(6), 38-9

Thibault, M., Boude, J.P. and Prevost, E. (1984) La production mondiale de saumons (1952-1982) et le commerce extérieur français des saumons (1956-1983), INRA Dept Hydrobiologie. Bulletin Scientific Fiche, 15, 19-21

Vibert, R. (1980) Primauté des décisions politiques dans l'épuisement ou le développement de la ressource saumon. Saumons, 34, 7-13

Chapter Five

THE STATUS OF EXPLOITATION OF SALMON IN ENGLAND
AND WALES

G.S. Harris
Awdurdod Dŵr Cymru/Welsh Water Authority

5.1 NATURE AND DISTRIBUTION OF THE RESOURCE

Despite a long history of environmental degradation during and
subsequent to the Industrial Revolution which eradicated or
otherwise drastically reduced migratory fish stocks in many
once-productive rivers (Grimble, 1913; Netboy, 1968), the
distribution of salmon and sea trout in England and Wales is still
extensive with some 40 'major' and many other 'minor' rivers
supporting exploited stocks in varying abundance (Figure 5.1).
     With the notable exception of the Wye, Dee, Severn, Usk,
Eden and Avon, where a significant proportion of the rod catch is
taken before May in any year, the salmon rivers are for the most
part 'summer' or 'autumn' fisheries where the run consists
predominantly of grilse. Many of the rivers are small spate
streams less than 80 km long. In a typical season about 25 per
cent of the total rod catch for England and Wales is taken in the
Wye.
     Sea trout have a somewhat wider distribution than salmon
and the main rod fisheries are located on the west coast, with
roughly 50 per cent of the total rod catch being reported from
Wales.
     Catch records exist for most major salmon fisheries from
the early 1950s in relation to both the commercial and sport
fisheries and for some of the more prestigious rivers (such as the
Wye) the record goes back to the beginning of the century.
Interest in sea trout is of relatively recent origin. Few detailed
records exist prior to the 1950s and it was not until the late
1970s that records, comparable to those for salmon, were
collected generally for sea trout throughout the area.
     The historical catch record for salmon in England and Wales
for the period 1952-1985 is shown in Figure 5.2. A breakdown of
catches for 1984 by administrative regions is given in Table 5.1
along with relevant supplementary information.

Figure 5.1: Principal salmon and sea trout rivers in England and Wales within the area of each Water Authority

Figure 5.2: Declared catch of salmon and grilse in England and Wales 1952-1985, (a) Rods, (b) All commercial instruments, (c) Northumbrian Drift Nets

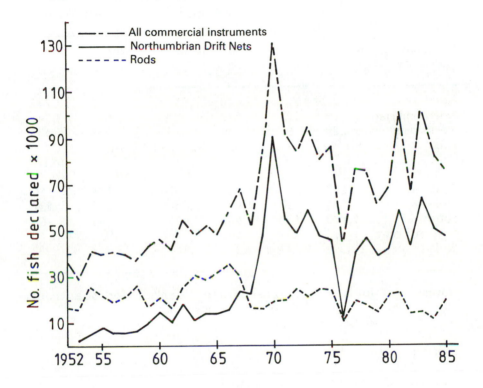

Whereas the salmon is generally more important in sustaining the commercial and recreational fisheries overall, in some regions the sea trout may be of equal or even greater importance, especially in the sport fishery.

The total reported catch for salmon (and grilse) in England and Wales over the period 1975-1984 has been roughly 3-7 per cent of the known catch in 'home waters' of the various 'producing' countries (Russell, Potter and Duckett, 1985).

Table 5.1: Number of licences issued and declared rod and commercial catch in each Regional Water Authority area in 1984

| Regional Authority | No. Licences Issued | | Declared Fish Catch | | | |
| | | | Salmon | | Sea Trout | |
	Rods	Nets	Rods	Nets	Rods	Nets
Northwest	(9,218)[a]	275	2,400	7,957	4,446	10,742
Northumbrian	2,219	121	572	50,685	1,010	43,541
Yorkshire	229	64	39	8,610	256	20,535
Anglia	NIL[b]	113	NIL	NO DATA[c]	NIL	NO DATA[c]
Thames	NIL[b]	NIL	6	NIL	NIL	NIL
Southern	273	4	960	157	156	156
Wessex	635	22	1,045	1,034	1,265	527
Southwest	5,123	100	1,652	7,455	5,041	9,261
Severn Trent	1,566	209	545	3,376	NIL	NIL
Welsh	19,371	177	3,802	3,947	18,386	10,937

Notes: a, Includes 3,962 separate licences for sea trout.
         b, Included with non-migratory trout.
         c, Return not collected.

Source: MAFF, 1985.

5.2   ADMINISTRATIVE STRUCTURE

There is a long history of fisheries administration in England and Wales which goes back over 120 years. The statutory agencies currently charged with the management and administration of the fisheries for salmon, sea trout and other inland water fish are the ten Regional Water Authorities (RWAs). These agencies were created in 1974 and have many different functions which include water supply and treatment, effluent treatment and disposal, land drainage and flood protection, pollution control, navigation, water-based recreation and fisheries. They are responsible for the integrated management of the water-cycle 'from source to sea' and as such are unique (Okun, 1977). The geographical areas of the RWAs are shown in Figure 5.1.

5.3   FISHERIES FUNCTIONS

The fisheries functions of the RWAs are defined by the Salmon and Freshwater Fisheries Act 1975. This imposes a statutory duty to 'maintain, improve and develop' the fisheries for salmon, trout, freshwater fish and eels within their respective areas.

Understandably, the ten RWAs have different levels of commitment to the discharge of their fisheries functions in accordance with the obvious regional differences in the nature and distribution of the fisheries, local pressures and perceptions of need. Only four regions have significant migratory fisheries and two have no exploitable stocks of salmon or sea trout. Nevertheless, the overall commitment to the management of the salmon and sea trout fisheries is substantial. Out of a total revenue expenditure on fisheries in 1981/82 of £7.3 million, roughly £3.2 million (45 per cent) was devoted to the conservation and improvement of migratory fish stocks (Bielby, 1983).

5.4   FISHERIES LEGISLATION

The principal statute relating to the management of inland water fisheries in England and Wales is the Salmon and Freshwater Fisheries Act 1975. This is not a recent statute however, but a consolidation of earlier legislation going back to the original Salmon Fishery Acts of the 1860s. The 1975 Act, like the preceding legislation provides for the conservation and regulation of the fisheries. Its sole purpose is to protect the resource from the deleterious effects of environmental degradation and overexploitation and so maintain an adequate spawning population.

The Salmon Fishery Act 1865 was significant in that it introduced the concept of licensing as a basis for the regulation of the fisheries. Thus, for over a century every person fishing for salmon and sea trout in England and Wales has had to obtain a licence to fish with rod-and-line or a commercial instrument and, as a consequence, historical data on fishing effort and the pattern of change within the fisheries is available.

In general terms, the means to regulate the fisheries and the nature and extent of their exploitation are provided by the 1975 Act or by the local orders and byelaws made under it. That differences exist in the orders and byelaws made by the RWAs reflects the inherent differences in the abundance, composition and pattern of runs within individual rivers and to some extent the tradition of fishing practice.

5.5 NATURE AND EXTENT OF EXPLOITATION

Salmon and sea trout are exploited both legally and illegally by a variety of means. For the present purpose, legal fishing is defined as using a licensed instrument (rod-and-line or commercial fishing gear) in accordance with the statutory requirements, while illegal fishing is defined as using any unlicensed instrument or using a licensed instrument in contravention of the statutory controls.

Commercial Fishing

There is a long history of control over the commercial fishery for salmon and sea trout in England and Wales which began with the 1860 Acts and has been progressively strengthened by subsequent legislation. Prior to the 1860s salmon and sea trout were exploited at all stages within the rivers and in tidal waters on a wholesale and indiscriminate basis by a multitude of methods (over the entire year). Such legislation that did exist was wholly inadequate and rarely, if ever, enforced.

Today, every commercial fishing instrument must be licensed and the public right that exists in tidal waters has been progressively derogated so that the maximum number of net licenses that may be issued within a particular locality is now 'fixed' under the terms of a local Net Limitation Order (NLO) in most tidal waters throughout England and Wales. In addition to stipulating the number of licences available, the NLO also defines the type(s) of net that may be used and the area in which it must operate. The length and timing of the annual close season for fishing is defined by the primary legislation and a 48 hour close time during each week of the fishing season is also stipulated. Only certain constraints may be applied by local byelaws but, when used in concert, the limited byelaw-making powers can be impressive. Thus, the length of the weekly close time may be extended and the length of the fishing season further curtailed. Byelaws may specify, among other things, the design, construction, materials and dimensions, method of use and, for nets, mesh size.

In 1985 a total of 1,015 commercial fishing instruments were licensed to fish in England and Wales. The two most common types of fishing gear are seine nets and drift nets. Although a few drift nets operated in very restricted locations in the southwest and northwest of England and in Wales, the main drift net fisheries are off the northeast coast of England. It is relevant to point out that many traditional and ancient commercial fishing instruments still operate in some areas and form part of the folk-culture heritage of the region. Coracle net-fishing is unique to Wales whereas stop-boat fishing, lave netting and putts and putchers are peculiar to the Severn

Estuary. Haaf netting is restricted to certain rivers in the northwest of England. (Many of the instruments are not particularly effective.)

Although a few instruments, mostly ancient traps, are still licensed to operate in non-tidal waters, this is very unusual and virtually all fishing now takes place in estuarial or coastal waters. A few fisheries may be privately owned or subject to ancient rights of privilege.

The history of commercial fishing for salmon and sea trout in Wales and parts of England has been described in great detail by Jenkins (1974) and was, clearly, very important in social and economic terms in many rural communities at the turn of the century. Only one study of the modern social and economic importance of commercial fishing has been undertaken in Great Britain. This was carried out in Wales in the late 1970s and, while fairly primitive by conventional standards, showed that commercial fishing did not contribute significantly to the incomes of the netsmen other than in just one or two (sic) instances. Harris (1983) suggests that the casual nature of the Welsh fishery and its limited value in providing a livelihood for the participants may be fairly typical for much of England other than for the northeast coast drift net fishery.

The history of commercial fishing in England and Wales since the 1860s can be characterised by three trends: (a) a progressive reduction in the number of licences issued, (b) increasing legislative controls over the nature and extent of the fishery, and (c) a general decline in its local importance to the community. Comprehensive data to illustrate the reduction in commercial fishing effort is only readily available for Wales (Harris, 1980) where the number of commercial licences issued in 1895 was 509 compared with 263 in 1921 and 175 in 1975.

Sport Fishing

Rod-and-line fishing is also subject to various restrictions intended to prevent over-exploitation. However, unlike commercial fishing, there is at present no statutory limitation on the number of licences available to anglers in any season and the rod fishery is not subject to a weekly close time during the fishing season. This contrasts with the situation in Scotland where angling for salmon and sea trout is prohibited on a Sunday. (Scottish fisheries legislation and administration is significantly different from that applying in England and Wales.)

It should be noted that virtually all of the rod fisheries in England and Wales are privately owned and that the owners or occupiers of such fisheries invariably impose restrictions on the anglers granted access to the fishing which supplement the constraints imposed by the 1975 Act and the local byelaws made by each RWA. Thus, while the 1975 Act does little more than define the period and minimum duration of the annual close season for rod-and-line fishing, the local byelaws and the restrictions imposed by the fishery owners or occupiers can be so complex that concise summary is impossible here. The local restrictions intended to prevent over-exploitation by anglers may vary extensively within and between different river systems and regions. Some fisheries may be fly-only while others may allow spinning and bait fishing also. Where spinning and, more particularly, bait fishing are permitted it is likely that certain baits, (maggot, prawn, shrimp) may be banned altogether or at certain times within the season, or that spinning and worm fishing may be proscribed in certain areas or otherwise, permitted only when the water conditions are unsuitable for fly-fishing. Some fisheries may impose bag limits and size limits on anglers in respect of sea trout but such constraints are not normally applicable to salmon. There is no statutory provision which makes it illegal for anglers to sell their catch during the angling season; but some fishery owners may impose such a constraint within the local regulations applying to their fishings.

Harris (1980) notes that there has been a dramatic increase in the number of anglers obtaining licences to fish for salmon and sea trout over the last century. Whole data for England and Wales are not available but within Wales licence sales have increased from 627 in 1868 to 2,174 in 1921 and to 30,200 in 1975. During the more recent period 1952-1967 the number of rod licences for salmon and sea trout issued in England and Wales increased from c. 16,000 to c. 36,000 (Anon, 1974).

The total number of rod licences issued for salmon and sea trout in 1984 was roughly 36,000 in England and Wales (Russell, et al., 1985) but the true figure, while probably not very much greater, is masked by anomalies in the licence structures operated by two of the ten RWAs and the existence of a separate rod licence for sea trout in two other regions. The different rod licence structures operated by the RWAs and their several predecessor organisations make it difficult, if not impossible, to highlight overall trends but, from the evidence available, it appears that rod licence sales peaked in early 1970s and then exhibited a fairly general decline. This decline has been substantial in some regions such as Wales where the number of rod licences issued has fallen from 22,409 in 1978 to 14,412 in 1984. There is no simple explanation for this decrease in angler effort. It is, in part, thought to be a manifestation of the current economic recession and, in major part, a crisis of confidence among the angling community resulting from the perceived decline in the quality of the sport provided by the fisheries (Harris, 1985).

Few studies on the social and economic importance of sport

fishing have been undertaken in England and Wales. Those relating to salmon and sea trout so far commissioned have been carried out primarily in Wales. Radford (1980) estimates the (gross expenditure) value of the Wye salmon fishery at roughly £1 million annually in terms of the expenditure generated by anglers in pursuit of their sport. The gross expenditure value of the entire Welsh sport fishery for salmon and sea trout has been estimated at about £10-15 million annually (Harris, 1983) and is thought to exceed the socio-economic value of the commercial fishery by a significant extent in most regions of England and Wales.

Illegal Fishing

It is now widely accepted that the most important single threat to the future status and wellbeing of the salmon and sea trout fisheries of England and Wales is the dramatic increase in the extent of illegal fishing over recent years which, in some areas, has reached epidemic proportions (Anon, 1983). 'Crooks don't keep books' and it is, therefore, impossible to quantify the level of exploitation by the illegal fishery in precise terms. Nevertheless few would challenge the general statement that the illegal catch may now exceed the legal catch on many Welsh and English rivers!

That salmon and sea trout have always been taken illegally in England and Wales is an accepted fact, but until the early 1970s the level of illegal exploitation was generally considered to be relatively insignificant in terms of overall catches and the problem was held to be under control by the enforcement agencies. For most areas this was probably so during the 1950s and 1960s when the illegal fishing was mainly confined to freshwater and entailed the use of traditional methods – such as the 'snatch', 'snare' and 'spear' – and was practised by a few individuals whose motivation was as much for pleasure as for profit.

The situation has changed significantly over the last 15 years. Illegal fishing has now become big business in many regions and the problem has extended down river into the estuaries and coastal zone; with the use of nylon monofilament gill nets being widespread. The local poacher has now been joined by well-organised, highly mobile and extremely efficient gangs who, motivated solely by profit, have become increasingly violent in their attitude to the enforcement agencies.

It is difficult to explain the recent escalation in the nature and extent of the illegal fishery, but it is likely the main causes include: (a) the high level of unemployment in many rural communities, (b) the ready availability of cheap, efficient, nylon nets that are easily transported and concealed, (c) the high price obtainable for salmon and sea trout, (d) the reluctance of the courts to impose penalties that serve as a deterrent and (e) a general awareness of the inadequacy of the legislation in combating illegal fishing and, particularly, of the loophole that allows the public right of fishing for sea fish to be used as a pretext for taking salmon and sea trout in tidal waters. Fortunately the threat posed by the illegal fishery has been recognised by the Government and new legislation is currently before Parliament which may, ultimately, prove effective in combating the problem.

5.6   INTERPRETATION OF DATA

In the absence of more reliable information about the abundance of adult stocks based either on direct counts obtained from electronic fish counters and, preferably, fixed trapping stations or otherwise from indirect assessments derived from carefully structured parr census surveys, our appreciation of the past and present status and wellbeing of our fisheries depends very largely on the historical record of catches.

This is particularly so in England and Wales where only one trapping station exists and where fish counters have generally been installed at very few sites (where they have proved to be unreliable). The traditional and widespread practice of redd counting as an indicator of stock abundance has now been discontinued in most regions and while parr census surveys are now increasingly commonplace their alternative use as a predictive and retrospective measure of adult stock abundance has yet to be validated. Thus, for England and Wales as a whole, the historical record of catches forms the sole basis for management of the resource in a local, regional, national and international context, and will continue to do so in the forseeable future. Therefore, it is fundamentally important that the interpretation of that record, and the proposition that 'catch' and 'stock' are directly linked, should be the subject of careful scrutiny and appraisal.

In order to be meaningful, either as an acceptable measure of catch in any year or as an indicator of trends over a period of years, several basic criteria must be fulfilled by the catch records:

(1)   The record should be as complete and as accurate as possible.
(2)   The number of participants producing the catch should be known.
(3)   The proportion of participants submitting a record of catch should be known.

(4) The record should be consistent in terms of its completeness and accuracy from year to year.

Clearly, when seeking to identify trends, it is of paramount importance to compare 'like-with-like'.

Critical analysis of the 30-year catch record for England and Wales (Figure 5.2) leads to the general conclusion that its interpretation as either a measure of true catch in any year or as an indicator of trends between years is so fraught with difficulties that it has little value in any practical context. In this respect, the problems of interpretation are many and may be summarised here under six main headings.

Variable Inputs

In the early 1950s catch data for minor rivers and some major rivers was not included in the record for either the rod fishery or the commercial fishery - or both - for the simple reason that it was not collected. The completeness of the record has steadily improved over the period but it was not until the mid 1970s that comprehensive data for all rod and net fisheries in Wales was included in the 'official' record for England and Wales and while the overall record has become progressively more complete over the period it is still not universal in its coverage of all licensed fisheries.

Thus the record in the latter third of the period is far more comprehensive and stable in terms of the fisheries included than in the earlier period. But, when viewed in its totality it is evident that the total catch has been steadily inflated by the progressive inclusion of catches from fisheries previously omitted from the record or from new fisheries that have been developed (e.g. the Rheidol and Ystwyth).

Although, by contrast, few fisheries have ceased to be included in the record because they have been closed, it is relevant to note that in 1985 the commercial salmon fishery on the Wye ceased operation and, as a consequence, the commercial record will hereafter be depressed by up to 2,500 salmon annually.

Changes in methods of collecting data

Although the Salmon and Freshwater Fisheries Act 1923 made it possible for the regional fisheries authorities to make a byelaw requiring 'persons taking salmon and trout to make a return', few appear to have done so and the collection of catches from the rod fishery and the commercial fishery appears to have been based largely either on the 'diary records' maintained by enforcement staff or on a few rivers, such as the Wye, by a return collected and submitted by the private owners of the individual fisheries. It was not until after the <u>Salmon and Freshwater Fisheries Act</u> 1972, which reiterated and expanded the 50-year-old power to make a byelaw requiring the mandatory submission of a return of catch '... in such form, giving such particulars and at such times as may be specified ...' and, importantly, including provision to require the submission of a 'nil' return of catch, that catch returns submitted directly by each individual participants became mandatory in most regions. Thus, over the period 1952-1986 the methods of collecting catch data have changed significantly.

In the early part of the period the record of catches was compiled by enforcement staff on the basis of 'diary records' which incorporated fish which were either 'seen' or 'heard' to have been caught (or both!). In some areas these diary records were incorporated without 'correction' into the 'official' statistics while in other areas they were adjusted by an arbitrary 'weighting factor' depending on whether it was judged to have been a 'good', 'typical' or 'poor' season. It is important to stress that these diary records were not based on 'creel-census' techniques of structured random sampling and that they were inevitably subject to extreme variations in the effort applied to the collection of the data within and between years in any 'accounting' area.

In the latter part of the record there has been a marked shift towards basing the official catch record on mandatory returns submitted directly by each participant (angler or commercial fishermen) to the extent that this now represents the principal method of compiling the record throughout England and Wales - with the notable exception of the Wye where (for good reasons) the record continues to be based on the return submitted by the owners of each individual rod fishery.

While the compilation of catch records from the mandatory returns submitted directly to the RWAs by each licensed instrument (rod or commercial) is now common practice throughout England and Wales, it is relevant to note that this approach is (a) relatively recent in practice, (b) subject to different levels of commitment, and (c) not yet wholly universal in its application.

Of the ten RWAs in England and Wales, nine require the submission of individual returns from each and every participant. In some regions special catch return forms are provided (which may be fairly basic or quite complex in the nature of the information required), but in some regions the return is submitted in a form 'convenient to the respondent'. Some regions do not require the submission of a return from commercial fishermen until after the end of the season; while other regions require a monthly catch return. One region which has no significant rod fisheries for salmon and sea trout but which issues (currently) 113 commercial licences (c. 10 per cent of the total for England and Wales) does not yet require the submission of a catch return from the individual fishermen. Of the nine regions where catch returns are mandatory, some accept the initial response for the purpose of recording the declared catch, while others may issue one, two or three 'reminder notices' to non-respondent fishermen in an attempt to improve the accuracy of data.

The issue of formal reminder notices to non-respondent anglers has been standard procedure in Wales since 1976, but is a more recent innovation in many other regions. The 'official' digest of fishery statistics for 1984 (Russell et al., 1985) shows that whereas some regions report a 100 per cent response rate for anglers, the return in other regions may be as low as 18-23 per cent from anglers in the absence of reminder notices. (The regional response rate from the commerical fishery is not shown).

The issue of reminder notices to recalcitrant anglers has seen a dramatic improvement in the accuracy of the overall catch record for Wales (Table 5.2). The issue of one reminder notice has served to double the initial response rate from 27-33 per cent to 55-66 per cent of all licensed anglers and improved the record of catch for salmon by between 9 per cent (1,015 fish) and 27 per cent (1,673 fish). Although the issue of a second reminder in 1983 added only 5 per cent to the overall response rate (which increased from 61 to 66 per cent as a result) it added a further 505 salmon (8 per cent of the declaration) to the 'official' record. (Data for sea trout are not shown in Table 5.2, but doubling the response rate improved the declared catch by 77 per cent in 1979!)

It is relevant to stress the sensitivity of the record to 'improvements' in the collection of data. The introduction of a catch-reminder system in Wales for anglers in 1976 resulted in the total declared rod catch for England and Wales being increased by roughly 10 per cent in subsequent years! A parallel response by most of the other RWAs in more recent years has, therefore, improved the accuracy of the historical record by a significant - but unknown - extent and these continued improvements in the completeness and accuracy of the record have, clearly, done much to mask any underlying trends over the period.

Accuracy of returns

In addition to the proportion of participants making a return, the accuracy of the recorded catch compiled on the basis of returns submitted by anglers and commercial fishermen depends very much on the honesty, integrity and powers of recall of the participants. It also depends in large part on the understanding and awareness of the participants of the importance of 'good'

Table 5.2: Improvements in the catch return data for salmon obtained from anglers by the issue of postal reminder notices to non-respondents [Welsh Water Authority Area]

Year	Licences issued (rods)	Voluntary return			1st Reminder			2nd Reminder			Total for year		
		%	Salmon Catch	Catch/Return	%	Salmon Catch	Catch/Return	%	Salmon Catch	Catch/Return	%	Salmon Catch	Catch/Return
					(+)	(+)	(+)	(+)	(+)	(+)			
1976	18,790	28	4,109	0.78	22	838	0.20				60	4,947	0.44
1977	21,354	27	9,108	1.58	36	1,262	0.16				63	10,371	0.77
1978	22,402	26	8,807	1.51	32	1,015	0.14				58	9,822	0.76
1979	22,532	27	5,128	0.86	28	1,375	0.22				55	6,503	0.54
1980	22,171	28	8,509	1.48	33	1,163	0.17				61	9,672	0.77
1981	21,497	33	9,726	1.37	27	1,015	0.17				60	10,742	0.83
1982	19,368	28	4,785	0.85	38	1,502	0.20				66	6,287	0.47
1983	16,768	28	4,123	0.88	33	1,673	0.30	5	505	0.60	66	6,301	0.57
1984	14,412	33	2,394	0.60	26	800	0.26				59	3,194	0.45

catch records in relation to the future management and regulation of the fisheries.

Although 95-100 per cent of all commercial fishermen now submit a return after the end of the season, it is widely held that the catches declared by the commercial fishermen are deliberately falsified and grossly under-represent the true catch. The evidence for this is largely circumstantial and anecdotal (Anon, 1974). Nevertheless, the fact that it is a view shared by most fishery managers and their enforcement staff is significant and generates the suspicion that the catch record for the commercial fishery should be viewed with extreme caution - if it is to be used at all. This view has been reinforced by a study undertaken in Wales during 1985 where the number of salmon and sea trout observed to have been taken by selected commercial fishermen was carefully recorded by enforcement staff and then, with equal care, cross-checked against their monthly return of catch. This study, which was admittedly limited in its scope, duration and focus, suggested that while some netsmen did submit a reasonably true return of catch, some others did not and that the true catch for the sample in question was 300-400 per cent greater than that actually declared to the Authority!

The reasons why commercial fishermen should wish to under declare their catches are to some extent, understandable: (1) the commercial fishermen are often in direct competition with each other, (2) the commercial fishery is in competition with the rod fishery, and under increasing pressure from sport fishermen and tourist-related interests for its reduction and ultimate abolition, (c) the wish to evade tax and any other financial liability assessed on income (= catch), and (d) the perceived link between the cost of a licence and the profitability of the fishery.

Other than perhaps for the perceived link with the cost of a rod licence, there appears to be no good reason why anglers should deliberately underdeclare their catches. This may well occur, but it is believed that the majority of anglers submit an accurate catch return - within the limits of recall - at the end of the season. That the catch return is incomplete in some regions because of a variable response rate is, however a serious constraint on the use of the data for management purposes.

Changes in effort

Over the period of the record (Figure 5.2) there have been two significant changes in the overall fishing effort within both the rod fishery and the commercial fishery.

Rod licence sales show that the number of anglers participating in the fisheries increased by roughly 81 per cent over the 20-year period 1952-1972 (Anon, 1974). No strictly comparable data are available since 1972 but it is probable that this trend has continued and that the total rod fishing effort has approximately trebled by the end of the period when compared with the start of the period.

While rod licences sales provide one measure of effort it is important to note also that each individual angler may fish more frequently than hitherto and that spinning and worming are now widely practised on many fisheries where fly-only fishing once prevailed. Thus an appreciation of the level of effort within the rod fishery must take into account changes in the number of participants, the frequency of fishing and the methods employed. It is also important to note that very significant changes can occur at a regional and local level. For example, the increase in rod licence sales over the period has been 100 per cent on the Usk, 400 per cent on the Wye and 1,400 per cent within Wales as a whole.

Throughout most of England and Wales the commercial fishing effort has been relatively stable over the period in terms of the number of participants. Net Limitation Orders have been introduced and periodically reviewed for most significant fisheries and where changes have occurred these have been of minor or very local significance only. There is however one very notable exception in Northumbria where the dramatic increase in the drift net fishery during the 1950s and 1960s inflated the catch statistics for England and Wales by a very significant extent (Figure 5.2). In 1953 the catch declared by the 58 licensed drift nets was 2,006 salmon - or 6 per cent of the commercial catch for England and Wales. By 1970 the number of licensed nets had increased to 218 with a declared catch of 90,587 salmon - or 72 per cent of the commercial catch for England and Wales. The fishery is now subject to a limitation of 121 nets and in 1984 the catch had dropped to 50,685 fish - representing 61 per cent of the total declared commercial catch for England and Wales.

The effect of this expanded fishery and the increase and decrease in licensed fishing effort over the period has grossly distorted the historical record. The fact that 94 per cent of the salmon have been shown to be of Scottish origin (Mills, 1983) further adds to the confusion.

Extraneous factors

Interpretation of the catch record for any year also requires an appreciation of several factors that may affect the catch. Changes in the length of the fishing season, further restrictions on the use of bait, prohibitions on the use of monofilament nets, and increases in fishing licence duties may variously affect the efficiency of fishing, the total fishing effort and, hence, the total catch in various ways at a local or regional level. An awareness of such factors and their impact may be important in translating the record for management purposes.

One very important factor is the weather and water conditions that prevailed during the fishing season. Prolonged low water conditions may favour commercial fishing in some situations and are not conducive to angling success, whereas the frequency of effort may be severely curtailed by inclement weather in some commercial fisheries (e.g. drift net fisheries). Similarly, sustained high river flows during the season may be detrimental to certain types of commercial fishing gear (especially estuarial seine nets) and favour angling success (especially spinning and bait fishing).

England and Wales were affected by an unprecedented drought in 1976. Some regions had virtually no detectable rainfall from February until September. It is not surprising that the salmon catch slumped markedly for both rods and nets during that season (Figure 5.2).

Catch/stock relationships

The widespread assumption that catch data can be used as an indicator of stock abundance has never been critically examined for any river in England and Wales. No information whatsoever on either the numerical abundance of the runs of fish or the rate of exploitation by the commercial and sport fisheries exists for any single river system and, as a consequence, the use of catch data as an index of stock abundance are - notwithstanding the obvious inadequacies of the catch record itself - fraught with unsubstantiated assumptions and, therefore, risks.

That a high level of stock will result in a high catch (and vice versa) is to be accepted, but for practical purposes it is the precise relationship between the level of (available) stock and the catch that is important and it is to be anticipated that catch will depend as much on the prevailing water and weather conditions over the fishing season as on the abundance of the available 'in-season' stock. It will also depend to an appreciable extent on the number of participants (= crude effort) and on the various fishing methods practised within the fishery. The rate of exploitation within the sport fishery is expected to vary enormously in relation to river discharges depending on whether or not spinning and bait fishing are (a) prohibited altogether, (b) permitted without constraint, or (c) allowed only when river conditions are unsuitable for fly fishing. Chadwick (1985) discusses the link between catch and stock for Canadian rivers and concludes that exploitation rates may be more constant among years than not. However he, like others, records that water discharges can 'greatly influence catchability'.

The only good data on rod catches in relation to stock

abundance over a long period of time on any river in the British Isles are from the Salmon Research Trust of Ireland (a fly-only, lake fishery for salmon and sea trout) (Ann. Reps. SRTI, 1970-1983). This has established that, at best, the link between stock abundance and catch is extremely tenuous and of little practical value in a management context. For example, comparing one year with another, doubling the stock resulted in a fourfold increase in catch whereas a 50 per cent reduction in stocks resulted in a 50 per cent increase in catches in selected years.

In England and Wales relatively few salmon are now taken on the fly compared with the situation that prevailed in the period up to the 1950s. A comparison of success for various methods of angling (Anon, 1974) shows that less than 10 per cent of salmon are taken on the fly. Contemporary data for Wales establishes that whereas about 50 per cent of the declared catch of sea trout is taken by fly fishing, almost 90 per cent of the declared catch of salmon is taken by worm fishing or spinning. Gee and Milner (1980) suggest that the rate of angler exploitation on (early-run) salmon heavier than 9.1 kg may approach 100 per cent while later running (grilse) salmon less than 3.2 kg are under-exploited.

The need to give priority to establishing the rate of exploitation on salmon (and sea trout) stocks by rods and nets and the link between stocks and catches is clearly of paramount importance throughout England and Wales as a whole. Until the nature of the relationship has been defined - on as many systems as is practicable - the assumed link between stock and catch cannot - and should - not be used for practical management purposes.

It is to be noted that any link between catch and stock can only apply to the stock entering the fishery during the fishing season. There is some evidence to suggest that on many river systems in England and Wales a major part of the stock now enters the fishery after the end of the fishing season. This component of the total stock is not available for legal exploitation and, despite the fact that it may provide a very significant contribution to regeneration of the resource, it will not be reflected within the commercial or sport fishing catch record. Thus not only is the relationship between catch and 'available' (= in-season) stock to be questioned but also the relationship between catch and total stock has yet to be determined and, moreover, the nature of the relationship (if any) established in the context of its management implications.

5.7   CONCLUSIONS AND COMMENTS

Despite the fact that catch records have been laboriously collected and faithfully published for England and Wales for over 30 years, it is considered that the historical record is so incomplete, inaccurate and inconsistent over the period such that the past and present status of exploitation of the resource cannot be stated with any practical meaning. All that can be said with any confidence is that the catches shown in any year represent but an unknown proportion of the true catch and are to be accepted as minimum values only. It can be said also that the declared catches for the latter part of the record are better minimum values than those presented for the earlier part of the record. However, whether or not the true catch in any year is 50 per cent or 500 per cent higher than that declared is a matter for speculation, grave concern and immediate action!

That other commentators (Chadwick, 1985; Hansen, 1985) have variously expressed concern about the accuracy and value of the historical catch record in other countries provides no comfort. In the absence of other meaningful data on the status and wellbeing of the resource in England and Wales it is inevitable that, for the forseeable future, decisions concerning the management and regulation of the fisheries will continue to be based principally on catch statistics. It is, therefore, of paramount importance that the catch record should be as complete, accurate and consistent as possible within the obvious constraints that exist. That this has not been so in the past is regrettable. That it should be so in the future is imperative!

One of the first requirements must be to validate the accuracy of the various methods used for collecting catch data. Evidence from Wales has shown that while the returns submitted by owners were more accurate than those submitted to the Water Authority by individual anglers, this approach was not practicable for other river systems within the region and that, while diary records maintained by enforcement staff in one area produced a greater reported catch than that submitted by anglers returns, this did not apply in most other areas. Most administrative areas have now adopted direct reporting by individual participants and it is thought that this approach represents the best practical option. There is clearly wide variation in response rate achieved from anglers within and between regions and obvious regional differences in the information required and in the manner in which it is collected. Whether or not a more uniform approach between areas is necessary or otherwise desirable merits further consideration. What is important, however, is that the future record should be presented alongside some proper measure of effort.

It remains to be established if the number of licences issued is in fact a useful index of effort or if a standard catch return form - such as now used in Wales - which requires details of fishing effort and catch to be entered on a daily basis for the commercial fishery and a breakdown of catches with a statement of the number of fishing trips made during the currency of the licence (including those when nothing was caught) for the rod fishery - should be adopted generally.

That the initial response rate to the submission of mandatory catch returns by anglers is so low is cause for concern and there is clearly a need for a concerted effort to educate all participants in the importance of submitting accurate catch returns. In parallel with efforts to improve the accuracy of the catch record, priority should be given to establishing the relationship between stock and catch and to testing the assumption that catches provide an acceptable indicator of the status of the fishery.

Bearing in mind the considerable variation in the nature and extent of exploitation on individual river systems and the differences in the characteristics of the fisheries themselves, it is appropriate that careful thought be given to the adoption of a suitable number of 'index rivers' (Chadwick, 1985) for the purposes of identifying trends in a regional and national context. In this respect the benefits of 'good' information from a small number of carefully selected fisheries rather than 'poor' information from every fishery are apparent and this approach recognises the plain fact that the financial, manpower and other resources available to the regulatory agencies are limited and subjected to many other demands.

Mention has been made of legislative proposals currently before Parliament to combat the threat posed by the dramatic escalation in illegal fishing in recent years. That the Government has recognised that the need is not an increased commitment to the prohibitively costly enforcement of inadequate legislation but better legislation to make the sale of illegally caught fish as difficult as possible is heartening. The principal regret by many is that the widely supported proposals promoted by the RWAs (Anon, 1983) for the introduction of a modified, Canadian-type, tagging scheme as a simple, cheap, effective and practical means of regulating the sale of fish have not received Ministerial support. It is noteworthy in the present context that one of the most significant incidental advantages of such a scheme was that, with certain amendments, it provided an unprecedented opportunity for the collection of catch records which were likely to be more accurate and complete than ever likely to be achieved by present methods.

REFERENCES

Anon (1974) Taking Stock. A Report to the Association of River Authorities. Association of River Authorities March 1974, 48pp.

Anon (1983) Salmon Conservation - A New Approach. A report of the Salmon Sales Group of the National Water Council. National Water Council July 1983, 31pp.

Bielby, G.H. (1983) The role of the Water Authorities in the conservation of salmon and sea trout. European Parliament Committee on Agriculture. Working Group on Fisheries. Working Document PE 85.263, 6pp.

Chadwick, E.M.P. (1985) Fundamental research problems in the management of Atlantic Salmon, Salmo salar L., in Atlantic Canada. In: Williams, R.W. et al. (eds) The Scientific Basis of Inland Fisheries Management. Journal of Fisheries Biology, 27. Suppl. A, 9-25

Gee, A.S. and Milner, N.J. (1980) Analysis of 70-year catch statistics for Atlantic Salmon (Salmo salar) in the River Wye and its implications for management of stocks. Journal of Applied Ecology, 17, 41-57

Grimble, A. (1913) The Salmon Rivers of England and Wales. Kegan, Tench, Trubner, London, 310pp.

Hansen, L.P. (1985) The data on salmon catches available for analysis in Norway. In: D. Jenkins and W.M. Shearer (eds) The Status of the Atlantic Salmon in Scotland ITE Symposium No. 15, Institute of Terrestial Ecology, Abbots Ripton, pp. 79-83

Harris, G.S. (1980) Ecological constraints on future salmon stocks in England and Wales. In: A.E.J. Went (ed.), Atlantic salmon: Its future, Proc. 2nd Int. Atlantic Salmon Symp. Edinburgh 1978, Fishing News Books, Farnham, pp. 82-97

Harris, G.S. (1983) The social and economic importance of salmon and sea trout in England and Wales. European Parliament. Committee on Agriculture. Working Group on Fisheries. Working Document PE 85.418, 8pp.

Harris, G.S. (ed.) (1985) Report of the working group on Welsh salmon and sea trout fisheries. Welsh Water Authority, October 1985, 154pp.

Jenkins, J.G. (1974) Nets and coracles. David and Charles (Holdings), London, 335pp.

Mills, D.H. (1983) Problems and solutions in the management of open seas fisheries. Atlantic Salmon Trust, Farnham, 23pp.

Netboy, A. (1968) The Atlantic Salmon: a vanishing species. Faber and Faber, London, 377pp.

Okun, D.A. (1977) Regionalisation of water management. A revolution in England and Wales. Applied Science Publishers, London, 377pp.

Radford, A.F. (1980) Economic survey of the River Wye recreational salmon fishery. Portsmouth Polytechnic Marine Resources Research Unit. Research Paper Series No. 20, 94pp.

Russell, I.C., Potter, E.C.E. and Duckett, L. (1985) Salmon and migratory trout fisheries statistics for England and Wales, 1984. Ministry of Agriculture, Fisheries and Food. Directorate of Fisheries Research. Fisheries Research Data Report No. 7, 16pp.

Chapter Six

STATUS OF EXPLOITATION OF ATLANTIC SALMON IN SCOTLAND

R.B. Williamson
Inspector of Salmon and Freshwater Fisheries for Scotland

6.1 INTRODUCTION

Salmon have been exploited in Scotland since the earliest times and it is clear from Boece's History of Scotland, 1527, that they were considered an important resource and that local fishermen had a good working knowledge of the biology of the fish long before the scientific discoveries of later centuries (Neill, 1946). The many salmon laws passed in the twelth to seventeenth centuries also give some indication of the value of the fish. In economic terms, exploitation of salmon may be less important now than then but the increased recreational use of the resource offsets the decline in the relative importance of the annual production of fish.

This paper on the status of exploitation of salmon in Scotland includes descriptions of the ownership of the fishing rights, the permitted methods of fishing and other controls on exploitation, as well as commenting on the catches made over the last 30 years.

6.2 DISTRIBUTION OF SALMON

Atlantic salmon occur in all the river systems in Scotland except some small streams that run direct to the sea. Within these systems the distribution is, in some cases, limited by natural or artificial obstructions or, in the industrial central belt, by pollution. The effect of direct industrial pollution on the distribution of salmon is however decreasing. The return of salmon in small but significant numbers to the River Clyde in recent years is just an example of what is also happening in smaller and less well known previously polluted waters.

Information on the distribution of salmon in Scottish rivers has recently been collected and published on a map at a scale of

1:625000 (Gardiner and Egglishaw, 1986). During the survey it was found that the great majority of streams over 10 m wide are free of obstructions such as waterfalls and contain salmon but that there is scope for artificial stocking in many streams less than 10 m wide where impassable waterfalls are commoner (Anon, 1986).

In the larger east coast rivers salmon come in from the sea in every month of the year but in smaller rivers, and especially on the west coast, runs may be limited to the summer and autumn.

6.3   OWNERSHIP OF FISHING RIGHTS

In Scotland wild salmon do not belong to any individual person until they are caught but there is no public right of fishing for such salmon. The fishing right at any place, whether in a river or on the sea coast, belongs in each case to some person who has the exclusive right to fish for salmon at that site. These rights are heritable property and can be bought, sold or leased independently of the adjacent land. There are exceptions in Orkney and Shetland where some relics of Norse udal law persist and salmon rights may go with the land. The right to fish for salmon includes the right to fish for sea trout which are for all legal purposes treated as if they were salmon.

Salmon fishing rights, whether by rod or by net, are thus owned by a wide variety of corporations, public companies, family businesses, clubs, syndicates and private individuals. These owners may run their fisheries themselves, employing such persons as necessary, or may let them out to others. The Crown Estate, which under feudal law originally held all salmon fishings, still owns a significant amount on some parts of the Scottish coast (about 30 per cent of the coastal fishings in total, Hunt (1978)) and lets these to tenant fishermen on standard 9-year leases.

The fact that the salmon fishings in rivers and on the coast are a private heritable right, and the way in which ownership is distributed, has a significant effect on the way in which exploitation has developed and is controlled.

6.4   ADMINISTRATION

Each salmon fishery is directly administered by its owner and/or occupier under laws made by central government to ensure the maintenance of the resource as a whole for general benefit. The country is divided into 108 salmon fishery districts on a river catchment basis and the law provides that the owners in any district may set up a local Salmon Fishery Board for the protection and development of salmon fisheries in their district. The Boards may appoint officers to enforce the salmon fishery law and some also operate hatcheries for re-stocking purposes. They are self-financing and have a power to levy a rate on the proprietors in their districts.

6.5   METHODS OF FISHING

Scottish salmon fisheries law distinguishes between the methods allowed in rivers (including their estuaries), on the sea coast and in the open sea.

Rivers and estuaries

Since earliest times, salmon fishing in rivers has, for conservation reasons, been limited to the use of rod-and-line and a restricted form of beach seining called net-and-coble. In net-and-coble fishing the net is not allowed to lie stationary relative to either the water or the land but must be both shot and hauled speedily to surround the fish; the net must not be operated as a static trap or be used as a trawl or to obstruct the movement of salmon (except momentarily as it is shot). In theory, salmon nets can be used well upstream in rivers and in inland lochs, and in former times they were. Now, owners find rod fishing to be more appropriate and/or profitable and, in practice, netting is restricted to the lower reaches of rivers, mostly in the tidal waters of the estuary.

Salmon cruives, consisting of a weir with inset box-traps, are no longer used in Scotland although rights to operate such devices still exist at a few places.

Sea coast

On the coast all known methods of fishing for salmon are prohibited except the use of fixed traps called bag-nets and stake-nets, usually set within 200 m of the shore, or beach seine nets. There are some places where salmon may be caught by rod-and-line in the sea but that method, though lawful, is not much used. A stake-net is shown in Figure 6.1. The bag-net is of basically similar shape but is designed to be set floating at the surface over deeper water and is often used off rocky shores.

Open sea

Fishing for salmon in the open sea is a fairly recent development. In Scotland it started with the development of a significant drift net fishery on the east coast in the early 1960s (see Table 6.3).

Figure 6.1: A salmon stake net

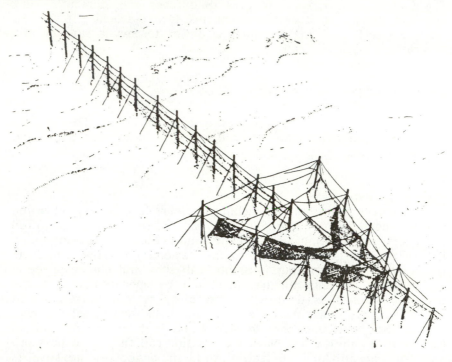

This was considered undesirable on conservation and fishery management grounds and drift netting for salmon was prohibited at the end of 1962. Since then all other methods thought of as being possible for fishing for salmon in the open sea, i.e. other gill nets, trawl nets, seine nets, trolling or use of long-lines, have been prohibited.

6.6   CLOSE TIMES FOR FISHING

There has been a weekly close time for salmon fishing in Scotland since at least the twelfth or early thirteenth century and an annual close time since 1424. The present weekly close time for fishing with nets is from noon on Saturday to 6 am on the following Monday. The annual close time is fixed at 168 days but the starting date varies from district to district depending partly on the timing of different runs of salmon. In general the close time runs from late August to mid-February but in some rivers, e.g. the Tweed, net fishing continues to mid-September. Angling is allowed to continue after the nets are off and, in some districts, to start before they come on. The extra time for angling varies from district to district and extends the season to the end of September or later. The weekly close time for angling is Sunday.

It is proposed, in a Bill at present before Parliament, that both the weekly and annual close times may be increased by the Secretary of State for Scotland after consultation with interested parties.

6.7   FISHING EFFORT

Amount of effort
The most recent analysis of numbers of persons employed directly in the salmon net fishery, i.e. in actually operating the nets, is based on information submitted with the annual catch returns for 1984 and it indicates that between March and August inclusive the average number of men employed monthly was 744. Between May and August, the most productive part of the season, the average number employed was 962 and in the peak month of July it was 1,144 (Hansard, 1986). The number of net-and-coble crews peaked at 186 in July and averaged 164 over the period June-August; the number of pockets (traps) operated in bag nets and stake nets during the same year averaged 1,721 between May and August, August being the peak month with 1,943 (see Table 6.1).

A much larger, but unknown, number of persons fish for salmon by rod-and-line but, by the nature of things, most of them each for a very much shorter time. An attempt was made in 1982 to make an estimate of the number of rod-days of fishing let by salmon fishery proprietors during the previous year. This was in connection with a study on the economic value of sport salmon fishing (TRRU, 1984) and produced an estimate of 373,703 rod-days for Scotland as a whole, but the estimate is subject to a number of qualifications and is of limited value. It does, however, appear that the intensity of angling (measured by the number of rod-days fished) has increased over the past decade and longer. The letting of previously private beats and the letting of fishing by the day rather than by the week (or in earlier times months) has contributed to this increase as has the development of time-share ownership of salmon rod fishings whereby an individual acquires an exclusive right, in perpetuity, to fish a particular salmon beat for one named week each year.

Control of fishing effort
Control of fishing effort in Scotland is indirect, through limitation of the permitted methods of fishing. Because the right to fish at any point is privately owned and belongs to the owner to the exclusion of others, control of method does give rise to a

Table 6.1: Number of crews and men employed at .the method 'net and coble' and the number of traps in operation and men employed at the method 'fixed engines' in 1984

Month	Net and Coble				Fixed Engines			
	No. of Crews		No. of men empoyed		No. of traps		No. of men employed	
	Max.	Min.	Max.	Min.	Max.	Min.	Max.	Min.
February	20	20	100	96	88	73	94	87
March	27	23	129	106	297	219	142	119
April	37	34	165	146	602	433	231	216
May	75	61	267	215	1,397	1,160	348	326
June	141	115	484	385	1,662	1,410	475	442
July	186	165	634	538	1,881	1,648	510	484
August	165	147	591	509	1,943	1,648	479	444
September	31	28	128	109	500	471	122	119

partially self-regulating limit on the amount of effort. The methods permitted (see Section 6.5) are inherently inefficient by modern standards, requiring significant labour and expensive equipment. As new more efficient methods developed they have been successively declared unlawful either by the courts, which maintain a restrictive definition of what is meant by net-and-coble, or by statutory regulation, for example the prohibition of the use of drift nets and other gill nets for salmon fishing. Because a fishermen has the exclusive right to fish his part of the shore or river he can decide how many nets to operate without fear of competition within that area. Given that a unit of netting effort will tend to catch fish in proportion to the abundance of the stock during the period fished, and given that each individual net is relatively expensive to operate and to some extent competes with others in the same area, there is an economic force which gives rise to a decrease in the number of nets operating at times of low abundance and an increase in periods of high abundance. An example of this effect is shown by the decrease in the number of coastal nets operated in the spring months during recent years (see Table 6.2).

Table 6.2: Number of traps (pockets) in bag nets and stake nets reported operating in the spring months (February to April) in the years 1978 to 1985 (figures missing for year 1982)

	1978	1979	1980	1981	1982	1983	1984	1985
February	281	197	386	313	–	97	88	82
March	908	1,062	1,071	727	–	507	297	147
April	1,216	1,654	1,538	880	–	927	602	463

Because salmon fisheries are saleable, salmon netsmen or companies can acquire fishing rights over a sufficiently large area to enable them to operate on an efficient basis. A good historical example of that is the arrangement whereby the various salmon net fisheries in the lower reaches of the River Tay were brought under single management at the turn of the century. This resulted in a decrease in the netting effort but, according to contemporary accounts, an increase in the value of both the net fisheries and the upstream rod fishings (Elgin Report, 1902).

The system of ownership also allows rod-fishing interests to buy out netting stations and limit their operation for the purpose of benefiting the rod fishings upstream as was done on the River

Dee between Banchory and Aberdeen in the years between 1870 and 1910 (Calderwood, 1921). More recently, an Atlantic Salmon Conservation Trust has been set up in Scotland specifically to purchase coastal netting stations for the benefit of salmon management on a river-by-river basis. On the River Spey there has been an arrangement for the last 2 years whereby the net-and-coble effort is reduced in certain circumstances and annual payments are made to the netting proprietors by the upstream interests.

The system is of course not claimed to be perfect as a means of fine control but it does have significant advantages over direct control of fishing effort in home-water fisheries where the size of the exploitable stock cannot be predicted from year to year; and it can continue to work provided that exploitation outside home waters is also controlled.

6.8 RESTOCKING

The salmon fishery is based on naturally regenerating stocks but there is some enhancement of those stocks by the planting of eggs and fry in areas not accessible to adult salmon or, more commonly, where the natural stock is thought (not always correctly) to be less than the nursery area can support. In a recent survey (R. Gardiner, personal communication) it was estimated that some 13.7 million salmon eggs and fry were used for re-stocking in 1983. There is also an increasing amount of stocking with parr and smolts, partly as a result of the development of commercial salmon farming and the consequent availability of surplus parr from smolt rearing units geared to producing one-year-old smolts. The survey resulted in estimates of at least 202,500 parr and 64,000 smolts used in 1983.

6.9 SALMON CATCHES

General
Comprehensive salmon catch statistics for rods and nets have been collected by the Department of Agriculture and Fisheries for Scotland since 1952. Summaries are now published annually (Anon, 1984a, 1985 a,b) and a compendium for the years 1952-1981 has also been published in two volumes (Anon, 1983, 1984b). The all-Scotland catch by each of the three methods and in total, for the years 1952-1984 are given in Tables 6.4 - 6.7. These statistics do not include the drift-net catches of 1960-1962 because those were made by sea fishermen outside the ambit of the salmons' statistics scheme. Estimates for that fishery are given in the first Hunter Committee Report (Anon, 1963) and are summarised in Table 6.3.

Table 6.3: Estimated catch of salmon and grilse in the Scottish salmon drift-net fishery from its inception in 1960 until it was prohibited at end of 1962

Area	Estimated catch (numbers)		
	(1960)	(1961)	(1962)
Tweed	9,000	14,000	30,000
Arbroath/Montrose	-	12,000	37,000
Aberdeen/Stonehaven	-	)	25,000
		)About	
Peterhead/Fraserburgh	-	)2,000	18,000
		)	
Moray Firth	-	)	5,000
Total	9,000	28,000	115,000

Consideration of the value of salmon catch statistics and the significance of relationships between the size of catch and stock are the subject of other papers at this symposium. Catch statistics are used here as an indication of the status of exploitation and, in that context, changes in reported catch are of interest in themselves independently of their relationship to the size or state of the stock from which they are taken.

Whole-Scotland catch

Nets
Tables 6.5 and 6.6 show that catches by the two netting methods, net-and-coble (used mainly in estuaries) and fixed engines (used on the open coast) have fluctuated together over the whole period and have more-or-less equal shares of the total catch. On a whole-country basis they can therefore be added and considered as a single net catch. This is shown in Figure 6.2. The data show no continuous trend over the 34 years and can best be described as a period of high catches during 1962-1975 flanked by two periods of lower catches 1952-1961 and 1976-1985. The average catch during the second of these periods is significantly less than that of the first.

Table 6.4: Number and weight of salmon and grilse caught by rod-and-line in Scotland each year from 1952 to 1984

	Numbers					Weight (lb)				
	Salmon			Grilse	Salmon +Grilse	Salmon			Grilse	Salmon +Grilse
	Jan-Apr	May-Dec	Annual			Jan-Apr	May-Dec	Annual		
1952	12,483	22,481	34,964	6,133	41,097	133,519	230,932	364,451	30,268	394,719
1953	16,219	27,501	43,720	6,383	50,103	160,056	255,753	415,809	32,521	448,330
1954	19,860	34,429	54,289	4,872	59,161	197,480	336,048	533,528	26,356	559,884
1955	17,039	29,453	46,492	4,147	50,639	171,977	291,748	463,625	21,057	484,682
1956	15,556	32,907	48,463	7,686	56,149	160,298	323,785	484,083	39,779	523,862
1957	26,539	37,922	64,461	9,645	74,106	248,498	342,366	590,864	49,814	640,678
1958	20,881	40,606	61,487	9,462	70,949	204,618	380,222	564,840	47,359	632,199
1959	20,975	26,460	47,435	3,605	51,040	203,119	255,092	458,211	17,219	475,430
1960	23,567	31,319	54,886	6,458	61,344	226,241	294,652	520,893	32,805	553,698
1961	18,556	29,667	48,223	6,720	54,943	178,291	274,212	452,503	33,819	486,322
1962	14,645	46,882	61,527	10,602	72,129	157,018	451,432	608,450	57,895	666,335
1963	23,068	49,099	72,167	9,239	81,406	226,638	484,432	711,070	48,523	759,593
1964	16,809	52,143	68,952	13,005	81,957	157,579	469,496	627,075	68,209	695,284
1965	19,080	49,456	68,536	9,693	78,229	196,746	455,003	651,749	52,026	703,775
1966	16,374	46,771	63,145	8,719	71,864	166,086	448,440	614,526	44,425	658,951
1967	15,278	47,699	62,977	14,819	77,796	165,577	445,961	611,538	80,837	692,375
1968	14,215	31,563	45,778	6,838	52,616	135,085	287,368	422,453	36,144	458,597
1969	10,147	33,481	43,628	8,441	52,069	103,709	307,287	410,996	47,459	458,455
1970	11,724	37,353	49,077	13,752	62,829	114,408	329,442	443,850	74,038	517,888

Table 6.4 (Cont'd)

Year	Numbers					Weight (lb)				
	Salmon Jan-Apr	May-Dec	Annual	Grilse	Salmon +Grilse	Salmon Jan-Apr	May-Dec	Annual	Grilse	Salmon +Grilse
1971	10,233	31,214	41,447	8,078	49,525	96,461	272,705	369,166	44,442	413,608
1972	14,301	38,594	52,895	6,751	59,646	140,249	358,056	498,305	36,693	534,998
1973	10,492	48,042	58,534	7,772	66,306	116,941	475,797	592,738	44,981	637,719
1974	8,363	41,981	50,344	9,576	59,920	90,681	395,748	486,429	53,721	540,149
1975	14,885	46,136	61,021	7,231	68,252	159,807	459,944	619,751	41,798	661,549
1976	9,370	31,485	40,855	7,819	48,674	96,880	286,572	383,452	43,086	426,538
1977	11,674	44,109	55,783	8,784	64,567	109,779	412,325	522,104	48,414	570,518
1978	19,876	46,250	66,126	11,270	77,396	205,757	450,226	655,983	62,701	718,684
1979	13,756	52,165	65,921	15,244	81,165	132,490	479,355	611,845	83,062	694,907
1980	18,170	41,167	59,337	11,167	70,504	181,127	412,083	593,210	55,859	649,069
1981	9,257	41,749	51,006	12,287	63,293	96,030	411,690	507,720	67,170	574,890
1982	7,157	40,758	47,915	16,839	64,754	75,458	384,349	459,807	86,097	545,904
1983	7,528	44,433	51,961	14,109	66,070	72,637	430,517	503,154	76,843	579,997
1984	4,054	34,076	43,130	15,543	58,673	88,110	321,711	409,821	79,215	489,036

Table 6.5: Number and weight of salmon and grilse caught by net-and-coble in Scotland each year from 1952 to 1984

	Numbers					Weight (lb)				
	Salmon			Grilse	Salmon +Grilse	Salmon			Grilse	Salmon +Grilse
	Jan-Apr	May-Dec	Annual			Jan-Apr	May-Dec	Annual		
1952	40,217	53,116	93,333	61,025	154,358	404,418	686,407	1,090,830	298,084	1,388,910
1953	30,510	42,889	73,399	51,498	124,897	292,889	509,203	802,092	281,561	1,083,650
1954	59,998	50,925	110,923	45,567	156,490	564,809	598,265	1,163,070	254,103	1,417,180
1955	36,043	71,818	107,861	51,528	159,389	337,609	844,190	1,181,800	256,811	1,438,610
1956	27,658	44,337	71,995	45,959	117,954	270,085	500,373	770,458	234,274	1,004,730
1957	32,027	53,752	85,779	78,089	163,868	275,546	570,468	846,014	403,149	1,249,160
1958	30,886	56,125	87,011	74,857	161,868	282,859	622,229	905,088	386,419	1,291,518
1959	55,885	64,118	120,003	49,690	169,693	489,845	746,602	1,236,450	269,023	1,505,470
1960	27,120	59,945	87,065	90,198	177,263	249,528	706,136	955,664	493,179	1,448,840
1961	23,562	46,877	70,439	70,141	140,580	213,847	523,578	737,425	374,111	1,111,540
1962	15,754	72,190	87,944	140,985	228,929	158,902	803,995	962,897	817,364	1,780,260
1963	38,499	70,103	108,602	62,554	171,156	361,836	846,455	1,208,290	349,307	1,557,600
1964	20,333	78,880	99,213	117,252	216,465	186,068	861,849	1,047,920	653,931	1,701,050
1965	22,491	61,627	84,118	97,324	181,442	203,871	684,817	888,688	563,054	1,451,740
1966	21,650	64,237	85,887	95,430	181,317	205,140	721,808	926,948	545,227	1,472,180
1967	20,528	99,296	119,824	155,889	275,713	191,159	1,080,728	1,271,880	906,593	2,178,470
1968	19,234	78,474	97,708	94,810	192,518	174,072	904,214	1,078,290	542,401	1,620,690
1969	10,944	97,667	108,611	169,557	278,168	111,073	1,065,270	1,176,350	1,039,600	2,215,950
1970	12,875	64,441	77,316	110,164	187,480	117,676	707,263	824,939	640,610	1,465,550

Table 6.5 (Cont'd)

	Numbers					Weight (lb)				
	Salmon			Grilse	Salmon +Grilse	Salmon			Grilse	Salmon +Grilse
	Jan-Apr	May-Dec	Annual			Jan-Apr	May-Dec	Annual		
1971	10,636	57,276	67,912	117,839	185,751	94,858	598,352	693,210	702,870	1,396,080
1972	11,374	88,539	99,913	101,003	200,916	111,007	951,100	1,062,110	635,116	1,697,220
1973	13,335	89,914	103,249	109,438	212,687	143,456	1,004,470	1,147,930	692,950	1,840,880
1974	8,301	70,543	78,844	120,933	199,777	92,399	782,525	874,924	734,556	1,609,480
1975	15,435	71,124	86,559	94,463	181,022	164,306	798,761	963,067	587,345	1,550,410
1976	8,105	34,489	42,594	71,401	113,995	83,521	364,892	448,413	418,314	866,727
1977	7,030	38,325	45,355	82,373	127,728	65,735	418,204	483,939	485,102	969,041
1978	12,388	41,243	53,631	81,854	135,485	128,697	472,502	601,199	481,994	1,083,190
1979	6,003	30,471	36,474	80,946	117,420	54,674	341,123	395,797	456,616	852,413
1980	15,223	43,895	59,118	47,157	106,275	144,655	534,489	679,144	242,509	921,653
1981	11,599	48,808	60,407	58,549	118,956	116,957	546,487	663,444	330,991	994,435
1982	6,812	33,758	40,570	76,023	116,593	70,872	361,724	432,596	403,298	835,894
1983	4,738	49,537	54,275	82,029	136,304	42,098	513,829	555,927	471,358	1,027,290
1984	2,382	30,624	33,006	68,412	101,418	21,613	346,682	368,295	364,998	733,293

Table 6.6: Number and weight of salmon and grilse caught by fixed-engine in Scotland each year from 1952 to 1984

| | Numbers | | | | | Weight (lb) | | | | |
| | Salmon | | | Grilse | Salmon +Grilse | Salmon | | | Grilse | Salmon +Grilse |
	Jan-Apr	May-Dec	Annual			Jan-Apr	May-Dec	Annual		
1952	45,895	60,662	106,557	83,203	189,760	443,468	702,412	1,145,880	396,998	1,542,880
1953	35,849	58,244	94,093	83,140	177,233	335,336	641,589	976,925	441,343	1,418,270
1954	34,495	55,228	89,723	65,775	155,498	342,618	603,834	946,452	352,917	1,299,370
1955	30,295	68,969	99,264	79,465	178,729	295,595	724,531	1,020,130	396,055	1,416,180
1956	19,882	59,310	79,192	62,165	141,357	196,150	618,745	814,895	304,417	1,119,310
1957	16,091	53,294	69,385	107,500	176,885	144,931	540,746	685,677	547,597	1,233,270
1958	13,302	62,222	75,524	115,206	190,730	135,001	679,003	814,004	585,463	1,399,470
1959	32,295	69,468	101,763	61,763	163,526	291,713	750,397	1,042,110	313,330	1,355,448
1960	12,410	50,109	62,519	96,863	159,382	117,749	545,122	662,871	515,091	1,177,960
1961	14,754	45,147	59,901	78,712	138,613	135,380	463,248	598,628	416,013	1,014,640
1962	10,548	52,555	63,103	128,773	191,876	106,990	557,492	664,482	720,170	1,384,650
1963	29,963	60,991	90,954	97,467	188,421	286,679	678,134	964,813	520,721	1,485,530
1964	16,242	83,436	99,678	152,425	252,103	146,381	846,808	993,189	813,234	1,806,420
1965	16,211	56,690	72,901	106,144	179,045	156,117	602,718	758,835	596,605	1,355,440
1966	12,170	62,534	74,704	110,118	184,822	119,878	651,762	771,640	613,082	1,384,720
1967	5,748	72,978	78,726	172,455	251,181	57,416	776,743	834,159	961,508	1,795,678
1968	9,757	61,558	71,315	117,633	188,948	92,559	658,063	750,622	648,958	1,399,580
1969	5,004	54,762	59,766	172,255	232,021	50,387	560,266	610,653	1,024,610	1,635,270
1970	5,458	36,590	42,048	116,679	158,727	51,178	387,617	438,795	645,649	1,084,440

Table 6.6 (Cont'd)

	Numbers					Weight (lb)				
	Salmon			Grilse	Salmon +Grilse	Salmon			Grilse	Salmon +Grilse
	Jan–Apr	May–Dec	Annual			Jan–Apr	May–Dec	Annual		
1971	4,554	47,688	52,242	136,243	188,485	41,039	480,660	521,699	801,181	1,322,880
1972	6,060	59,155	65,215	143,711	208,926	59,091	612,677	671,768	902,576	1,574,340
1973	9,839	66,298	76,137	175,880	252,017	101,606	709,736	811,342	1,131,960	1,943,300
1974	6,398	52,771	59,169	158,907	218,076	68,252	580,495	648,747	967,335	1,616,080
1975	12,621	47,777	60,398	120,670	181,068	132,022	504,257	636,279	724,546	1,360,830
1976	4,460	26,673	31,133	109,272	140,405	45,521	272,995	318,516	634,753	953,269
1977	3,950	33,899	37,849	103,107	140,956	37,641	364,409	402,050	615,735	1,017,790
1978	11,403	31,794	43,197	111,346	154,543	114,977	348,963	463,940	650,888	1,114,830
1979	5,444	24,670	30,114	91,046	121,160	48,940	260,667	309,607	514,969	824,576
1980	9,631	44,502	54,133	63,117	117,250	93,385	510,715	604,100	325,729	929,829
1981	8,287	55,021	63,308	79,902	143,210	80,708	608,425	689,133	459,788	1,148,920
1982	3,340	36,417	39,757	115,199	154,956	30,620	390,105	420,725	604,803	1,025,530
1983	1,523	38,202	39,725	113,479	153,204	13,620	407,842	421,462	662,989	1,084,450
1984	1,342	29,735	31,077	129,124	160,201	12,598	320,313	332,911	676,927	1,009,838

Table 6.7: Number and weight of salmon and grilse caught by all methods in Scotland each year from 1952 to 1984

| | Numbers | | | | | Weight (lb) | | | | |
Year	Salmon Jan–Apr	Salmon May–Dec	Annual	Grilse	Salmon +Grilse	Salmon Jan–Apr	Salmon May–Dec	Annual	Grilse	Salmon +Grilse
1952	98,595	136,259	234,854	150,361	385,215	981,405	1,619,750	2,601,160	725,350	3,326,518
1953	82,578	128,634	211,212	141,021	352,233	788,281	1,406,550	2,194,838	755,425	2,950,250
1954	114,353	140,582	254,935	116,214	371,149	1,104,910	1,538,150	2,643,050	633,376	3,276,430
1955	83,377	170,240	253,617	135,140	388,757	805,081	1,860,470	2,665,550	673,923	3,339,470
1956	63,096	136,554	199,650	115,810	315,460	626,533	1,442,900	2,069,440	578,470	2,647,910
1957	74,657	144,968	219,625	195,234	414,859	668,975	1,453,580	2,122,560	1,000,560	3,123,120
1958	65,069	158,953	224,022	199,525	423,547	622,478	1,681,450	2,303,930	1,019,240	3,323,170
1959	109,155	160,046	269,201	115,058	384,259	984,677	1,752,090	2,736,770	599,572	3,336,340
1960	63,097	141,373	204,470	193,519	397,989	593,518	1,545,910	2,139,430	1,041,080	3,180,500
1961	56,872	121,691	178,563	155,573	334,136	527,518	1,261,040	1,788,560	823,943	2,612,500
1962	40,947	171,627	212,574	280,360	492,934	422,910	1,812,920	2,235,830	1,595,420	3,831,250
1963	91,530	180,193	271,723	169,260	440,983	875,153	2,009,020	2,884,170	918,551	3,802,730
1964	53,384	214,459	267,843	282,682	550,525	490,028	2,178,150	2,668,180	1,535,370	4,203,560
1965	57,782	167,773	225,555	213,161	438,716	556,734	1,742,540	2,299,270	1,211,690	3,510,960
1966	50,194	173,542	223,736	214,267	438,003	491,104	1,822,010	2,313,110	1,202,730	3,515,850
1967	41,554	219,973	261,527	343,163	604,690	414,152	2,303,420	2,717,570	1,948,940	4,666,510
1968	43,206	171,595	214,801	219,281	434,082	401,716	1,849,650	2,251,360	1,227,500	3,478,860
1969	26,095	185,910	212,005	350,253	562,258	265,169	1,932,830	2,198,000	2,111,670	4,309,670
1970	30,057	138,384	168,441	240,595	409,036	283,262	1,424,320	1,707,500	1,360,300	3,067,880

Table 6.7 (Cont'd)

	Numbers					Weight (lb)				
	Salmon Jan-Apr	Salmon May-Dec	Annual	Grilse	Salmon +Grilse	Salmon Jan-Apr	Salmon May-Dec	Annual	Grilse	Salmon +Grilse
1971	25,423	136,178	161,601	262,160	423,761	232,358	1,351,720	1,584,080	1,548,490	3,132,570
1972	31,735	186,288	218,023	251,465	469,488	310,347	1,921,830	2,232,180	1,574,390	3,806,570
1973	33,666	204,254	237,920	293,090	531,010	362,003	2,190,010	2,552,010	1,869,890	4,421,900
1974	23,062	165,295	188,357	289,416	477,773	251,332	1,758,770	2,010,100	1,755,610	3,765,710
1975	42,941	165,037	207,978	222,364	430,342	456,135	1,762,960	2,219,100	1,353,690	3,572,790
1976	21,935	92,647	114,582	188,492	303,074	225,922	924,459	1,150,380	1,096,150	2,246,530
1977	22,654	116,333	138,987	194,264	333,251	213,155	1,194,940	1,408,090	1,149,250	2,557,340
1978	43,667	119,287	162,954	204,470	367,424	449,431	1,271,690	1,721,120	1,195,580	2,916,710
1979	25,203	107,306	132,509	187,236	319,745	236,104	1,081,150	1,317,250	1,054,650	2,371,900
1980	43,024	129,564	172,588	121,441	294,029	419,167	1,457,290	1,876,450	624,097	2,500,550
1981	29,143	145,578	174,721	150,738	325,459	293,695	1,566,600	1,860,300	857,949	2,718,250
1982	17,309	110,933	128,242	208,061	336,303	176,958	1,136,180	1,313,130	1,094,200	2,407,330
1983	13,789	132,172	145,961	209,617	355,578	128,355	1,352,190	1,480,540	1,211,190	2,691,730
1984	12,778	94,435	107,213	213,079	320,292	122,321	988,706	1,111,027	1,121,140	2,232,167

Figure 6.2: Number of salmon and grilse caught by net (net-and-coble and fixed-engine), 1951-1985

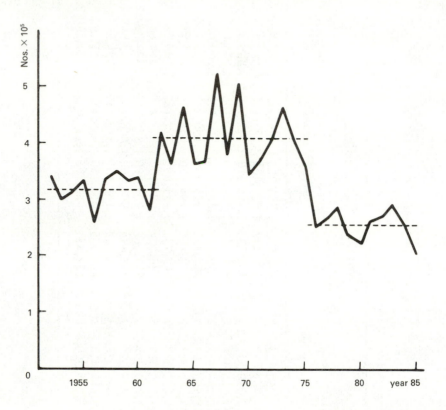

The distribution of catch within the season has also changed with time. The decline of catch of spring fish (taken as those fish caught during the period January-April) is most marked (see Figure 6.3). The decline in abundance of spring fish has led to a reduction in the number of nets that are operated early in the season (see Table 6.2) and that has contributed further to the decrease in reported landings.

Rod and line
Rod catches have held much steadier and do not closely reflect the fluctuation in net catch. Indeed it is notable that the rod

Figure 6.3: Number of salmon caught by net (broken line) and rod (solid line) during the spring (January-April) each year 1952-1985

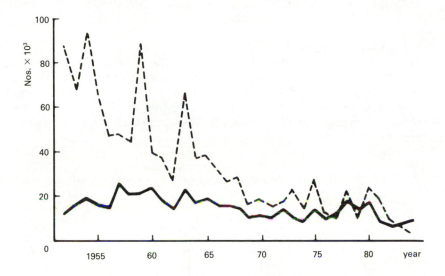

Figure 6.4: Number of salmon and grilse caught by rod-and-line each year, 1952-1985

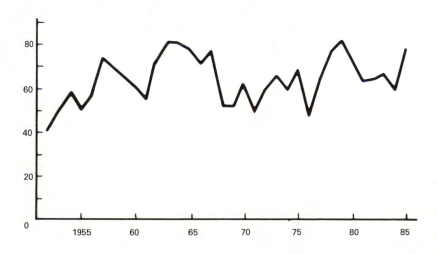

catch has been above average for most of the last 10 years (see Figure 6.4). However, as with the nets, the numbers of fish taken during the spring has dropped, though not by so much (Figure 6.3) and this has an effect on the value of the salmon fishing.

Salmon/Grilse ratio.
The ratio of the number of salmon to the number of grilse in the reported catch has changed in the past 30 years. In the early years salmon were dominant but since 1967 it has been the grilse (Figure 6.5). However, with the exception of 1984, the statistics show salmon as still providing the greatest of the catch by weight (Table 6.7).

Figure 6.5: Number of salmon and grilse caught by all methods each year, 1952-1985

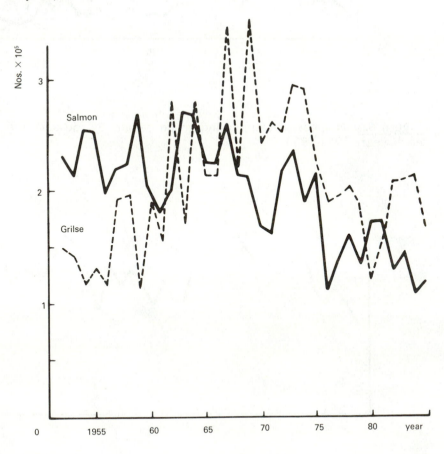

Determination of the salmon/grilse ratio is subject to a variable error due to changes in the average weight of grilse from year to year and the fact that those reporting the catches distinguish salmon from grilse on the basis of weight rather than on years spent in the sea. In years when average grilse weights are high many more than usual are misreported as salmon. An attempt was made to correct this error for the years 1962-1976 when grilse were numerous and the fish were large, and this showed that the proportion of grilse in those years was significantly higher than the published statistics show (Mitchell, 1978).

Changes in the grilse to salmon ratio have occurred before in Scotland; George (1982) describes many fisheries where catches of spring salmon were low and those of grilse high during the early years of this century before an increase in the spring run during the 1930s. The cycle, if it is a cycle, is however of sufficiently long wave-length that a present decline is not always viewed in the context of previous increases.

Nets/rods catch ratio

Catch by rod is a relatively small part of the total but has been a significantly greater proportion in the years since 1977, averaging about 21 per cent and being consistently at or above the maximum of the years before 1977 (Figure 6.6). This change results mainly from the decrease in the size of the net catch since 1975 but is compounded by above average rod catches in recent years.

The same trend is seen in both the grilse and the salmon catch but the number of rod-caught grilse is a very small part of the total grilse catch: only 2-5 per cent in the years 1952-1978 and between 6 and 11 per cent since then. There is significant misreporting of grilse as salmon (see above) and probably to a greater extent amongst anglers than netsmen but it is not suggested this is the whole explanation for the apparently lower catchability of grilse.

Regional differences

The status of exploitation of salmon varies from region to region, and within region from district to district, both in the fishing methods used and in changes in the size of catch taken.

For example, in the East Region more than 90 per cent of the net catch is taken by net-and-coble whereas in the adjacent North East Region less than 30 per cent is taken by that method. In the Moray Firth Region the rod fishermen have taken more than 30 per cent of the catch in several recent years but in the North East Region only 15 per cent or less. There is not space in

Figure 6.6: Number of salmon and grilse caught by rod-and-line expressed as a percentage of the total catch each year 1952-1985

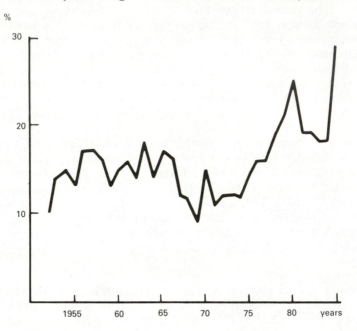

this paper to describe all the regional differences but, as an illustration, Figure 6.7 shows the differences in total catch in four regions over the period 1952-1984. The catch taken in each region by each method for each year is given by the Department of Agriculture and Fisheries for Scotland (Anon, 1985b) and will be updated in the Statistical Bulletin giving the 1985 catches (in press).

Egglishaw, Gardiner and Foster (1986) in seeking to relate a decline in salmon catch with afforestation in upland areas, noted that in some districts the total catch in a recent decade (1972-1981) was between 50 and 90 per cent of that made in either of the two preceding decades (which he termed for the purposes of his paper, 'decline'), in others it was less than 50 per cent ('serious decline') while in the rest it was more than 90 per cent of either of the previous decades ('no decline'). The 'serious decline' areas are all relatively small districts on the west coast and, in terms of total numbers, the decrease in catch there is less important than the less serious 'decline' in the large and more productive east coast areas such as the Tweed. In the first half of the 1970s, the East Region (Tweed to Tay) provided between one-third and one-half of the whole Scottish catch but in recent years only about one-fifth of it. Clearly changes in this

Figure 6.7: Number of salmon and grilse caught by all methods in four regions (East, Moray Firth, North and Solway) in years 1952 to 1984 to illustrate difficulties between regions

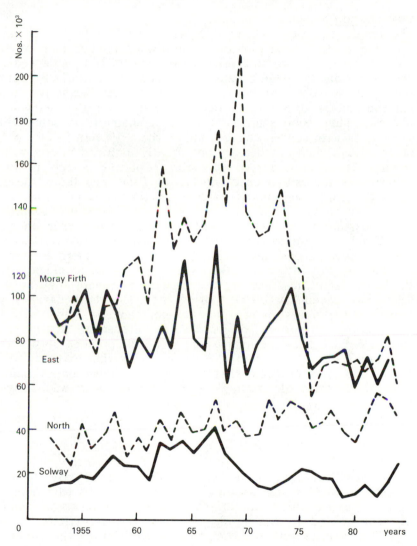

region, and in the large North East and Moray Firth regions have a significant effect on the overall Scottish catch.

Undeclared catches

The catch statistics used in this paper are based on information reported to the Department by the owner or occupier of each fishery in response to an annual questionnaire. It is accepted that despite a substantial enforcement effort there is unlawful salmon fishing in Scottish waters. The size of the catches is not known but the amount of salmon taken has varied over the years and is likely to have been significant in some districts in some years, sometimes with sudden changes (as in the drift-net fishery of the early 1960s). Such catches are not usually included in any official return. It is considered that the reported catch probably understates the actual catch of lawful fishermen (both rods and nets) but that the amount of bias in this does not vary much from year to year; and that the unlawful catch has sometimes been substantial and has varied markedly from year to year.

Catches of salmon of Scottish origin made in other countries, e.g. England, Faroes, Greenland, Ireland and Northern Ireland, do not strictly fall into the category of undeclared landings but, to the extent that they are not ascribed to Scotland, they must be taken into account when comparing recent Scottish catches with those of the years before the development of these so-called 'interceptory' fisheries. The increase in catches of Scottish-origin salmon outside Scotland may balance the differences between the low catches since the mid 1970s and the higher catches of the 1950s, and it is probable that the total exploitation of Scottish salmon, in terms of catch, is as great now as it was 30 years ago.

6.10 SOME CONCLUSIONS

Exploitation of salmon of Scottish origin gives rise to catch levels that are as high as they were in the 1950s (the beginning of the present statistical series). However, in terms of numbers of fish landed in Scotland, catches are significantly less.

The proportion of the total catch that is taken by rod-and-line has increased over the past 10 years, mainly as a result of the decrease in the catch by nets. It is likely that that position will be maintained unless there is a significant increase in the number of salmon reaching the Scottish coast during the fishing season, because salmon net fishing by lawful methods becomes uneconomic when salmon are less abundant. This is particularly so now that the development of salmon farming has put an end to the high prices that used to prevail during periods of scarcity of wild fish.

The trend towards a relatively higher rod catch may also be reinforced by the current Salmon Bill which, if enacted as it stands, will provide a framework for stricter controls on netting by traditional methods.

The Scottish system of private ownership of all salmon fishing rights and the resultant regime of marketable fishing rights, coupled with central control of fishing methods, close-times etc, provides a mechanism for controlling the intensity of fishing that has worked over many centuries of active and sometimes intensive exploitation. Although there are improvements that can be made in detailed arrangements, the regime provides a sound basis for controlled management of the resource in Scotland.

6.11 SUMMARY

Salmon are distributed widely in Scotland and, in the larger rivers, run in from the sea over a long season. There is some artificial stocking but the fishery is based on a naturally regenerating stock.

Salmon fishing rights, in rivers and the sea, are privately owned and may be bought, sold and leased like land.

Salmon fishing methods are restricted to net-and-coble or rod-and-line in inland waters, and to those methods and bag nets or stake-nets on the coast.

Catches in Scotland have been lower in the period since 1976 than formerly but the catch of Scottish origin stock has not declined in the same way and is as high as in the 1950s.

Grilse have provided the majority of the total catch since 1967 but salmon still predominate in the rod catch.

The rod catch has constituted about 20 per cent of the total catch over the past 8 years compared with less than 15 per cent in the period before that.

REFERENCES

Anon. (1963) Scottish salmon and trout fisheries (First Report of Lord Hunter's Committee). Cmnd 2096, HMSO, Edinburgh

Anon. (1983) Scottish salmon catch statistics 1952-1981 (East, Northeast, Moray Firth and Solway Regions). Department of Agriculture and Fisheries for Scotland, Edinburgh

Anon. (1984a) Scottish salmon and sea trout catches: 1982. Statistical Bulletin No 1/1984, Department of Agriculture and Fisheries for Scotland, Edinburgh

Anon. (1984b) Scottish salmon catch statistics 1952-1981 (North Northwest, West, Clyde Coast, Outer Hebrides, Orkney and Shetland Regions). Department of Agriculture and Fisheries for Scotland, Edinburgh

Anon. (1985a) Scottish salmon and sea trout catches: 1983. Statistical Bulletin No 1/1985, Department of Agriculture and Fisheries for Scotland, Edinburgh

Anon. (1985b) Scottish salmon and sea trout catches: 1984. Statistical Bulletin No 2/1985, Department of Agriculture and Fisheries for Scotland, Edinburgh

Anon. (1986) Triennial review of research 1982-1984; Freshwater Fisheries Laboratory. Department of Agriculture and Fisheries, Edinburgh

Calderwood, W.L. (1921) The salmon rivers and lochs of Scotland. Edward Arnold, London

Egglishaw, H., Gardiner, R. and Foster, J. (1986) Salmon catch decline and forestry in Scotland. Scottish Geographical Magazine, 102, 57-61

Elgin Report (1902) Report of the Commissioner on Salmon Fisheries, (Lord Elgin's Commission). HMSO, London

Gardiner, R. and Egglishaw, H. (1986) A map of the distribution in Scottish rivers of the Atlantic salmon (Salmo salar L.) Department of Agriculture and Fisheries for Scotland, Edinburgh

George, A.F. (1982) Cyclical variations in the return migration of Scottish salmon by sea age c. 1790 to 1976. Thesis submitted for Degree of MPhil, Open University

Hansard (1986) House of Commons Official Report, Standing Committee D, 22 May 1986, col 283

Hunt, D.T. (1978) The salmon and the crown. The Salmon Net, XI, Salmon Net Fishing Association of Scotland, Aberdeen

Mitchell, K.A. (1978) Changes in the reported catch of summer salmon in the Scottish salmon net fishery, 1952-1976. ICES/CM, 1978, M23

Neill, R.M. (1946) Early salmon biology. Aberdeen University Review, 31, 243-7

TRRU (1984) A study of the economic value of sporting salmon fishing in three areas of Scotland Tourism and Recreation Research Unit, University of Edinburgh

Chapter Seven

HARVEST AND RECENT MANAGEMENT OF ATLANTIC SALMON IN CANADA

T.L. Marshall
Fisheries Research Branch, Department of Fisheries and Oceans,
P O Box 550, Halifax, Nova Scotia, Canada B3J 2S7

7.1   INTRODUCTION

Harvesting the Atlantic salmon resource in Atlantic Canada (Figure 7.1) is widespread, although less so now than just a few years ago. People in all five eastern provinces, Newfoundland and Labrador, Quebec, New Brunswick (NB), Nova Scotia (NS), and Prince Edward Island (PEI) (the latter three referred to collectively as the Maritime Provinces) participate. Major harvesters are the commercial and sport fishermen; lesser harvesters include native peoples, commercial fishermen taking other fish species and the poacher.

The first harvesting of Atlantic salmon in eastern Canada probably started about 8600 BC - the first dating of Amerindians in the vicinity. The Vikings, who briefly colonised areas of insular Newfoundland, possibly Labrador and the Maritime Provinces between 995 AD and 1014 AD and the Basques, fishing the Labrador Sea between 1400 and 1420, were the first Europeans to indulge in the North American Atlantic salmon resource (Dunfield, 1985). Commercial fishing in Newfoundland and Labrador is known since at least 1583 (Taylor, 1985); river fisheries operated by Arcadians of Nova Scotia and New Brunswick date from the mid 1600s (Dunfield, 1985). Angling was introduced by British army officers frequenting Nova Scotia and New Brunswick in the 1770s. Licences to fish commercially and laws to protect salmon from overfishing, obstructions and other hazards were in operation by 1880.

Each province has developed patterns and harvest methods which are somewhat different. Differences are in part due to tradition, economics and proprietary rights of each province to license angling in non-tidal waters, and of Quebec to license its commercial fishery. This paper summarises recent catch and effort statistics collected variously by federal (Newfoundland and Labrador, New Brunswick, Nova Scotia, and Prince Edward Island)

Figure 7.1: Atlantic Canada, showing location of provinces, regions and rivers/drainages mentioned in text. Numbers identify the following rivers: 1. Eagle; 2. St Genevieve; 3. Exploits; 4. Conne; 5. George; 6. Koksoak; 7. St Augustin; 8. Olomane; 9. Natashquan; 10. Mingan; 11. Moisie; 12. Betsiamites; 13. Escoumins; 14. Jupiter; 15. Matane; 16. Cascapédia; 17. Richibucto; 18. Middle; 19. St Mary's; 20. LaHave. Shaded areas denote drainage of the top five rivers for recreational landings

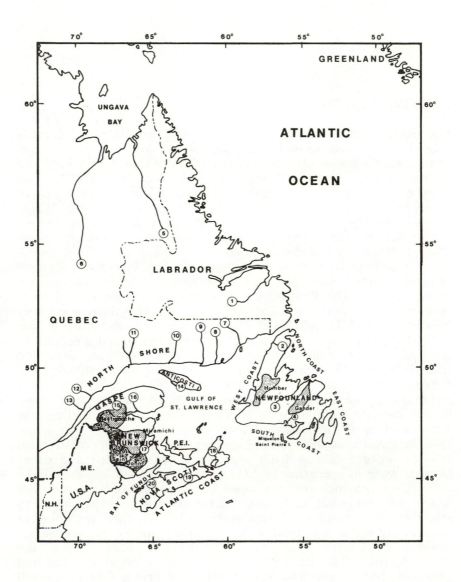

and provincial (Quebec) fisheries officials and reviews major elements of the most recent management plans designed to reduce exploitation and increase spawning escapement.

7.2   TOTAL LANDINGS

Table 7.1: Nominal catch of commercial[a] and recreational Atlantic salmon (tonnes round fresh weight) taken in Canada, 1960-1985

	Salmon (tonnes)		
Year	Small[b]	Large	Total
1960	-	-	1,636
61	-	-	1,583
62	-	-	1,719
63	-	-	1,861
64	-	-	2,069
65	-	-	2,116
66	-	-	2,369
67	-	-	2,863
68	-	-	2,111
69	-	-	2,202
1970	1,562	761	2,323
71	1,482	510	1,992
72	1,201	558	1,759
73	1,651	783	2,434
74	1,589	950	2,539
75	1,573	912	2,485
76	1,721	785	2,506
77	1,883	662	2,545
78	1,225	320	1,545
79	705	582	1,287
1980	1,763	917	2,680
81	1,619	818	2,437
82	1,082	716	1,798
83	903	530	1,434
84	645	467	1,112
85	526	574	1,100

Notes: a, Includes estimates of some local sales and by-catch.
          b, Includes some 1SW fish < 2.7 kg which are not necessarily destined to be 'grilse' (mature 1SW fish).

Source: Anon. (1986a).

Nominal sport plus commercial catches in Canada for the period 1960-1985 (Table 7.1; Anon., 1986a) have averaged 2,020 t/yr. The 1970-1985 landings averaged 2,000 t/yr of which 1,320 t/yr (66 per cent) was of large or multi-sea-winter (MSW) salmon. Landings of 1,112 t and 1,100 t in 1984 and 1985, respectively, are the lowest of the 26-year data set (Table 7.1). Data presented later indicate that since 1970, about 90 per cent of the salmon by weight were landed in the commercial fisheries.

Commercial Landings, 1910-1985
Total annual Canadian landings by commercial fishermen, 1910-1985, have ranged from a peak of 6,101 t in 1930 to lows of 815 t and 842 t in 1984 and 1985, respectively (Figure 7.2). The latter low values contrast with annual means of 4,201, 2,925, 1,852, 2,027 and 1,894 t for the 1930s, 1940s, 1950s, 1960s and 1970s respectively. Even the addition to Canadian landings of 50 per cent of the West Greenland harvest in the previous year (assumed on the basis of Anon. (1986a) to be of Canadian origin) underscores the smaller commercial harvests of Canadian stocks during the period 1960-1985, mean of 2,469 t, relative to the mean of 3,739 t for the 1920-1945 period.

Figure 7.2: Commercial landings of Atlantic salmon in provinces of Atlantic Canada, 1910-1985 (Anon. 1978; updates by S.F. O'Neill, Y. Côté, and T.R. Porter personal communication). Potential total includes 50% of nominal catches at West Greenland (Anon. 1986a) in year i-1 and total Canadian catches in year i

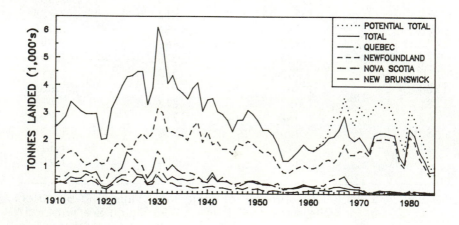

Commercial landings reflect regional differences in both stock strength and in restrictive measures to enhance escapement/angling. Both factors have contributed to the dramatic decline for Quebec, New Brunswick, Nova Scotia, and Prince Edward Island (Figure 7.2) from an annual mean of 1,854 t, 1925-1940, (45 per cent of total Canadian landings) to one of 213 t, 1970-1985 (12 per cent of the total). Decline from 2,265 to 1,533 t/yr for the same periods in the Province of Newfoundland and Labrador has been less dramatic.

Recreational fisheries

Although documentation of the recreational harvest of Atlantic salmon for many rivers of Atlantic Canada dates back to the early 1900s, annual totals of ISW and MSW fish for some provinces are not available before 1965. The total number of fish retained, including black salmon or kelts in New Brunswick (25 per cent of NB retention), has remained fairly constant and averaged 98,064 (coeff. of variation 0.20) fish (between 1965 and 1985, Figure 7.3). Over the period 1970-1983, during which time

Figure 7.3: Numbers of large salmon and grilse retained in recreational fisheries of the provinces of Atlantic Canada 1965-1985. Values for New Brunswick include black salmon

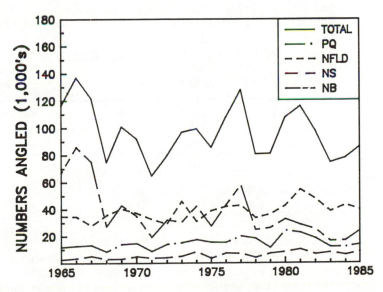

Source: O'Neill, Bernard and Singer (1985), updates by S.F. O'Neill, Y. Côté and T.R. Porter (personal communication)

the data for each province are most reliable, the Province of Newfoundland and Labrador has provided 46 per cent of the total Canadian bright salmon. New Brunswick, Quebec and Nova Scotia provided 28, 19 and 7 per cent of the total, respectively, while Prince Edward Island with an average of only 18 fish/year has provided less than 1 per cent.

Each province has its own angler licence system. Each system is different with respect to age limits affecting who is licensed and to duration for which the licence is valid (days, weeks, season). The latest provinces to issue angling licences specific to salmon were Quebec in 1976 and Nova Scotia and PEI in 1983. Licence sales for Newfoundland and Labrador, New Brunswick and Quebec, 1976-1985, which were 89 per cent of the total of all provinces, 1983-1985, averaged 56,430 (Table 7.2).

Table 7.2: Sales of resident and non-resident angling licences for Atlantic salmon by the provinces of Atlantic Canada, 1976-1985

	Provinces					
Year	NFLD	NB	PQ	NS	PEI	Total
1976	20,103	18,757	11,854	–	–	50,714
1977	21,061	21,252	14,283	–	–	56,596
1978	19,349	21,141	15,380	–	–	55,870
1979	21,481	21,721	12,772	–	–	55,974
1980	18,372	20,633	17,521	–	–	56,526
1981	20,554	21,916	20,450	–	–	62,920
1982	22,554	22,535	21,239	–	–	66,328
1983	21,859	22,397	18,775	7,576	321	63,031[b]
1984	19,866	16,119	12,401	5,790	68	48,386[b]
1985	18,476[a]	17,778	13,347	5,825	109	49,601[b]

Notes: a, Preliminary.
          b, Exclusive of NS and PEI.

Source: Anon. (1986b).

Non-resident licences averaged 22 per cent of the total in New Brunswick, 16 per cent in Quebec and 8 per cent in Newfoundland and Labrador. Licence sales in every province, 1984-1985, were less than those for the period 1981-1983, which can in part be attributed to reduced sizes of runs and in the Maritime Provinces, to prohibition of the retention of salmon > 63 cm in fork length.

Native fisheries

Native fisheries for Atlantic salmon are officially conducted from occupied Reserve properties in Quebec, New Brunswick and Nova Scotia (Table 7.3). Micmacs settled on non-reserve lands near Conne River, Newfoundland, (Figure 7.1) acquired a licence to

Table 7.3: Location and quotas of active Indian/Inuit food fisheries of Atlantic Canada 1986. (Most are additionally constrained by limitation on seasons and number and length of nets.)

Province	Stock/River	Band/community	Quota
Newfound-land and Labrador	Conne	Conne River	1,200 1SW fish
Quebec	Restigouche	Cross Point	6,995 kg
	Cascapédia	Maria	909 kg
	Escoumins	Escoumins	57 salmon
	Betsiamites	Betsiamites	200 salmon
	Moisie	Sept.-Iles/ Malioténam	350 salmon
	Mingan	Mingan	0 salmon (Band restraint)
	Natashquan	Point-Parent	1,500 salmon
	Olomane	LaRomaine	120 salmon
	St Augustin	St Augustin	250 salmon
	Koksoak	Kuujjuaq	8,000 kg
	George	Kangiqsualujjuaq	4,000 kg
New Brunswick	Restigouche	Eel River Bar	no quota
	Richibucto	Big Cove	no quota
	Miramichi	Red Bank	no quota
		Eel Ground	no quota
		Burnt Church	no quota
	St John	Kingsclear	900 salmon
		Oromocto	150 salmon
Nova Scotia	Middle	Wagmatcook	100 salmon

fish salmon in saltwater adjacent to the river outflow in 1986. The Naskuapi and Montagnais Indian Bands of Quebec and Inuits of Labrador are also permitted to fish for food in the freshwaters of Labrador but most efforts are towards species other than salmon. Inuits of northern Labrador participate in licensed commercial fisheries under the same regulations as non-native fishermen. Inuit commercial and food fisheries also exist in Ungava Bay, Quebec. All Indian fisheries are related to aboriginal rights to the resource and are for food purposes only.

All food fisheries are licensed and restricted by quotas, numbers of nets, net length, or seasons. Some Bands, such as Wagmatcook in Nova Scotia and Conne River in insular Newfoundland, operate within agreed limits. However, many Bands, particularly on the Saint John, Miramichi and Restigouche rivers (Figure 7.1 and Table 7.3), defy documentation of catch levels and are widely believed to exceed permissible fishing effort and/or quotas. While the impact of the uncontrolled Indian fisheries on stocks of affected rivers is significant, the total number of fish harvested in all food fisheries would not rival removals by recreational fisheries on a national basis. Most food fisheries do, however, remove MSW salmon which, in the Maritime Provinces, have recently been made illegal for anyone else to retain.

7.2   PROVINCIAL PERSPECTIVES

Commercial fisheries

Newfoundland and Labrador.
Over the period 1970-1985, the Province of Newfoundland and Labrador accounted for 88 per cent of the total Canadian commercial landings. Regional data for the period 1974-1985 indicate that Labrador, followed by the north coast contributed 40 per cent and 24 per cent of the average annual landings of 1,578 t (Figure 7.4). The east, south and west coasts followed with respective proportions of 16, 15 and 5 per cent. Large salmon constituted 64 per cent of the total landings with Labrador accounting for 47 per cent and the west coast for only 3 per cent of the total large salmon landings. Low landings in 1984 and 1985 relative to the previous all-time low landings of 1978 and 1979 are evident in Figure 7.4.

Estimates of total provincial fishing effort, in terms of licensed gill nets, 50 fathoms in length, are shown in Figure 7.4. The data indicate a peak in effort in 1975 associated with the implementation of a new salmon licensing policy providing licences to previously unlicensed partners of licensed salmon fishermen. Licensed effort has been reduced by over 40 per cent between 1975 and 1985 and by 35 per cent between 1978/79 and 1985.

Figure 7.4: Landings (100s of tonnes) of small (< 2.7 kg) and large (> 2.7 kg) Atlantic salmon and licensed effort (1,000s of units of fixed gear 50 fathoms in length) in five regions of the Province of Newfoundland and Labrador 1974-1985

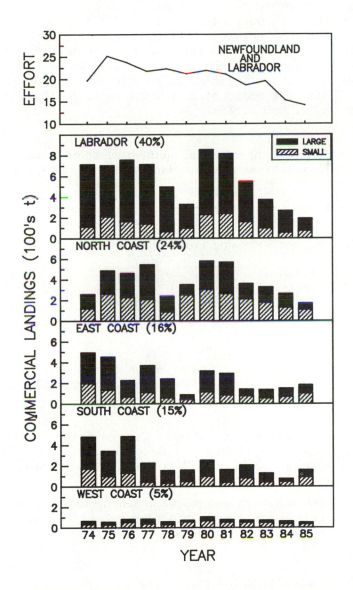

Source: O'Connell, Dempson, Reddin and Ash (1986); T.R. Porter, D.G. Reddin, and J.L. Peppar personal communication.

Reductions in licensed effort and in catch can in part be attributed to major changes in licensing and duration of commercial salmon seasons (Table 7.4; Anon., 1985a). Significant changes in 1984 included a 3-week delay from that of 1980 in the opening date of the fishery in most statistical areas of the province, complete closure of Area J_2 (most westerly area of south coast), a voluntary buy-back of salmon fishing licences (27 per cent reduction between 1983 and 1985) and prohibition of retention of any salmon captured in non-salmon gear. In 1985 there was a mandatory buy-back of approximately 700 salmon licences from part-time commercial fishermen. In total, there was a 27 per cent reduction in licensed fishing effort between 1983 and 1985 (Anon., 1986a).

Table 7.4: Major changes to management of commercial fisheries for Atlantic salmon in Newfoundland and Labrador[a], 1975-1985

1975

New Salmon Licensing Policy implemented

Main features:
(1) Freeze on new entrants
(2) Policy on attrition introduced
(3) Strict transfer policy introduced

1976

Licensing policy modified to eliminate from the fishery persons employed full-time in jobs other than the fishery

1978

Reduced fishing season in area Cape St Gregory south to Cape Ray from 15 May - 31 December to 1 June - 10 July; and in area Cape Ray to Pass Island, season reduced from 15 May - 31 December to 20 May - 10 July

Changes in herring and mackerel fishing season to reduce salmon by-catch-closed period: herring, 15 June to 31 July; mackerel, 1 July to 31 July

1979

To reduce salmon by-catch:

(1) mesh size in cod leaders increased to 177 mm;
(2) monofilament prohibited in cod traps

Table 7.4 (Cont'd)

1981

Commercial salmon season changed from 15 May - 31
December to 18 May - 31 December for all areas except
Area J which remained 20 May - 10 July and Areas K and
L which remained 1 June - 10 July
Closure of Bay of Islands to cod traps
Closure of area outside two nautical miles off Port aux
Basques to commercial salmon fishing

1982

Fourteen separate management zones to be implemented
(includes Gulf area of Newfoundland). This will result in
more specific localised management plans on a
zone-by-zone basis if necessary

1983

Implement a programme to standardise amount of fishing
gear per licensed fisherman such that full-time fishermen
are limited to 200 fathoms and part-time fishermen limited
to 50-100 fathoms. The programme was brought in over
two years. In 1983 all part-time fishermen who were
previously licensed for more than 100 fathoms were
reduced to 100 fathoms and full-time fishermen who were
licensed for 300 fathoms were reduced to 200 fathoms and
those fishermen who had been licensed for more than 300
fathoms were reduced to 300 fathoms in 1983 and to 200
fathoms in 1984. Full-time fishermen who had been
licensed for less than 200 fathoms had their licensed gear
increased to 200 fathoms in 1983

1984

Area J_2 closed to salmon fishing
Transfer of licences restricted to immediate family
members
It became illegal to retain salmon captured incidentally in
non-salmon commercial gear
Voluntary buy-back of fishing licences
No transfer of part-time licences

1985

Mandatory buy-back of part-time commercial salmon
licences (approx. 700)

Table 7.4 (Cont'd)

1975 - 1985	Open seasons (summary)	
1975 - 1977	All areas A to O	15 May - 31 December
1978 - 1980	Areas A to I; M to O	15 May - 31 December
	Area J	20 May - 10 July
	Areas K, L	1 June - 10 July
1981 - 1983	Areas A to I; M to O	18 May - 31 December
	Area J	20 May - 10 July
	Areas K, L	1 June - 10 July
1984 - 1985	Areas A to I; M to O	5 June - 31 December
	Areas J_1, K, L	5 June - 10 July
	Area J_2	Closed

Notes: a, where Labrador = Area O; north coast = Areas A, B; east coast = Areas C, D, E, F; south coast = Areas G, H, I, J_1, J_2, and west coast = Areas K, L, M, N.

Source: Table 22 in Anon. (1985a).

Quebec.
Quebec commercial landings, 1960-1985, which between 1970 and 1985 made up only 6 per cent of the total Canadian landings, are illustrated in Figure 7.5 (Côté, personal communication). Most of the landings were MSW fish taken in trap and set gill nets of 5.5 in. (140 mm) minimum mesh size. Reduced provincial landings can at least be partially attributed to District closures, quotas, effort reduction from approximately 360 licenced fishermen in 1971 to 222 in 1985 and stock size.

The effects of reduced effort are particularly evident in Gaspé region (60 per cent of the provincial landings during the 1960s) where landings in 1972-1983 averaged only 7 per cent (10 t) of those of the 1960s (134 t). In 1984 and 1985 all Districts of Gaspé were closed to salmon fishing. The remaining salmon licences in Gaspé, 1985, were only 25 per cent of those in 1971.

Landings on the north shore, averaging 91 t in the 1960s and 84 t, 1972-1983, were free of District closures and quotas. The 1984 and 1985 landings (66 t average) were affected by the closure of two of the most westerly Districts and a formulated TAC in the remaining seven Districts. The small commercial fishery on Anticosti Island was closed in 1966.

Figure 7.5: Landings (100s of tonnes) of Atlantic salmon in North Shore and Gaspé regions of Quebec[a] and Gulf, Atlantic Coast and Bay of Fundy regions of the Maritime Provinces[b]

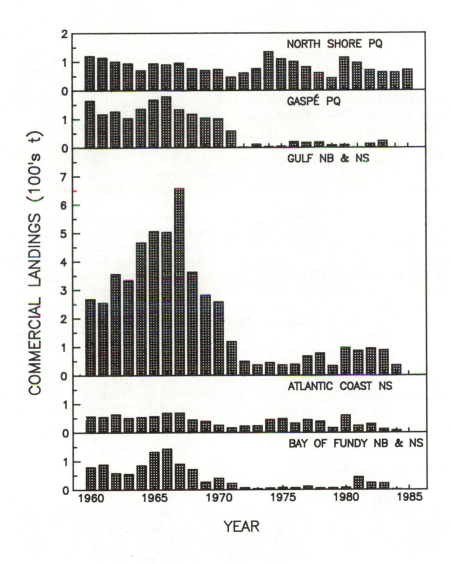

Sources: a, Y. Côté (personal communication).
b, Atlantic Salmon Commercial Catch Statistics, Maritime Region, annual series beginning 1967; update and postdate by S.F. O'Neil (personal communication).

Data for commercial/food fisheries (50:50) of Ungava Bay are incomplete. However, estimates for 1981-1983 and 1984-1985 averaged about 10 t and 8 t, respectively (Côté, personal communication).

Maritime Provinces.
Landings from the Maritime Provinces, averaging 6 per cent of Canadian landings between 1970 and 1985, can be lumped into those from the Gulf of St Lawrence, Atlantic and Bay of Fundy coasts. The Gulf of St Lawrence fishery was the largest of the region (Figure 7.5) comprising two-thirds (90 t) of the average landings 1969-1984. Atlantic and Bay of Fundy coasts comprised two-(30 t) and one-third (15 t), respectively, of the remainder.

Traditional salmon gear in the region has comprised trap, drift and gill nets; all, until 1981, of 5.5 in. (140 mm) minimum mesh size. This mesh size largely excluded 1SW fish grilse (grilsie) (illegal in New Brunswick prior to 1981) from the statistics.

As in Quebec, declining landings in the Maritime Provinces can, in large part, be attributed to management activities which: closed the salmon fisheries of the Restigouche, Miramichi (Gulf, NB) and Saint John (Bay of Fundy, NB) rivers and their outflows, 1972-1980 and 1984-1985 (the Restigouche was open for just two weeks in 1984); enacted variously abbreviated seasons and quotas for the aforementioned fisheries 1981-1983; dramatically reduced length of the open seasons in Nova Scotia, 1982 to 1984; reduced 1,147 eligible salmon licences in 1969 (approximately 50 per cent trap net and 25 per cent of each of the other gear types) to 478 in 1984; and closed, with cash compensation, all commercial fisheries for salmon in 1985. By-catch, which became illegal in 1984 and which had contributed substantial numbers of 1SW fish to the statistics, has not been acknowledged in the statistics since that legislation.

Recreational fisheries

Newfoundland and Labrador.
Differing in many respects from the commercial fishery, the recreational harvest of the province comprised less than one-half (46 per cent) of the number of angled salmon retained in eastern Canada, 1970-1983. Provincial data for the period 1970-1985 are shown in Figure 7.6.

The west coast, followed by the north and south coasts, contributed 35, 24 and 22 per cent (1970-1985 average of 13,705, 9,523 and 8,834 salmon), respectively. Labrador and the east coast provided only 12 and 7 per cent (4,740 and 2,945 salmon), respectively.

Figure 7.6: Numbers of bright 1SW and MSW Atlantic salmon retained and catch per unit effort (fish/rod day) in the sport fishery of various regions of the Province of Newfoundland and Labrador 1970-1985

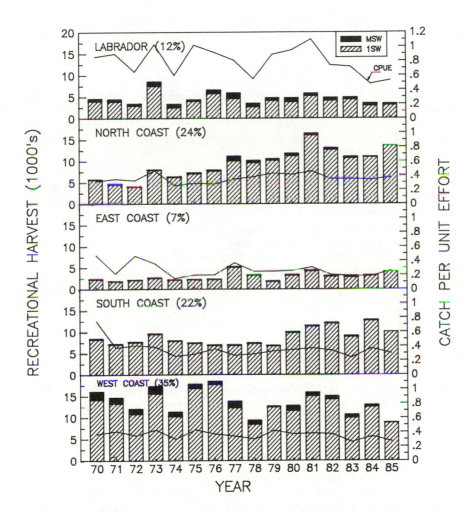

Sources: Ash and Tucker (1984); Moores, Penney and Tucker, (1978); Moores and Tucker (1979, 1980); Moores and Ash (1984); update by T.R. Porter (personal communication).

Unlike the provincial commercial landings which were 64 per cent large salmon, sport catches consisted of only 5 per cent MSW fish. These sport catches better reflect the proportional production of 1SW and MSW fish in this province's rivers. Nearly

131

one-half of these MSW fish were taken from west coast rivers; one-third were from Labrador.

Angling effort has averaged about 108,900 rod days/year, 1970-1985. With only minor variations, Labrador and east coast efforts correspond with the level of removals from each region of the Province. Average annual CPUE values (fish/rod day) 1970-1985 (Figure 7.6) were a maximum in Labrador, 0.79, and a minimum on the east coast, 0.25. Average CPUEs for the other three regions ranged from 0.34 to 0.36.

Based on average annual numbers of fish caught for the period 1970-1985, top-producing rivers (Figure 7.1) were the Humber (3,704), Gander (2,660), Conne (2,128), Eagle (1,842), Exploits (1,468) and St Genevieve (1,372). Data for the period 1974-1985 reflecting recent enhancement activities on the Exploits River would elevate landings from that system above those of the Eagle River.

Recent changes in the regulation of the Newfoundland and Labrador sport fishery have not severely hampered the angling community. Open seasons have not changed significantly in recent years with opening dates in 1985 ranging from 1 June to 1 July and closing dates ranging from 2 to 15 September. Possession limits have been four fish (two-day limits); there was no season bag limit nor tagging required. Closure of nine small rivers in 1978 and requirement in insular Newfoundland beginning in 1984 and fully implemented in 1985, to release MSW salmon had little impact on provincial totals composed mostly of 1SW fish.

Quebec.
Sport data for the Province of Quebec, which in total represented 19 per cent of the 1970-1983 Canadian retention, can be divided into the Ungava, north shore, Gaspé and Anticosti regions. The Gaspé and north shore regions accounted for 42 per cent (6,975 fish) and 35 per cent (5,861 fish) respectively of the 1970-1982 average annual provincial landings (Figure 7.7). Ungava accounted for 14 per cent (2,251 fish) and Anticosti the remaining 9 per cent (1,553 fish).

In contrast to Newfoundland, MSW fish averaged 74 per cent of the 1970-1982 sport landing. Regional extremes of MSW composition were Anticosti (69 per cent MSW) and Ungava (92 per cent MSW) (Figure 7.7).

Angling effort has averaged an estimated 34,500 rod days/year, 1970-1982. Efforts in 1984 and 1985 were about 38,400 and 40,500 rod days and represent declines of about 15 per cent from 1981 and 1982 which, in turn, were the second and third highest since 1970. Approximately 50 per cent and 40 per cent of the total provincial effort are reported from Gaspé and the north shore, respectively. Corresponding CPUE values (Figure 7.7) for

Figure 7.7: Numbers of 1SW and MSW Atlantic salmon retained and catch per unit effort (fish/rod day) in the sport fishery of various regions of the Province of Quebec, 1966-1985

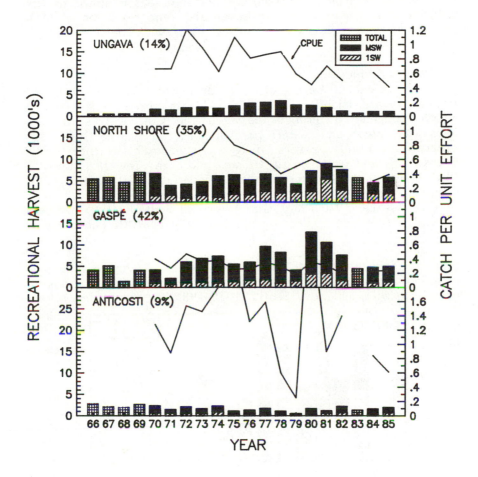

Source: Côté (personal communication).

Gaspé and north shore averaged 0.3 and 0.6. Average CPUE values for Anticosti and Ungava were 1.2 and 0.6. Recent CPUE values are the lowest since 1978 and 1979 (Figure 7.7).

Based on average annual catch, 1970-1982 best yields of salmon have come from the Quebec portion of the Restigouche (2,420 fish), George (1,307 fish), Jupiter (1,078 fish), Matane (1,001 fish) and Moisie (925 fish) rivers (Figure 7.1).

The most significant changes to management measures of recent years were taken in 1984 when opening dates in most zones of Gaspé and north shore were delayed by ten days, about

20 (25 per cent) of rivers with a low spawning escapement on the north shore and Gaspé were closed to angling, tagging of retained fish became mandatory, daily and season bag limits were reduced (e.g. Gaspé - from two fish/day and no season limit to one fish/day and a season limit of seven fish) and retention of MSW salmon in waters boundary to New Brunswick became illegal.

Maritime Provinces.

The recreational fisheries of New Brunswick, Nova Scotia and Prince Edward Island, can be divided into three major regions which recognise differences between the Gulf waters of New Brunswick (north and eastern portion bordering on the Gulf of St Lawrence), Fundy waters of New Brunswick (southern New Brunswick), and Nova Scotia (if we ignore the few fish taken in PEI). In total, they represent about 35 per cent of the 1970-1983 Canadian retention of bright salmon. Of the salmon harvested in New Brunswick (1970-1983), Gulf rivers accounted for 85 per cent (annual average of 20,065) of the fish (Figure 7.8). Fundy rivers

Figure 7.8: Numbers of bright 1SW and MSW Atlantic salmon retained and catch per unit effort (fish/rod day) in regions of New Brunswick and Nova Scotia, 1970-1985

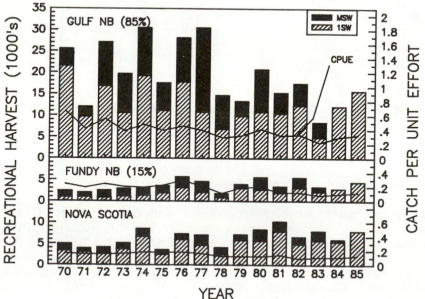

Sources: O'Neil and Swetnam (1984); O'Neil et al. (1985, 1986); Swetnam and O'Neil (1984); update by S.F. O'Neil (personal communication).

of New Brunswick accounted for the remaining 15 per cent (3,585 fish). Nova Scotia rivers yielded an average of 6,352 salmon or 21 per cent of the total for the Maritime Provinces.

MSW salmon were most dominant, 1970-1983, in Fundy, New Brunswick (48 per cent). Gulf rivers yielded a catch consisting of 39 per cent MSW fish, while catches for Nova Scotia rivers were, on average, 27 per cent MSW fish. In 1984 and 1985 the retention of MSW fish was banned in the Maritime Provinces. However, yields of 1SW fish were above average in all three major regions (Figure 7.8), particularly due to MSW catch-and-release regulations.

Annual effort for Nova Scotia and New Brunswick, 1970-1983, averaged 117,200 rod days; 61 per cent of the effort was in New Brunswick and 70 per cent (50,000 rod days) of that was in Gulf NB rivers. Recent effort, 1980-1983, averaged 47,300, 35,800 and 71,600 rod days in Gulf NB, Fundy NB and Nova Scotia rivers, respectively. The 1984-1985 average dropped to respective values of 43,900, 33,200, and 62,400 rod days. CPUE values have, in general, declined over the period of the data (Figure 7.8) to 1980-1983 averages of 0.33, 0.13 and 0.12 fish/rod day for Gulf NB, Fundy NB, and Nova Scotia, respectively. Similar values for 1984 and 1985 do not, however, reflect MSW fish caught and released.

River systems leading the recreational landings 1970-1982, were the Miramichi (average of 15,331 fish) and Restigouche (4,827 fish) in Gulf NB, Saint John (2,861 fish) in Fundy NB and the LaHave and St Mary's (1,039 and 811 fish, respectively) in Nova Scotia (Figure 7.1).

The most significant management measure applied to the sport fishery of the Maritime Provinces occurred in 1984 when the retention of MSW salmon (> 63 cm) was prohibited. Open seasons for bright salmon vary according to river system and the few recent changes have been minor in nature. Bag limits in Nova Scotia and New Brunswick in the 1980s stood at ten for the season, six fish in possession and two fish for a daily limit. These limits included black salmon (kelts) taken in a 15 April - 15 May season in New Brunswick. Daily and possession limits for salmon in the few rivers of PEI were one fish each; the season limit was five fish. Tagging programmes to minimise illegal harvest and marketing of salmon were introduced to New Brunswick in 1981, to Nova Scotia in 1983 and to PEI in 1985.

Leading salmon rivers.
Trends of annual Atlantic salmon landings and CPUE for the five top-producing sport fishing rivers (Figure 7.1) in Canada appear in Figure 7.9. Average annual landings for 1970-1982 place the Miramichi River NB, (15,331 bright fish) well ahead of the

Figure 7.9: Numbers of bright 1SW and MSW Atlantic salmon retained and catch per unit effort (fish/rod day) on the Miramichi, Restigouche, Humber, Saint John and Gander rivers, Atlantic Canada, 1970-1985

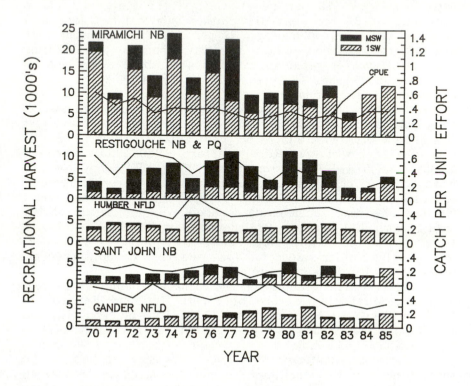

Sources: As detailed in Figures 7.6, 7.7 and 7.8.

Restigouche River (7,247; NB and Quebec combined) and the Humber River, Newfoundland (3,915). The Saint John River, NB, and Gander River, Newfoundland followed at 2,861 and 2,683 respectively.

The Humber and Gander rivers, insular Newfoundland, yielded mostly 1SW fish (95 and 93 per cent respectively); 70 per cent of Miramichi landings, 47 per cent of those of the Saint John and 32 per cent of those of the Restigouche were 1SW fish.

The impact of regulations prohibiting the retention of MSW fish in 1984 and 1985 is particularly evident on the three major NB rivers.

7.3   OVERVIEW FOR 1986

In general, 1986 sport and commercial regulations for salmon in Atlantic Canada remained, with a few exceptions, much as they were in 1985. One exception was the 15 October closure of the commercial fishing season in all areas of Newfoundland which had previously closed on 31 December. The intent was to reduce the impact of the small north and east coast autumn fishery on MSW stocks believed to be returning to mainland rivers of Canada and USA. Additionally, the opening date of the northern area of the west coast of Newfoundland was delayed by five days to diminish fishery impact on outbound kelts. A season limit of 15 fish was also placed on the recreational fishery of Newfoundland and Labrador.

Early indications are that landings of small and large salmon in the commercial fishery of Newfoundland and Labrador in 1986 increased from 1984 and 1985 levels, but are below pre-1983 landings. Recreational fisheries and counts at fishways in Newfoundland and Labrador are similar to previous years. Counts of 1SW fish in particular, in most index rivers of the Maritime Provinces, are better than the average of 1983-1985. Indications are that both 1SW and MSW fish abundance in rivers of Gaspé, Quebec, are about double that of 1985.

7.4   DISCUSSION

Over both the long (Figure 7.2) and short-term (last 5 years), commercial landings, which formerly represented about 90 per cent or more of the landed weight of Canadian removals of Atlantic salmon have declined. Recent declines would appear to be the result of reduced stock abundance and regulations which have diminished fishing effort.

Total numbers removed by sport fishermen, 1965-1985, (Figure 7.3) have, despite annual variation, been fairly consistent. The proportionate utilisation of the resource by sport and commercial interests in 1984-1985, as illustrated by the ratio of tonnes landed by all sport fisheries to tonnes landed by all commercial salmon fishermen was 1:5. The comparable ratio for 1970-1971, when landings in the Newfoundland and Labrador fishery were double those of 1984-1985, was 1:10.

Regionally, reduction in landings in Labrador and the north coast of Newfoundland made the most impact on the 1983-1985 decline in Canadian commercial landings. Within the Newfoundland recreational harvest, totals for the north and south coasts (Figure 7.6) show an upward trend while west coast Newfoundland and Gulf NB reveal a downward trend. Most evident, however, is the absence of MSW removals in the Maritime Provinces, 1984-1985

(Figure 7.8) and north and west coasts of Newfoundland, 1985, and the associated reduction in licensed effort since 1983 (Table 7.2).

Evidence prepared by the Anadromous, Catadromous, Freshwater Fishes Subcommittee of the Canadian Atlantic Fisheries Scientific Advisory Committee (CAFSAC) which indicated recent declines in stock abundance was reviewed by the Working Group on North Atlantic Salmon (ICES). The Working Group noted low return rates for the 1983 smolt class returning as 1SW fish (wild origins) to Western Arm Brook, insular Newfoundland, and as 1SW fish (hatchery origin) returning to the Saint John River, NB (Anon., 1985b). It was also noted for the Saint John River that 1984 was the third consecutive year in which reduced marine survival had been observed for this river (Anon., 1985b). Significant correlations of 1SW and MSW return rates for the Saint John River salmon with recorded Canadian catches further suggested that these return rates were an index of marine survival for Canadian stocks. Low abundance of 1SW stocks in some rivers was also partially attributed to low egg depositions in 1978 and 1979 (Anon., 1985b).

In 1986 the Working Group on North Atlantic Salmon also assessed the impact of management measures in Canada by reviewing the average landings of MSW salmon caught in commercial fisheries which were closed in 1985 and recreational fisheries in which the retention of MSW salmon was prohibited in 1985 (Anon., 1986a). The Working Group estimated that the Canadian restrictions reduced the harvest of MSW salmon by 22 per cent and 1SW fish by 2 per cent. The Working Group also reasoned, on the basis of 1981-1983 landings, that the delay of the opening of the commercial season in Newfoundland to 5 June affected 11 per cent and 1 per cent of the MSW and 1SW salmon, respectively. However, because of the potential for subsequent fishing mortality of the Newfoundland affected fish, the Working Group could not quantify the reduced harvest (Anon., 1986a). Data were also insufficient to quantify the impact of the 27 per cent reduction in licensed fishing effort in Newfoundland and Labrador. However, an indication of the impact of management changes in homewaters was provided by Randall, Chadwick and Pickard (1986a) and Randall, Chadwick and Schofield (1986b) who compared the proportions of MSW spawners to total MSW returns in both the Miramichi and Restigouche rivers, NB, 1983 and 1985. Proportions of spawner/return for the Miramichi increased from 0.10 to 0.93, and for the Restigouche, from 0.11 to 0.70 (Anon., 1986b).

In 1986, cut-backs in the commercial season in Newfoundland further reduced fishing effort. Also, the mandatory buy-back of commercial salmon licences possessed by part-time fishermen in Newfoundland and Labrador, coupled with a 1986 announcement to buy back idle commercial salmon licences in the Maritime Provinces and an ongoing policy forbidding the transfer of salmon licences, will ensure further reductions in commercial fishing effort and opportunity for increased spawning escapements. Apparent improvements in stock abundance in 1986 must in part be credited to sacrifices made by Canadian commercial and sport fishermen to increase spawning escapement in the early 1980s. Increased sacrifices by both groups, 1984 through 1986 in particular, should contribute to increased stock in the late 1980s and early 1990s.

7.5   SUMMARY

Commercial landings of Atlantic salmon in Canada, about 90 per cent of all landings by weight, have generally declined from an average of 4,200 t/yr in the 1930s, 1900 t/yr in the 1970s to 830 t/yr in 1984-1985. The annual number of salmon caught between 1965 and 1983 by recreational fishermen has averaged, with little variation, about 98,000 fish. Landings by sport fishermen as well as licence sales were below average in 1984 and 1985.

During the past 15 years, Newfoundland and Labrador accounted for 88 per cent of the total commercial landings - nearly two-thirds by weight being MSW fish. Management measures in Newfoundland and Labrador to increase salmon returns through their North American range have in recent years reduced commercial licensed fishing effort by as much as 40 per cent. These reductions were the result of complete closures of fishing areas along the southwest coast and voluntary and mandatory buy-back of salmon licences. In addition, commercial fishing seasons were shortened. Quebec, which over the same period accounted for 6 per cent of the total Canadian commercial landings, has also undergone shortening of open seasons, implementation of TACs and a number of closures. The Maritime Provinces, also averaging 6 per cent of the 15-year total commercial landings, have experienced various closures culminating in 1985 by the closure of all licensed salmon fisheries.

Of the total Canadian recreational landings, 1970-1983, 46, 35 and 19 per cent were taken in Newfoundland and Labrador, the Maritime Provinces and Quebec, respectively. Respective MSW components for the same period averaged 5, 37 and 74 per cent. Recent management measures to increase escapement of MSW spawners from recreational fisheries have included: the closure of a number of small rivers and in 1984, a ban on the retention of MSW in insular Newfoundland; reduction of bag limits, mandatory tagging of creeled fish and prohibition since 1984 of the retention of MSW fish in the Maritime Provinces; and closure of about 20 rivers, reduction of bag limits and mandatory tagging of creeled fish in Quebec.

Largest yields to the sport fishery in order of decreasing abundance have come from the Miramichi, Restigouche, Humber, Saint John and Gander rivers.

Improvements in stock abundance in 1986 can be credited in part to improved spawning escapements in 1980 and 1981 and sacrifices by Canadian sport and commercial fishermen. Increased sacrifices by both groups through 1986 should ensure a better opportunity for increasing stock abundance into the early 1990s.

REFERENCES

Anon. (1978) Harvesting the resource subcommittee report. Atlantic Salmon Review. Fisheries Marine Service Mimeograph, revised, 1980, 106 pp.

Anon. (1985a) Report of meeting of the working group on North Atlantic salmon, St Andrews, NB, Canada, 18-20 September 1984. ICES, Doc. C.M. 1985/Assess:5, 56 pp.

Anon. (1985b) Report of meeting of working group on North Atlantic salmon, Copenhagen, 18-26 March 1985. ICES, Doc. C.M. 1985/Assess:11, 67 pp.

Anon. (1986a) Report of meeting of working group on North Atlantic salmon, Copenhagen, 17-26 March 1986. ICES, Doc. C.M. 1986/Assess:17, 88 pp.

Anon. (1986b) Report of special Federal-Provincial working group on Atlantic salmon (In prep).

Ash, E.G.M. and Tucker, R.J. (1984) Angled catch and effort data for Atlantic salmon in Newfoundland and Labrador, and for other fishes in Labrador, 1982. Canadian Data Report for Fisheries and Aquatic Sciences, 465, xv + 96 pp.

Dunfield, R.W. (1985) The Atlantic salmon in the history of North America. Canadian Special Publication on Fisheries and Aquatic Sciences, 80, 181 pp.

Moores, R.B. and Ash, E.G.M. (1984) Angled catch and effort data for Atlantic salmon in Newfoundland and Labrador, and for other fishes in Labrador, 1981. Canadian Data Report for Fisheries and Aquatic Sciences, 451, xv + 88 pp.

Moores, R.B. Penney, R.W. and Tucker, R.J. (1978) Atlantic salmon angled catch and effort data, Newfoundland and Labrador, 1953-77. Fisheries Marine Service Data Report, 84, 274 pp.

Moores, R.B. and Tucker, R.J. (1979) Atlantic salmon angled catch and effort data, Newfoundland and Labrador, 1978. Fisheries Marine Service Data Report, 147, 106 pp.

Moores, R.B. and Tucker, R.J. (1980) Atlantic salmon angled catch and effort data, Newfoundland and Labrador, 1979. Canadian Data Report for Fisheries and Aquatic Sciences, 212, xiv + 86 pp.

O'Connell, M.F., Dempson, J.B., Reddin, D.G. and Ash, E.G.M. (1986) Status of Atlantic salmon (Salmo salar L.) stocks of the Newfoundland Region, 1985. CAFSAC Research Document, 86/23. 60 pp.

O'Neil, S.F., Bernard, M. and Singer, J. (1985) 1984 Atlantic salmon sport catch statistics, Maritime Provinces (Redbook). Canadian Data Report for Fisheries and Aquatic Sciences, 530, v + 98 pp.

O'Neil, S.F., Bernard, M. and Singer, J. (1986) 1985 Atlantic salmon sport catch statistics, Maritime Provinces. Canadian Data Report for Fisheries and Aquatic Sciences, 600, v + 71 pp.

O'Neil, S.F. and Swetnam, D.A.B. (1984) Collation of Atlantic salmon sport catch statistics, Maritime Provinces, 1970-79. Canadian Data Report for Fisheries and Aquatic Sciences, 481, ix + 297 pp.

Randall, R.G., Chadwick, E.M.P. and Pickard, P.R. (1986a) Status of Atlantic salmon in the Restigouche River, 1985. CAFSAC Research Document, 86/1:22 pp.

Randall, R.G., Chadwick, E.M.P. and Schofield, E.J. (1986b) Status of Atlantic salmon in the Miramichi River, 1985. CAFSAC Research Document 86/2:23 pp.

Swetnam, D.A. and O'Neil, S.F. (1984) Collation of Atlantic salmon sport catch statistics, Maritime Provinces, 1980-83. Canadian Data Report for Fisheries and Aquatic Sciences, 450, ix + 194 pp.

Taylor, V.R. (1985) The early Atlantic salmon fishery in Newfoundland and Labrador. Canadian Special Publication on Fisheries and Aquatic Sciences, 76, 71 pp.

Chapter Eight

STATUS OF EXPLOITATION OF ATLANTIC SALMON IN NORWAY

Lars P. Hansen
Directorate for Nature Management, Fish Research Division,
Tungasletta 2, 7000 Trondheim, Norway

INTRODUCTION

In Norway, Atlantic salmon, Salmo salar L., are found in rivers along the coast from the border with the USSR southward to the border with Sweden. In recent years, several salmon populations have been wiped out by pollution, particularly in southern Norway where several important rivers now lack salmon because of the effects of acidification of the water (Jensen and Snekvik, 1972; Leivestad, Hendrey, Muniz and Snekvik, 1976). Another threat to several salmon populations in Norway is the very recent introduction of the fluke Gyrodactylus salaris to several rivers. This parasite kills salmon parr, and at present the estimated loss of salmon due to G. salaris is between 250 and 350 tonnes of salmon (Johnsen and Jensen, 1985).

At present there are between 400 and 500 rivers and streams in Norway populated by Atlantic salmon. The life history patterns and morphology vary considerably between populations and reflect the great variability between river systems.

Atlantic salmon is a very popular species in Norway both for commercial fishermen and sport fishermen, which has led to a very high exploitation of this species (Hansen, 1986). In this paper the present salmon fishery in Norwegian home waters and exploitation of some salmon stocks will be described. Finally some effects of the net fishery such as net damage and selective fishing will be discussed.

THE FISHERY

Catch statistics

In the past, fishing for salmon in Norway was carried out in rivers and estuaries. However, as gear which caught salmon efficiently in the fjords and coastal areas was developed, the

143

fishing intensity in these areas increased, and at present the fishery in these areas catch most of the salmon returning to Norway.

Systematic collection of data from the different salmon fisheries began in 1876. The results were, however, very poor; less than 100 rivers were included and the sea catch was very incomplete. Later, the statistics were improved, but there are still great problems in obtaining reliable figures. Even though the official statistics seriously underestimate the actual catch, it is generally accepted that the data describe the fluctuations in catches and the development of the fisheries (Anon., 1985a; Rødstøl and Gerhardsen, 1983).

The official figures of the sea catch, river catch and total catch since 1876 are presented in Figure 8.1. There is great

Figure 8.1: The reported catch of Atlantic salmon in Norway from 1876. Salmon caught in the northern Norwegian Sea by Norwegian fishermen are included

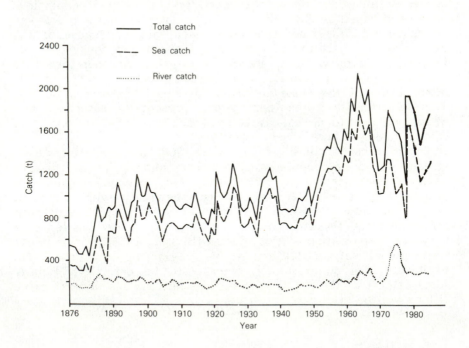

variation in catches between years. Initially the total catches increased, but from 1900 to about 1950 they were remarkably stable. From the beginning of the 1950s the total catch increased considerably and reached a peak in the middle of the 1960s. This increase was probably due to a number of factors such as improved catch statistics, increased abundance of salmon due to stock enhancement and an increased fishing effort. In spite of improved catch statistics the total catches declined towards the end of the 1960s.

This could have been caused by natural factors, but it is probably more likely that at least a part of the downward trend might be explained by the development of a long line interceptory fishery in the Norwegian Sea and a subsequent reduction in the numbers of salmon returning to home water. In addition there are also grounds for suggesting that there could be an overexploitation of salmon in home waters, and a subsequent reduction in smolt production. Effects of pollution, stream regulation, etc. are probably only of minor importance during the most recent period.

Description of the present fishery

Many different types of gear have been used to catch salmon. At present, drift nets, bag nets, bend nets, stationary lift nets and stake nets are the legal methods used in the commercial fishery in saltwater in Norway. There is also an expanding sport fishery with rods in the fjords. Rod-and-line fishing is the most common method in rivers. In a very limited number of rivers, nets and traps can be operated, but the regulatory measures of these fisheries follow those of the commercial sea fisheries.

In recent years, there has been a substantial change in the structure of the Norwegian home water fisheries (Figure 8.2). The

Figure 8.2: The reported catch of Atlantic salmon in Norway by different gear from 1960

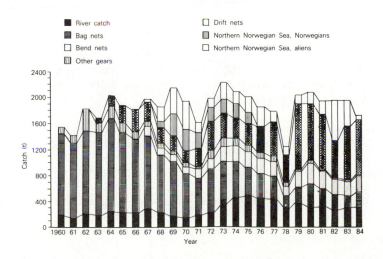

increase in the drift net and bend net fisheries has been very pronounced, and these methods have replaced the bag nets. The reported increase in catches by drift nets after 1978 was probably due partly to the introduction of a licensing system which required log books to be kept and the subsequent resulting improvement in catch statistics.

Drift nets are mostly manufactured from monofilament twine and can be operated between the baseline and the 12 mile limit. All other methods are used inside the baseline. Monofilament twine is used in bend nets, while the other nets are manufactured from spun nylon twine. The legal minimum mesh size is 58 mm knot to knot (116 mm stretched mesh) and most bag nets, lift nets and stake nets are of this size. The mesh size of bend nets can vary, and more than 70 per cent of drift nets have mesh sizes of 65-70 mm knot to knot.

Drift and bend net fishing is permitted during the period 1 June to 5 August, and bag nets and lift nets may operate from 15 May to 5 August. There is a weekly close time for all nets extending from 1800 h on Friday to 1800 h on the following Monday. A drift net fisherman also requires a licence to operate, and in 1984 a total of 627 licences was issued. The number of drift nets which can be fished per licence is restricted to 20, 35 and 50 in vessels, with 1, 2 and 3 fishermen respectively. Bend, bag, stake and lift net fisheries belong to the owners of the adjoining land, and at present do not require a licence. River fishing is carried out mainly with rod and line and with few exceptions, river fishing can be carried out from 1 June to 1 September. The fishery belongs to the owners of the river.

In 1984 there were 21,210 drift nets, 1,697 bag nets and 35 lift nets in operation in the Norwegian home water fishery (Anon., 1985a). There were no official figures for stake nets, but these are very few and restricted to a small area in south east Norway.

New regulations in the Norwegian home water fishery
On the 11 April 1986 the Norwegian Government decided to impose further regulations on the salmon fishery in Norwegian home water. Among these are:

(1)   Total ban on the drift net fishery (from 1989)
(2)   Total ban on the use of monofilament in salmon nets (from 1988)
(3)   Introduction of a licence scheme for anchored gear (from 1988)
(4)   Fishing for salmon in rivers with weak populations might be banned.

In addition, the Government also decided to shorten the fishing season for bag nets and lift nets.

EXPLOITATION

Norwegian salmon stocks are exploited in home waters and in the Faroese fishery. Some Norwegian salmon are also caught in the West Greenland fishery. There is also a small interceptory fishery on Norwegian salmon off the west coast of Sweden. For many years, Norwegian salmon stocks have been heavily exploited. Jensen (1979a, 1981) and Rosseland (1979) presented exploitation rates for River Eira and River Laerdalselv stocks. Autumn counts of spawning salmon in River Laerdalselv, West Norway, were combined with the number of salmon caught in the same year to calculate the rate of exploitation in the river. This rate was combined with the official statistics of salmon catches in the rivers and in the sea for the Sogn salmon Fisheries District to estimate the total rate of exploitation of River Laerdalselv's salmon stock. The same method but based on salmon catches and counts of salmon spawning redds was used in River Eira, West Norway. For both stocks the total rates of exploitation were apparently very high (0.80-0.97) (Tables 8.1 and 8.2). The method used is only approximate.

Table 8.1: Estimated rates of exploitation on the River Laerdalselv salmon stock. C_S = sea catch, C_R = river catch, μ_R = rate of exploitation in the river, μ_T = total rate of exploitation

Year	River catch (No.)	Spawners (No.)	μ_R	C_S	C_R	C_S:C_R	μ_T
1960	827	1,059	0.44	62,901[a]	14,246[a]	4.16	0.80
1962	1,434	1,888	0.43	91,172[a]	19,532[a]	4.67	0.81
1964	1,343	991	0.58	83,656[a]	16,659[a]	5.02	0.89
1965	1,640	1,285	0.56	85,059[a]	19,765[a]	4.30	0.87
1966	1,010	858	0.54	67,188	9,042	7.43	0.91
1968	1,385	645	0.68	42,304	9,594	4.41	0.92
1972	1,401	1,295	0.52	64,202	11,906	5.39	0.87
1974	2,519	1,579	0.61	53,352	23,615	2.26	0.84

Note: a, Salmon + sea trout.

Source: Rosseland (1979); Jensen (1981).

Table 8.2: Estimated rate of exploitation on the River Eira stock. The symbols used are as in Table 8.1

Year	River catch (No.)	Redds (No.)	Spawners (No.)	μ_R	C_S	C_R	$C_S:C_R$	μ_T
1964	125	93	93–186	0.57–0.40	41,291[a]	5,120[a]	8.06	0.92–0.86
1965	240	170	170–340	0.59–0.41	42,197[a]	7,905[a]	5.34	0.90–0.82
1966	199	119	119–238	0.63–0.46	41,429	6,279	6.60	0.93–0.87
1967	259	82	82–164	0.76–0.61	43,801	6,644	6.59	0.96–0.92
1968	251	55	55–110	0.82–0.70	42,880	10,645	4.03	0.96–0.92
1969	256	97	97–194	0.73–0.57	46,262	7,206	6.42	0.95–0.91
1970	295	62	62–124	0.83–0.70	34,960	6,794	5.15	0.97–0.93
1971	264	55	55–110	0.83–0.71	53,116	9,333	5.69	0.97–0.94
1972	316	98	98–196	0.76–0.62	79,556	8,060	9.87	0.97–0.95
1973	389	139	139–278	0.74–0.58	79,373	20,578	3.86	0.93–0.87
1974	292	179	179–358	0.62–0.45	53,255	15,919	3.35	0.88–0.78

Note: a, Salmon + sea trout.

Source: Jensen (1981)

Exploitation of the River Imsa stock

As part of a large sea ranching programme in the River Imsa hatchery-reared and wild salmon smolts have been Carlin-tagged (Carlin, 1955) and released each year since 1981.

River Imsa is situated near Stavanger in south-western Norway. The catchment area is 128 km^2, of which 12 per cent is lake surface. The river near the mouth is about 10 m wide, the yearly average water discharge is 4.5 m^3/s. A Wolf trap 100 m above the river mouth catches all descending fish larger than about 10 cm at water discharges less than 30 m^3/s. Under the Wolf trap an upstream trap is located to catch all ascending salmon.

Apart from Atlantic salmon, large populationss of brown trout, Salmo trutta L., Arctic charr, Salvelinus alpinus (L.), whitefish, Coregonus lavaretus (L.), Atlantic eels, Anguilla anguilla (L.) and three-spined stickleback, Gasterosteus aculeatus L. are present in the watercourse. Escaped rainbow trout, Salmo gardneri Richardson from nearby fishfarms also ascend the river to spawn, but no stock has been established. The salmon spawning areas are found in the river between the mouth and the first lake, a distance of only 1 km, so that the salmon stock is very small. Most naturally produced smolts are 2, with a few 3 years old. The spawning stock consists of grilse and a smaller component of 2- sea winter fish.

In May-June 1981-1984 wild smolts caught in the smolt trap were anaesthetised with chlorbutol, individually tagged with Carlin tags and released below the trap. During the same years Carlin-tagged 1+ and 2+ hatchery smolts from the River Imsa stock were released below the trap, near the river mouth.

Most recaptures were reported by commercial fishermen. The returns to River Imsa were monitored in the fish trap just above the river mouth.

The number of reported recaptures from the different experiments in different years are shown in Table 8.3. The Norwegian sea recaptures are mainly those reported from Faroese fishermen, although there are a few recaptures from the Danish long-line fishery, especially for the smolt releases in 1981 and 1982. Recaptures in the Norwegian home water fishery are those taken in saltwater.

Exploitation rates were estimated (Anon., 1985b), and the following assumptions and approximations were made:

(1) Tagged fish escaping home water fisheries return to River Imsa.
(2) The monthly instantaneous mortality rate was taken to be 0.01.
(3) Tagged and untagged fish were equally vulnerable to the gear.

149

Table 8.3: Number of salmon reported recaptured in the Norwegian Sea, home waters and the River Imsa trap from smolt taggings in River Imsa 1981-1984

		1 - SW				2 - SW	
Smolt type	No. tagged	Norw. Sea	Norw. home wat.	R. Imsa trap	Norw. Sea	Norw. home wat.	R. Imsa trap
Released 1981							
R. Imsa wild	3,214	0	244	66	34	59	9
R. Imsa 2+	5,819	5	235	114	36	34	6
Released 1982							
R. Imsa wild	736	0	17	5	7	4	1
R. Imsa 1+	5,581	0	36	1	12	15	1
R. Imsa 2+	8,501	17	249	25	55	25	4
Released 1983							
R. Imsa wild	1,287	0	71	31	7	8	1
R. Imsa 1+	5,861	0	11	1	1	1	0
R. Imsa 2+	6,052	4	80	12	7	5	0
Released 1984							
R. Imsa wild	936	0	41	30			
R. Imsa 1+	1,863	0	8	5			
R. Imsa 2+	7,445	13	137	48			

(4)   Non-catch fishing mortality was taken to be negligible.
(5)   The mean dates of capture in the Norwegian Sea, Norwegian home waters and River Imsa trap were taken to be 15 March, 15 July and 15 September respectively.
(6)   Tag reporting efficiency was assumed to range between 50 and 70 per cent in Norwegian home waters, and was estimated to be 75 per cent in the Norwegian Sea.

Adjusting for non-reported tags for natural mortality between the Norwegian Sea, Norwegian home waters and River Imsa, it was possible to produce estimates of the number of salmon of the different year classes available both to the Norwegian Sea and Norwegian home water fisheries, and hence to estimate exploitation rates.

Tables 8.4 and 8.5 give estimates of 1-SW and 2-SW salmon of the River Imsa stock available to the different fisheries and estimated exploitation rates in these fisheries for the two levels of tag reporting rate in the Norwegian home water fishery. In all groups, exploitation in the Norwegian Sea of salmon in their first sea-winter is zero for wild salmon and salmon released as 1+ smolts. There is a small exploitation of salmon released as 2+ smolts. The reason for this is likely to be that because 2+ smolts are bigger than wild 1+ smolts, they are larger and thus more vulnerable to long-lines during their first sea winter.

Home water exploitation is very high both for grilse and 2-SW salmon. The exploitation rate for 2-SW fish in the Norwegian Sea is smaller than for home water, but due to the fact that a higher number of 2-SW fish is available in the Norwegian Sea, the absolute catch of 2-SW fish in the Norwegian Sea is relatively high.

EFFECT OF FISHERY

Net damage

Net marks on salmon appear when the fish escape from different types of nets. Damage varies from small injuries on the dorsal fin to serious damage on the body, depending on how the salmon escaped from the net. With the introduction of nylon and monofilament material in salmon nets an increase in fishing effort with these nets occurred in Norway. During the 1960s and 1970s there was an increase in the proportion of net-marked salmon in Norwegian rivers. Therefore a systematic recording of this problem was initiated in 1978.

Jensen (1979b) reported a drop-out rate of 5.5 per cent by patrolling the nets set from a Norwegian driftnet vessel. However, this is a minimum estimate because of the relatively high probability that salmon entangled in the nets would drop out

Table 8.4: Estimated number of 1-SW and 2-SW salmon of the River Imsa stock available to the Norwegian Sea fishery and Norwegian home water fishery, and estimated exploitation rates. The number of salmon caught in the trap in the River Imsa is considered to be the total river escapement. The estimates are based on 75 and 50 per cent tag reporting rates in Norwegian Sea and Norwegian home waters respectively

Smolt tagged	No. tagged	1-SW Norw. Sea No. of fish avail.	1-SW Norw. Sea Expl. rate	1-SW Norw. home wat. No. of fish avail.	1-SW Norw. home wat. Expl. rate	1-SW Norw. home wat. No.in trap	2-SW Norw. Sea No of fish avail.	2-SW Norw. Sea Expl. rate	2-SW Norw. home wat. No. of fish avail.	2-SW Norw. home wat. Expl. rate	2-SW Norw. home wat. No. in trap
Released 1981 R.Imsa wild	3,214	776	0.00	555	0.88	66	177	0.25	127	0.93	9
R.Imsa 2+	5,819	757	0.01	586	0.80	114	125	0.38	74	0.92	6
Released 1982 R.Imsa wild	736	61	0.00	39	0.87	5	18	0.50	9	0.89	1
R.Imsa 1+	5,581	130	0.00	73	0.99	1	48	0.33	31	0.97	1
R.Imsa 2+	8,501	712	0.03	524	0.95	25	129	0.57	54	0.93	4
Released 1983 R.Imsa wild	1,287	211	0.00	174	0.82	31	27	0.33	17	0.94	1
R.Imsa 1+	5,861	27	0.00	23	0.96	1	3	0.31	2	1.00	0
R.Imsa 2+	6,052	205	0.02	172	0.93	12	19	0.47	10	1.00	0
Released 1984 R.Imsa wild	936	118	0.00	113	0.73	30					
R.Imsa 1+	1,863	22	0.00	21	0.76	5					
R.Imsa 2+	7,445	353	0.05	323	0.85	48					

Table 8.5: Estimated number of 1-SW and 2-SW salmon of the River Imsa stock available to the Norwegian Sea fishery and Norwegian home water fishery, and estimated exploitation rates. The number of salmon caught in the trap in River Imsa is considered to be the total river escapement. The estimates are based on 75 and 70 per cent tag reporting rates in Norwegian Sea and Norwegian home waters respectively

| | | 1-SW | | | | | 2-SW | | | | |
| | | Norw. Sea | | Norw. home wat. | | | Norw. Sea | | Norw. home wat. | | |
Smolt type	No. tagged	No. of fish avail.	Expl. rate	No. of fish avail.	Expl. rate	No. in trap	No. of fish avail.	Expl. rate	No. of fish avail.	Expl. rate	No. in trap
Released 1981 R.Imsa wild	3,214	592	0.00	416	0.84	66	142	0.32	93	0.90	9
R.Imsa 2+	5,819	596	0.01	452	0.74	114	105	0.46	55	0.89	6
Released 1982 R.Imsa wild	736	48	0.00	29	0.83	5	16	0.56	7	0.86	1
R.Imsa 1+	5,581	98	0.00	52	0.98	1	39	0.41	22	0.95	1
R.Imsa 2+	8,501	549	0.04	382	0.93	25	115	0.63	40	0.90	4
Released 1983 R.Imsa wild	1,287	163	0.00	133	0.76	31	22	0.41	12	0.92	1
R.Imsa 1+	5,861	20	0.00	17	0.94	1	2	0.50	1	1.00	0
R.Imsa 2+	6,052	154	0.03	126	0.90	12	16	0.56	7	1.00	0
Released 1984 R.Imsa wild	936	94	0.00	90	0.66	30					
R.Imsa 1+	1,863	17	0.00	16	0.69	5					
R.Imsa 2+	7,445	272	0.06	245	0.80	48					

a short time after contact with the nets. This is supported by Angelsen and Holm (1978) working with farmed salmon in an aquarium, who observed that salmon which are not able to escape from nets within a few seconds will be caught.

Table 8.6 shows the total net mark frequency on salmon from different rivers during the period 1978-1983. The frequency varies both between rivers and within size groups, one characteristic was that grilse showed a higher net mark frequency than multi sea-winter salmon, which could be explained by the selective nature of the nets. There was also a variation between years. Most salmon were, however, only slightly damaged.

Table 8.6: Total net mark frequency on Atlantic salmon from different Norwegian rivers 1978-1983 (N=number of salmon examined)

River	1978		1979		1980		1981		1982		1983	
	%	N	%	N	%	N	%	N	%	N	%	N
R.Komagelv	44	91	46	123	30	107	29	41	41	32	46	99
R.Børselv	51	90	51	81	52	62	47	90	38	86	39	46
R.Målselv	75	201	19	270	12	252	37	544	54	366	36	409
R.Ranaelv	57	260	41	152	25	174	41	29	30	54	33	45
R.Namsen	26	455	30	438	26	843	14	595	12	497	36	569
R.Surna	52	228	31	206	12	277	13	281	3	123	9	137
R.Etneelv	52	510	46	622	33	476	29	362	37	612	27	729

Sources: Hansen (1982); Gausen (1984).

Salmon from the River Ørsta stock showed the highest net mark frequency. The great majority of fish in this river are grilse with average weight between 1.2 and 2.1 kg. The net mark frequency on grilse during the period 1977-1981 varied between 73 and 91 per cent (Table 8.7). Considering that this represents the fraction of fish that are not caught by nets, the probability of contact with nets must be higher than the net mark frequency indicates. In addition, the river escapees are exploited by rods. The total exploitation rate on this stock must be too high, which is confirmed by a decline in catches in recent years (Table 8.7).

The main contributors to damaged salmon are drift nets and bend nets which can damage all size groups of salmon. The bag nets, however, only damage grilse which pass through the meshes.

Table 8.7: Net mark frequency and reported catch of Atlantic salmon in River Ørstaelv

Year	Net mark frequency (%)	No. of observations	Reported salmon catch (tonnes)
1970	–	–	11.2
1971	–	–	12.8
1972	–	–	5.6
1973	–	–	13.7
1974	–	–	13.3
1975	–	–	14.4
1976	–	–	6.4
1977	81	85	3.2
1978	84	425	2.7
1979	81	539	6.5
1980	73	331	3.0
1981	91	196	1.8
1982	–	–	1.1
1983	–	–	4.1
1984	–	–	2.8

In 1979 and 1980 net-damaged and undamaged salmon were caught in bag nets in a coastal area, in a fjord and in a river (Hansen, 1980; Hansen and Roald, 1981). In no cases were the differences in recapture rate between net-damaged and undamaged salmon significantly different (Table 8.8). There are, however, some uncertainties. We do not know whether net injuries affect the catchability of salmon. The composition of tagged salmon was a mixture of fish with old and fresh net injuries. This might hide a possible mortality a few hours or days after the salmon were injured. However, the recapture rate of salmon that were damaged by the bag nets when caught was not significantly smaller than the control group, although they had fresh wounds.

Ascending net-marked and undamaged salmon in the River Imsa trap were tagged and released above the trap in the period 1976-1981 (Hansen, 1982). When monitoring the descending kelts there was no significant difference in survival of net-marked and undamaged fish (Table 8.9). Most net-marked salmon had, however, relatively small injuries. Recently Berg, Abrahamsen and Berg (1986) reported from River Vefsna that badly damaged salmon do not spawn in normal spawning sites and that spawning success was poor.

Table 8.8: Recaptures of undamaged and net-damaged Atlantic salmon tagged and released on the coast (Kvaløya), in a fjord (Vefsnfjord) and in a river (River Vefsna). At Kvaløya and Vefsnfjord the salmon were caught in bagnets. In River Vefsna the fish were caught in a trap

| | 1979 | | 1980 | |
Group	Number tagged	% recapture	Number tagged	% recapture
Kvaløya undamaged	272	28.7	265	27.5
Kvaløya net damaged	71	28.2	65	23.1
Kvaløya net damaged when tagged	-	-	136	24.3
Vefsnfjord undamaged	309	42.1	226	37.6
Vefsnfjord net damaged	110	48.2	85	25.9
Vefsnfjord net damaged when tagged	-	-	82	30.5
River Vefsna undamaged	304	21.1	212	7.5
River Vefsna net damaged	758	24.8	146	11.0

Sources: Hansen (1980); Hansen and Roald (1981).

In 1980, net-damaged salmon were kept in net pens in fresh and salt-water (Table 8.10) (Hansen and Roald, 1981). There was a significantly increased mortality for both lightly damaged and badly damaged fish kept in saltwater when compared with the undamaged control groups. Damaged salmon kept in fresh water had a lower mortality. A similar experiment carried out in brackish water (5-10 per cent salinity) in 1979 showed no mortality at all (Roald, 1980). These experiments and the work of Bouck and Smith (1979) and Rosseland, Lea and Hansen (1982) suggest that salt-water is a very bad environment for salmon with fresh skin lesions. Badly damaged fish encounter osmotic problems which may lead to death if they do not enter a more isotonic environment. The reason why tagged damaged fish in a coastal area seemed to survive well, might be that the salmon move relatively fast into brackish water (Hansen and Roald,

Table 8.9: Number of ascending and descending net-marked and undamaged salmon in River Imsa

	With net marks			Without net marks		
	No. ascents	No. descents	Descents/ ascents (%)	No. ascents	No. descents	Descents/ ascents (%)
1976	12	9	75.0	86	59	68.6
1977	6	5	83.3	69	52	75.4
1978	18	8	44.4	45	29	64.4
1979	15	13	86.7	54	37	68.5
1980	32	19	59.4	43	30	69.8
Total	83	54	65.1	297	207	69.7

Source: Hansen and Roald (1981).

Table 8.10: Per cent mortality of undamaged and net-damaged Atlantic salmon kept in net pens in freshwater and seawater (salinity 19-24%) from the end of June to October/November 1980. n=number of fish at the start of the experiment

	Seawater	n	Freshwater	n
Undamaged	7.1	42	0.0	21
Some damaged	25.0	40	4.5	22
Seriously damaged	63.0	19	25.0	4

Source: Hansen and Roald (1981).

1981). A large part of the recaptures from the tagging experiments mentioned earlier, was reported from the inner coastal areas and fjords where water could be expected to be brackish. Ricker (1976), Ritter, Marshall and Reddin (1979) and Reddin (1980) suggested that mortality of salmon escaping from nets in the high seas is higher than for salmon escaping from coastal nets.

Selective fishing

Salmon gear is selective according to size, bodyshape, age at maturity, sex, seasonal return pattern etc. Ricker (1981) found in his analysis of body weights of different species of Pacific salmon that fish size decreased during the period 1950-1975. Although other variables could have affected the size decrease, Ricker concluded that at least a part of the decrease was caused by a size-selective fishery.

In the Miramichi River in Canada, Schaffer and Elson (1975) showed an increased proportion of grilse in the system. They suggested that the reason for this change was a size-selective fishery, e.g. very high exploitation of 2- and 3-sea winter fish. They also suggested a genetic change in the stock.

All fisheries on Norwegian salmon are selective. The total exploitation on grilse is smaller than on multi-sea winter salmon. Because the majority of the grilse are males, the gear harvesting Norwegian salmon is not only selective on size and age at maturity, but even sex selective, selecting for females.

The fishing season for salmon in Norwegian home water is short. Sea fishing is carried out between 15 May and 5 August and angling in rivers between 1 June and 1 September. Most salmon enter Norwegian rivers during this period, but in many rivers there is also a run after the fishing season has closed. This harvesting pattern will tend to exploit the early run more heavily than the late run. Salmon which enter the rivers in the autumn may be the major component in a spawning stock in river populations which are heavily exploited.

There is little direct evidence that a selective fishery has changed the genetic frequency in Norwegian salmon populations. However, in several Norwegian rivers it has been observed that salmon ascend later in the season compared with earlier years. This might be a result of selective fishing, but stock enhancement using late-returning fish as broodstock might also contribute to this phenomenon.

If we want to avoid possible undesirable effects of a heavy selective fishing pressure, it is important to make sure that there are sufficient numbers of spawners in the different populations, in such a way that the spawning stock is representative of the genetic variation in the population (Ryman and Ståhl, 1980). This is not the case for many Norwegian salmon stocks at present because of the very heavy mixed stock exploitation rates. The purpose of the drastic changes in the regulations of the Norwegian home water fishery from the 1988 and 1989 seasons is to change this present pattern.

Status of Exploitation of Atlantic Salmon in Norway

SUMMARY

At present fishing for salmon in Norwegian home water is carried out with drift nets, bag nets, bend nets, lift nets and stake nets in the sea, and mainly by rods in the rivers. Catch statistics have been collected since 1876, but reliability of the statistics is questionable. The rates of exploitation on many Norwegian salmon stocks are very high, and data are presented and discussed for River Laerdal, River Eira and River Imsa stocks. As a result of this heavy exploitation with nets, the frequency of net-marked salmon entering rivers is high especially for grilse, although most net-marked salmon are only slightly damaged. It is suggested that mortality of salmon that are net-damaged near or in brackish water is relatively small compared with salmon that escape from nets on the high seas. Some possible effects of selective fishing are also discussed.

REFERENCES

Anon. (1985a) Salmon and sea trout fisheries 1984. Central Bureau of Statistics of Norway, Oslo - Kongsvinger, 96 pp. (in Norwegian with English summary)

Anon. (1985b) Report of meeting of working group on North Atlantic salmon. ICES C.M. 1985/Assess, 11, 67 pp.

Angelsen, K.K. and Holm, M. (1978) Rapport om garnforsøk med laks, Fisk og Fiskestell, 8, 15-17. (in Norwegian)

Berg, M., Abrahamsen, B. and Berg, O.K. (1986) Spawning of injured compared to uninjured female Atlantic salmon, Salmo salar L. Aquaculture and Fisheries Management, 17, 195-9

Bouck, G.R. and Smith, S.D. (1979) Mortality of experimentally descaled smolts of coho salmon (Oncorhynchus kisutch) in fresh and salt water. Transactions of the American Fisheries Society, 108, 67-9

Carlin, B. (1955) Tagging of salmon smolts in the River Lagan. Report Institute of Freshwater Research Drottningholm, 36, 57-74

Gausen, D. (1984) Garnskaderegistreringer av laks og sjøørret 1983. Report Fiskekontoret, Direktoratet for vilt og ferskvannsfisk 1, 13 pp. (in Norwegian)

Hansen, L.P. (1980) Tagging and recapture of net-marked and undamaged Atlantic salmon in two sea localities and two rivers in Norway. ICES C.M. 1980/M:32, 8 pp.

Hansen, L.P. (1982) Registrering av garnskader på laks og sjøørret og merking av uskadet og garnskadet laks 1982. Report Fiskeforskningen, Direktoratet for vilt og ferskvannsfisk 3, 15 pp. (in Norwegian)

Hansen, L.P. (1986) The data on salmon catches available for analysis in Norway. In Jenkins D. and Shearer W.M. (eds), The status of the Atlantic salmon in Scotland. ITE symposium No. 15, Institute of Terrestrial Ecology, Abbots Ripton, pp. 79-83

Hansen, L.P. and Roald, S.E. (1981) Net mark registration and effects of damage caused by nets on Atlantic salmon Salmo salar L. in Norway 1980. ICES C.M. 1981/M:8, 20 pp.

Jensen, K.W. (1979a) Lakseundersøkelser i Eira. In T.B. Gunnerød and P. Mellquist (eds) Vassdragsreguleringers biologiske virkninger i magasiner og lakseelver. Norges Vassdrags og Elektrisitetsvesen, Direktoratet for vilt og ferskvannsfisk, pp. 165-71 (in Norwegian)

Jensen, K.W. (1979b) Experimental drift - netting for salmon in Norway 1978. ICES C.M. 1979/M:9, 4 pp.

Jensen, K.W. (1981) On the rate of exploitation of salmon from two Norwegian rivers. ICES C.M. 1981/M:11, 8 pp.

Jensen, K.W. and Snekvik, E. (1972) Low pH levels wipe out salmon and trout populations in southern Norway. Ambio, 1, 223-5

Johnsen, B.O. and Jensen, A.J. (1985) Parasitten Gyrodactylus salaris på laksunger i norske vassdrag. Report Regulering-sundersøkelsene, Direktoratet for vilt og ferskvannsfisk 12, 145 pp. (in Norwegian)

Leivestad, H., Hendrey, G., Muniz, I.P. and Snekvik, E. (1976) Effect of acid precipitation on freshwater organisms. In F.H. Braekke (ed.), Impact of acid precipitation on forest and freshwater ecosystems in Norway, SNSF project FR 6/76, Oslo-Ås, pp. 87-111

Reddin, D.G. (1980) Non-catch fishing mortality at West Green-land and in home water fisheries. ICES C.M. 1980/M:24

Ricker, W.E. (1976) Review of the rate of growth and mortality of Pacific salmon in salt water, and non-catch mortality caused by fishing. Journal of The Fisheries Research Board of Canada, 33, 1483-524

Ricker, W.E. (1981) Changes in the average size and average age of Pacific salmon. Canadian Journal of Fisheries and Aquatic Sciences, 38, 1636-56

Ritter, J.A., Marshall, T.L. and Reddin, D.G. (1979) A review of non-catch fishing mortality as it relates to Atlantic salmon (Salmo salar L.) fisheries. ICES C.M. 1979/M:25

Roald, S.O. (1980) Net marks on Atlantic salmon (Salmo salar) in Norwegian coastal areas. Preliminary report on gross, histological, serological and bacteriological signs. ICES, C.M. 1980/M:34

Rødstøl, R. and Gerhardsen, G.M. (1983) Økonomiske og distrikts-politiske aspekter av lakse- og sjøaurefisket i Norge. Fiskeriøkonomiske Skrifter, ser A, no 5. Fiskeriøkonomisk Institutt, Norges Handelshøyskole, Bergen. 209 pp. (in Norwegian)

Rosseland, L. (1979) Litt om bestand og beskatning av laksen fra Laerdalselva. In T.B. Gunnerød and P. Mellquist (eds), Vassdragsreguleringers biologiske virkninger i magasiner og lakseelver, Norges Vassdrags og Elektrisitetsvesen, Direk-toratet for vilt og ferskvannsfisk. pp. 174-86. (in Norwegian)

Rosseland, B.O., Lea, T.B. and Hansen, L.P. (1982) Physiological effects and survival of Carlintagged and descaled Atlantic salmon Salmo salar L. in different water salinities. ICES C.M. 1982/M:30, 23 pp.

Ryman, N and Ståhl, G. (1980) Genetic changes in hatchery stocks of brown trout (Salmo trutta). Canadian Journal of Fisheries and Aquatic Sciences, 37, 82-7

Schaffer, W.M. and Elson, P.F. (1975) The adaptive significance of variations in life history among local populations of Atlantic salmon in North America. Ecology, 56, 577-90

Chapter Nine

EXPLOITATION OF ATLANTIC SALMON IN ICELAND

Thór Gudjónsson
Institute of Freshwater Fisheries, Reykjavík, Iceland

INTRODUCTION

Iceland is an island in the North Atlantic situated between 63° 30' and 24° 32' W longitude and between 13° 30' and 24° 32' W longitude. It covers an area of 103,000 km^2. The climate is cool, temperate and oceanic with rapid changes. The summers are cool but the winters are relatively warm. There are only five species of fish living partly or wholly in freshwater native to the country. These are the Atlantic salmon (Salmo salar L.), the sea trout (Salmo trutta L.) as well as a land-locked variety, the brown trout, the sea charr (Salvelinus alpinus L.) and land-locked variety, the lake charr, the European eel (Auguilla anguilla L.) and the three spined stickleback (Gasterosteus aculeatus L.).

RIVERS

There are about 250 rivers in Iceland of which about 80 rivers and river systems hold salmon. There are three types of rivers with respect to their origin. These are spring-fed rivers, direct run-off rivers and glacial rivers. In some cases the rivers are mixtures of all three types. They vary with respect to types of riverbeds, discharge, variation in amount of flow, temperature of the water, ice cover and anchor ice. Since Iceland· is a mountainous country there are many hindrances in the rivers for migratory fishes. Salmon are found mostly in short streams and the lowland parts of the rivers and are generally most abundant in rivers which have their origin in lakes or flow through lakes.

The best salmon rivers with one exception are located in the western half of the country. The river system Ölfusá-Hvítá in the south and the river system of Hvítá in the west produce the largest catches which amount to about 40 per cent of the total yield in the country. The salmon rivers in the south and the west

Figure 9.1: Distribution of salmon-producing streams in Iceland and the average yearly catch 1971-1980 summarised for each of the seven districts

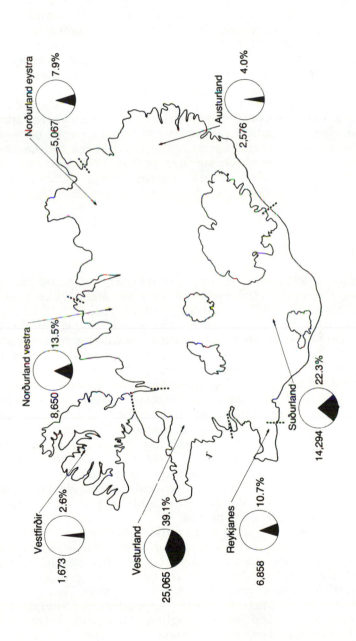

yield over 70 per cent of the catch and those of the north and the east less than 30 per cent (Figure 9.1).

On the eastern half of the country there are a few salmon rivers. These are located in the north and northeast districts. The Laxá in Thingeyjarsýsla district is by far the most productive river in these parts and one of the best salmon rivers in the country with respect to the number of salmon caught annually by anglers. It is also one of three rivers in the country yielding salmon with largest average weight and the one where most often the largest salmon are caught each year.

Northeast Iceland is the coldest part of the country and the sea off the coast there colder than elsewhere around the island. The fluctuation in salmon catches in rivers in these parts can be great, the ratio between the catches during the best years and the poorest ones during the period 1976-1985 was between 10:1 and 20:1, whereas the corresponding figures for larger salmon rivers elsewhere in the country were less than 4:1 during the same period.

CATCH STATISTICS

Official records of salmon catches are available since 1897. During the years 1897-1909 the average annual catch was 5,167 fish. From 1910-1950 the average catch rose to about 15,000 and further to about 64,000 during the years 1971-1980 the average total weight being about 240 metric tons. The record catch year was 1978 when 78,625 salmon were caught by rod and line and gill nets weighing 283 metric tons. Besides this 1,953 salmon returned to ocean ranching sites weighing about 7 metric tons. In 1979, 1981 and 1983 weather conditions were most unfavourable to the freshwater stages of the salmon and sea temperatures were unusually low causing a considerable drop in catches in general, especially in 1981, 1982 and 1984. Figure 9.2 shows the number of salmon caught from 1946-1986. At the lower right of the Figure the returns of salmon to ocean ranching sites are shown. The ratio between grilse and salmon varies most often between 50:50 to 60:40. There are only a few salmon that spend more than two years in the sea.

It is to be expected that the fishermen at first showed a certain reluctance to report accurately their catches since the records were also used in connection with taxation. But as time passed the catch statistics have improved considerably. During the last 40 years great emphasis has been put on obtaining accurate catch statistics with considerable success. Special logbooks are placed in all angling huts and lodges for daily recording of catches, and netsmen are cooperative in sending in their catch figures after each fishing season. The Institute of

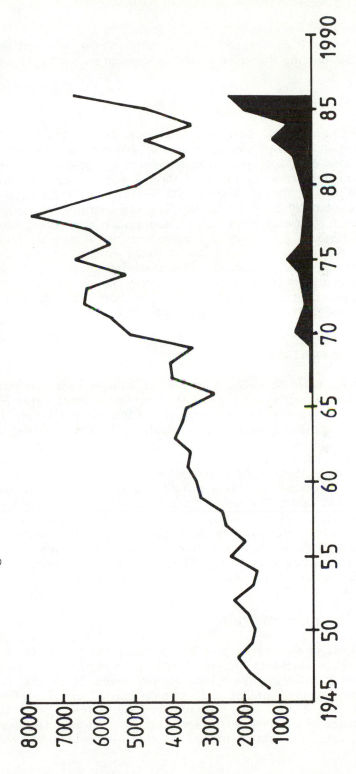

Figure 9.2: Number of Atlantic salmon caught by rod-and-line and gill nets in Iceland from 1946 to 1986. At the bottom right are the catches of ocean ranched salmon from 1966 to 1986

Source: Institute of Freshwater Fisheries.

Freshwater Fisheries collects the logbooks in the autumn each year for putting the information on computer.

FISHING METHODS

According to law, only two methods of catching salmon are allowed at present, i.e. angling and gill-netting. Earlier, salmon were also caught by seines, in traps and by spears. Salmon fishing is permitted for a maximum of three months in each salmon river during the period 20 May to 20 September. Netting is allowed only for half of each week from Tuesday morning to Friday night, and angling in each river for 12 hours each day during the period 7 a.m. and 10 p.m., with a limited number of rods in each river.
   Netting for salmon takes place mainly in three glacial rivers: the Thjórsá and the Ölfusá-Hvítá in the south and the Hvítá in the west. In most of the other rivers the salmon fishing is leased to anglers exclusively. The nets take annually about 30-40 per cent of the total river catches.

ADMINISTRATION

The freshwater fisheries including the salmon fishery comes under The Ministry of Agriculture. The Director of Freshwater Fisheries administers the fisheries under the Ministry and is the head of the Institute of Freshwater Fisheries. Locally the fishing associations, which are about 150 in number, manage the fishery for individual rivers and lakes, carry out enhancement programmes, hire bailiffs and often own angling huts and lodges. Fishing rights are privately owned and go with the land adjoining rivers and lakes. The owners are most often local farmers since almost all rivers and many lakes are in agricultural areas. The members of each fishing association are the owners or the tenants of farms on respective rivers. The fishing associations have functioned effectively. They play an important part in the administration of the salmon fishery in the country.

RATE OF EXPLOITATION

Iceland is in an unusual position to study the exploitation rate of the salmon stocks in individual rivers, since salmon fishing, with a minor exception, takes place in the rivers. There are, however, unknown numbers of Icelandic salmon caught off West Greenland and in the Faroe area, as indicated by a few Carlin tags and microtags which have been retrieved from salmon taken in these areas, and by a salmon tagged off West Greenland in 1972 which was caught in a West Iceland river the following year (Figure 9.3). Eight tags have been returned from West Greenland, one

Figure 9.3: Recoveries abroad of Atlantic salmon tagged and released as smolts in Iceland and recovery of a salmon tagged at West Greenland and caught in Iceland

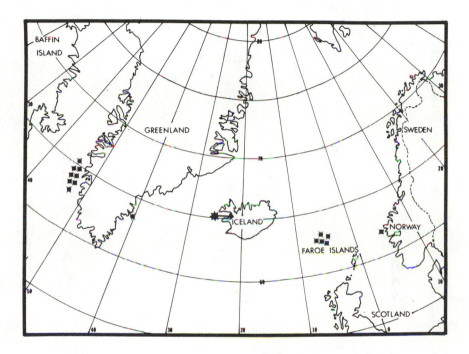

from East Greenland, five from the Faroes and one from West Norway. Since 1974 microtags have been used almost exclusively to tag salmon smolts in Iceland. It is not known how many micro-tagged Icelandic salmon may have been caught off West Greenland from 1975 until 1985 when scanning for microtags was started there. In this first year of scanning one Icelandic micro-tag was returned out of a total of 36 microtags retrieved.

A few attempts have been made to assess the exploitation rate of salmon in Icelandic rivers with different methods. In the Ellidaár direct counts were carried out at first and later by a mechanical fish counter. In the Úlfarsá, redds were counted. In the Nordurá a resistivity tube counter is located in a fish pass. In the Blanda and the Ölfusá-Hvítá the tag and recapture method was used.

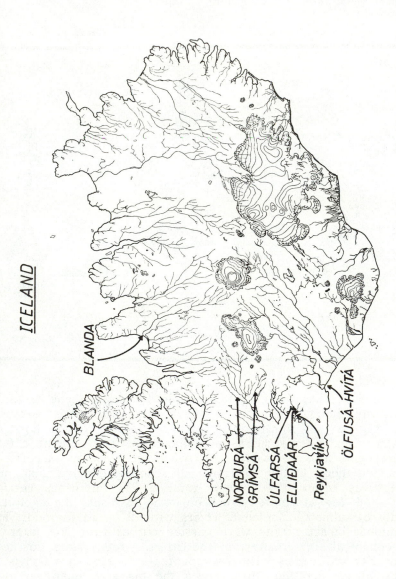

Figure 9.4: Map of Iceland showing the location of the salmon rivers discussed in this paper

168

The Ellidaár

The river Ellidaár is located within the City of Reykjavík in southwest Iceland (Figure 9.4). It is 6 km long and is the outlet of Lake Ellidavatn, which is 74 m above sea level. There is a hydro-electric power station on the river about 0.8 km above the estuary and a power dam about 1.5 km above the power station. There is a weir with a trap located just below the power station where the counts have been made since 1931. The fish were counted manually until 1960 when a mechanical fish counter was installed. There have been many changes made in the riverbed and in the natural flow pattern after the building of the power station and the power dam in 1921, causing degradation of natural conditions for salmon in the river system.

A study was made of the rate of exploitation of salmon in the Ellidaár from 1935 to 1976 based on total rod catch and total run (Table 9.1). Data were missing for a few years. Estimates

Table 9.1: The Ellidaár - average rate of exploitation of salmon caught by rod and line 1935-1976

Years	Total run	Total catch	Exploitation rate (%)
1935-1944	2,493	1,032	41.4
1945-1954	3,307	1,166	35.3
1955-1964	3,810	1,081	28.4
1965-1974	4,148	1,474	35.5
1975-1976	5,429	1,882	34.7
(1965-1976)	(5,234)	(1,851)	(35.4)

Average catch 1938-1976 34.6%

Source: Mathisen and Gudjónsson (1978)

were made for these years by methods discussed by Mundy, Alexandersdóttir and Eiriksdóttir (1978). The rod catch from 1935 to 1976 averaged 34.6 per cent of the run. During the period from 1935 to 1955 the average catch was 38 per cent. During the following decade it fell to 28 per cent, but from 1966 to 1976 it rose again to about 35 per cent. The range of the catch for individual years was from 18 to 58 per cent.

The Úlfarsá

The Úlfarsá is a small river located a few kilometres north of the Ellidaár. It is 10.6 km long and is the outlet of Lake Hafravatn which is 76 m above sea level. The average natural flow is about 800 l/s. Spawning takes place during the first half of November. Since the flow is small the redds are easily seen and counted except on a few occasions when redds were crowded at two places and some overcutting occurred there at high stock levels. In such cases, estimates of the number of redds were made. At spawning fords on the river there were a few too many redds at the same places every year, but on the other hand there were places where redds were observed on one or more occasions. Redds were counted for the entire river or parts there - of in the autumn of 1955, 1956, 1957, 1959, 1960 and 1963. It was estimated that one redd represented one female and one male. The rod catch for each year was added to the estimated number of salmon spawning in the river the same year. The result was that the average rod catch for the six years was 28.5 per cent of the run, varying from 14.1 per cent in 1959 to 46.2 per cent in 1957 (Table 9.2).

Table 9.2: The Úlfarsá - estimated exploitation rate of salmon based on redd counts and catch records

	1955	1956	1957	1959	1960	1963
Number of redds	495	140	53	269	117	205
Number of salmon	990	280	106	538	234	410
Catches	430	107	91	88	117	192
Stock size	1,420	387	197	626	351	602
Exploitation rate (%)	30.3	27.6	46.2	14.1	33.3	31.9

Average exploitation rate 28.6%

Source: Gudjónsson (1964).

The Nordurá

The Nordurá is a tributary of the glacial river Hvítá in Borgarfjördur district in West Iceland. Counting of salmon takes place through a resistivity tube counter in a fish pass at the waterfall Laxfoss, which is located about 16 km upstream from Nordurá-Hvítá confluence. About 3 km upstream from the Laxfoss is another waterfall, the Glanni, which salmon have been able to leap when the water level has been low. In 1985 a fish pass was built on one side of the waterfall. Above the Glanni salmon occupy about 27 km of river with good spawning grounds.

Table 9.3 shows the number of salmon counted at the Laxfoss for the years 1972-1985 with the exception of the year 1979 which registered an unbelievably high count, which is considered unreliable. Table 9.3 also lists the number of fish caught by rod and line above the fish counter as well as the total rod catch for the river.

Table 9.3: The Nordurá - counts of a resistivity tube counter in the fish pass at the Laxfoss waterfall and rod catches of salmon

Year	Counts	Rod catch above counter	Total rod catch in the river
1972	2,993	782	2,537
1973	3,189	816	2,322
1974	1,418	537	1,428
1975	3,993	632	2,132
1976	1,858	613	1,675
1977	1,894	532	1,470
1978	1,075	692	2,089
1980	3,722	838	1,583
1981	3,096	497	1,185
1982	3,524	592	1,455
1983	2,132	233	1,643
1984	323	216	856
1985	735	600	1,121
Total	29,952	7,580	21,496
Averages 1972-1985 except for 1979	2,304	583 25.3%	1,654

Source: Ágústsson (1973-1986).

The average count of salmon for the 13-year period amounted to 2,304 fish, the range being from 323 in 1984 to 3,993 in 1975. The average catch above the fish pass was 583 fish varying from 216 in 1984 to 838 in 1980. The average catch rate of fish above the counter was 25.3 per cent varying from 10.9 per cent in 1983 to 81.6 per cent in 1985.

Salmon of the Norduŕá origin are also caught by nets in the Hvítá. During the 13-year period in question 48.2 per cent of the total catch of salmon in the Hvítá river system was netted. Assuming that the mentioned percentage of Norduŕá salmon stock was caught in the Hvítá the catch should have amounted to 1,539 fish. The average annual rod catch for the same period was 1,654 salmon. The total catch of Norduŕá salmon would thus have been 3,193 fish. The average number of salmon left after the fishing season in the river above the counter was 1,721 fish or 57.4 salmon per kilometre above the counter on average. If one assumes that the same number of salmon per kilometre was left below the counter then there should have been 918 fish in that part of the river. The total salmon stock of Norduŕá should thus have been close to 5,832 fish. On the basis of the figures at hand it is estimated that 3,193 salmon or 54.8 per cent of the total stock were caught, 1,654 or 28.4 per cent by rod and line, 1,539 or 26.4 per cent by nets in the Hvítá, and 2,639 salmon or 45.2 per cent were left in the river after the fishing season.

Blanda
During the years 1982 to 1985 salmon in the glacial river Blanda in North Iceland were caught in a trap in a fish pass about 2 km above the estuary. The salmon were tagged with numbered 'spaghetti' tags before being released upstream. During this period 2,077 salmon were tagged, an average of 519 fish per year (Table 9.4). Salmon were able at a certain flow level to leap the cascades on one side of the fish pass. From the ratio of tagged to untagged fish caught above the fish pass it was estimated that 503 untagged salmon were caught. Adding them to the tagged ones the total run up above the fish pass amounted to 2,580 salmon or an average of 645 fish per year. The rod catch above the fish pass was 669 salmon or 26 per cent. The average catch varied from 21 per cent in 1982 to 31 per cent in 1983. Rod fishing in the Blanda is also carried out below the fish pass where a greater part of the total rod catches are made, amounting to 80 per cent on the average for the four years. When the rod catches for the whole river system are added up, they amount to 3,374 fish or 822 on average per year. The size of the run is estimated to be 5,198 fish or 1,300 per year on average. The total rod catch amounts to 65 per cent. The total catch varied from 82 per cent in 1982 to 55 per cent in 1985.

There is a great difference in exploitation rate above and below the fish pass. It may be explained by the difference in the kinds of bait used. Above the fish pass earthworms and artificial flies are used as bait, as in most Icelandic clear water rivers,

Table 9.4: The Blanda river system - counts and catches of salmon 1982 to 1985

	1982	1983	1984	1985	Total
Tagged and released at trap in fish pass	202	411	361	1,103	2,077
Estimate of untagged salmon above fish pass	59	87	88	269	503
Total number of salmon above fish pass	261	498	449	1,372	2,580
Rod catch above fish pass	55	153	131	330	669
Exploitation rate above fish pass (%)	21	31	29	24	26
Rod catch below fish pass	854	503	495	766	2,618
Exploitation rate below fish pass (%)	77	50	52	36	50
Net catch in upper Blanda	2	5	3	77	87
Total catch	911	661	629	1,173	3,374
Estimated run	1,115	1,001	944	2,138	5,198
Total exploitation rate (%)	82	66	67	55	65

Source: Gudbergsson and Gudjónsson (1986).

whereas large spoons are almost exclusively used in the glacial water below the fish pass.

The Ölfusá-Hvítá

The Ölfusá-Hvítá in South Iceland is one of the largest river systems in Iceland. The main river is of glacial origin, where netting for salmon takes place, while the tributaries are clear water streams with rod fishing only. Salmon can migrate as far as 90 km upstream in the main river. During the years 1960 to 1972, 924 clean salmon were trapped and tagged in the estuary of the Ölfusá-Hvítá. Tags from 333 salmon (36 per cent) were returned, varying from 18.3 per cent in 1961 to 49.1 per cent in 1969 (Table 9.5). Since the return of tags is expected to be incomplete, the actual catch of tagged salmon has been more likely close to 50 per cent.

Table 9.5: The Ölfusá-Hvitá - clean salmon tagged in the estuary and tag returns 1960 to 1972

	Number tagged	Tag returns	
1960-1964	255	74	29%
1965-1968	394	150	38%
1969-1972	275	109	40%
Total	924	333	36%

Source: Gudjónsson (1977).

DISCUSSION

The methods for assessing the exploitation rate of salmon in Icelandic rivers have certain shortcomings which render them inaccurate. The types and magnitude of inadequacies vary from river to river and the rate of exploitation varies from year to year depending on the size of runs and catchability during the fishing season. When the runs are at their smallest the percentage of the catches is highest, whereas when the runs are biggest the percentage of the catches is lowest. This can be seen from the Ellidaár data for 42 years, where the biggest run was 8.9 times larger than the smallest one, and the biggest annual catch was only 3.2 times larger than the smallest one.

It is seen from Table 9.1 that the salmon runs in the Ellidaár have increased during the period 1935 to 1976 from 2,493 for the first ten-year period to 5,234 for the 12-year period 1965 to 1976 and so have the rod catches from 1,032 fish to 1,851, respectively. The exploitation rate by rod fishing is about the same except for the period 1955-1964 when it is lower. The increase in the total run after 1965 may have been partly the result of a bag net fishery just outside the estuary being abandoned in 1964 allowing the salmon on their way to Ellidaár, which would otherwise have been caught in the bag net, to enter the river. The bag net average annual catch over the period 1947 to 1963 amounted to about 500 salmon. Another possible factor leading to an increase in the run is that in 1965 a fish pass was built in an impassable dam at the outflow of Lake Ellidavatn, opening up the lake and its tributary, the Hólmsá, for spawning and for rearing of salmon parr. Then enhancement measures have also played a part in enlarging the run. The number of salmon caught increased more than twofold with advancing years and the angling effort was almost doubled during the 42 years covered by the study.

The counts of salmon through the fish pass in Laxfoss in the Nordurá will not furnish accurate information about the run up above. The counter is mounted in the fish pass, which is located in the middle of the waterfall. It is exposed to great fluctuations in the amount of flow. Floods will impede salmon ascending and they have on a few occasions damaged the counting mechanism, but such damages have been repaired as soon as the floods have receded. Floods may last for many days at a time in the summer or even be frequent and lasting through the whole summer as was the case in 1983. An attendant looks after the counter once or twice daily, except when floods prevent him from wading across to the counter. He reads the counter, measures the air and water temperature, notes down the level on the water gauge and records weather conditions.

Another source of inaccuracy in the counts is the fact that salmon can leap the waterfall to some extent when the river is low. It is suspected that this has happened in 1974, 1978, 1984 and 1985. If counts and rod catches above the counter for these four years are left out, the average counts for the remaining nine years are 2,933 salmon instead of 2,304 and the average rod catch is 615 fish or 21.8 per cent.

As mentioned previously, a fish pass was built in 1985 in the waterfall Glanni which is located about 3 km above the Laxfoss. In June 1986 a resistivity tube counter was installed in the fish pass. A total of 1,166 salmon were counted through and 280 salmon were caught in the river above or 24.0 per cent, which is close to the exploitation rate for the thirteen year's count in the Laxfoss. But it is possible that some salmon leapt the Glanni waterfall at low water levels as they did before the fish pass was built.

The flow of the glacial river Blanda varies a great deal. On hot days the snow and ice melt on the glacier Hofsjökull in the central highlands causing floods in the Blanda with increased turbidity of the waters, which can slow down the upstream migration of the salmon. The same may happen during or after heavy precipitation on the extensive catchment area of the river. Below the cascades and the fish pass is a narrow gorge. During floods the heavy current through the gorge acts as a barrier to upstream migration of the salmon. This happens when the flow rises above 55-60 m^3/s. which can last for many days at a time. During such conditions the salmon will congregate in one large pool below the gorge, where they are heavily fished for by anglers. In 1982, 94 per cent of the total catch of the Blanda and its tributary Svartá was taken there, 76 per cent in 1983, 79 per cent in 1984 and 65 per cent in 1985.

It is to be expected that reporting of tags in the Ölfusá-Hvítá river system was incomplete. In 1967 and 1970, 412 salmon were tagged out of the total of 924 for the 13-year period of tagging, or 45.6 per cent. When going through the records of returned tags and which fishermen had reported them, it appeared that those who returned tags had caught 64 per cent of the netted catch and part of the angling catch. Assuming that the tagged salmon were distributed throughout the river system, the average exploitation rate should have been above 51 per cent, while the tag returns were 36 per cent. Even 51 per cent might be too low since it is not certain that the fishermen who returned tags have sent in all of those they retrieved.

SUMMARY

In Iceland are 80 rivers and river systems which hold Atlantic salmon. Most of them are found in the western half of the country. The fishing takes place almost exclusively in the rivers and their estuaries by rod and line and by gill nets mainly in three glacial rivers. The average annual catch for the last twenty years has been about 55,000 salmon averaging about 190 metric tons including returns of ocean ranched salmon.

Attempts have been made to assess the exploitation rates of salmon in five rivers by counting through traps, by redd counts and by the tag and recapture method.

The average rod catches were most often between 25.3 per cent to 34.6 per cent, except for the Blanda where the rod catch was 64.9 per cent on average for the four year study period.

The average rod catch for 42 years in the Ellidaár is 34.6 per cent. The catches vary from 23 per cent in 1963 to 58 per cent in 1937. In the Úlfarsá the average catch for six years is 28.6 per cent. The catches vary from 14.1 per cent in 1959 to 46.2 per cent in 1957. In the Nordurá above the fish counter at the Laxfoss the average catch for thirteen years is 25.3 per cent. The catches vary from 10.9 per cent in 1984 to 81.6 per cent in 1985. The latter figure is expected to be unreliable. An estimate for the average rod catch for the whole river amounts to 28.4 per cent. The average rod catch for four years below the fish pass in the Blanda is 50 per cent. The catches vary from 36 per cent in 1985 to 77 per cent in 1982. Above the fish pass the average catch for the same number of years is 26 per cent. The catches vary from 21 per cent in 1982 to 31 per cent in 1983. When the rod catches below and above the fish pass are summed, the average catch was 65 per cent, the smallest one was 55 per cent in 1985 and biggest one was 82 per cent in 1982. In the Ölfusá-Hvítá river system the average tag returns were 36.0 per cent, varying from 18.3 per cent in 1961 to 49.1 per cent in 1969. Because of incomplete tag returns it is estimated that the exploitation rate has been 51 per cent on the average, ranging from 25 to 70 per cent. An estimate was also made of the number of salmon of the Nordurá stock, which was netted in the Hvítá. This amounted to 26.4 per cent.

REFERENCES

Ágústsson, I. (1973-1986) Laxatalningar i Laxfossi. Nordurá 1972-1985. Manuscripts, Reykjavik

Gudbergsson, G. and Gudjónsson, S. (1986) Rannsóknir á fiskistofnum Blöndu. Manuscript, Reykjavik, 40 pp.

Gudjónsson, Th. (1964) Ahrif vatnostoku úr Úlfarsá á veioi í ánni. Manuscript Reykjavík, 37 pp.

Gudjónsson, T. (1977) Recaptures of Atlantic salmon tagged at the estuary of the river Ölfusá - Hvítá, Iceland. International Council for the Exploration of the Sea, C.M. 1977/M:40, 6 pp.

Mathisen, O.A. and Gudjónsson, Th. (1978) Salmon management and ocean ranching in Iceland. In O.A. Mathisen (ed), Salmon and trout in Iceland. Journal of Agricultural Research, Iceland, 10(2), 156-74

Mundy, P.R., Alexandersdóttir, M. and Eiríksdóttir, G. (1978) Spawner-recruit relationship in Ellidaár. In O.A. Mathisen (ed), Salmon and trout in Iceland. Journal of Agricultural Research, Iceland, 10(2), 47-56

Chapter Ten

THE ATLANTIC SALMON IN THE RIVERS OF SPAIN WITH PARTICULAR REFERENCE TO CANTABRIA

Carlos Garcia de Leániz
Marine Laboratory, Victoria Road, Aberdeen AB9 8DB,
Scotland

and Juan José Martinez
Servicio de Montes, Caza y Conservación de la Naturaleza,
Calle Rodriguez 5, 1, 39071 Santander, Spain

10.1   INTRODUCTION

Historical Perspective

In the Iberian peninsula, comprising Spain and Portugal, salmon are only found in rivers of the north and northwest provinces entering the Cantabrian Sea and the North Atlantic. They are absent from rivers flowing into the Mediterranean sea (Figure 10.1).

Perhaps the first written mention of Atlantic salmon (Salmo salar, L.) in Spain appears in several writings in the seventh century (Jusué Mendicouague, 1953). Later, in the eleventh century, contracts dealing with the fishing rights for salmon became commonplace (Camino, 1940). Salmon harvesting at that time was the patrimony and privilege of feudal landowners, abbeys and noblemen who owned large estates. Entire beats of the rivers and estuaries were crossed by a series of double stockades or postas which effectively trapped upstream migrants. The remains of these postas may still be seen in some rivers.

Salmon catches during these times must have been similar to those elsewhere in Europe. In 1258 the Spanish King Alfonso el Sabio (Alfonso the Wise) established what must have been the first attempt to regulate this fishery by introducing close seasons (Netboy, 1974). A royal decree in 1435 by King Juan II stipulated large fines for poachers caught poisoning the waters, suggesting that by the fifteenth century pressures on the fishery were rapidly mounting. Salmon harvesting in this century was, in fact, cause for much litigation between noblemen (Escagedo, 1927).

By the end of the sixteenth century specific gear regulations were being introduced, thereby protecting the smolt runs and the spawning of adults. The administration of the fisheries was handed over to the various provinces and boroughs, which were now free to implement their own netting regulations.

180

Figure 10.1: Major salmon rivers of Spain

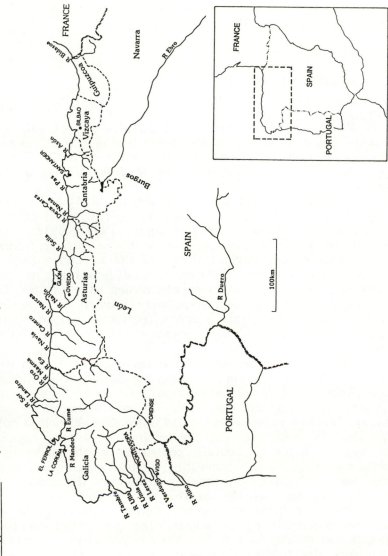

The Atlantic Salmon in Rivers of Spain

Salmon catches were still very large in these northern rivers. The historian Sanchez Reguart in his 'Diccionario Histórico' published in 1791-1795 estimated that 2,000 salmon were caught daily in the Principality of Asturias alone. The total Spanish salmon catch in the seventeenth and eighteenth centuries is said to have been between 8,000 and 10,000 fish per day which, according to Netboy (1974), must have amounted to 600,000-900,000 salmon annually, if daily catches are projected over a three month fishing season. Catches derived from historical accounts, however, should always be treated with some reservation although catches of similar magnitude have been reported elsewhere for other European and North American rivers (Dunfield, 1985).

By the end of the eighteenth century the monopoly on salmon stockades, traps and nets was finally lifted in most regions following a royal decree by King Carlos IV. Additional regulations stipulated that mill dams and lades be kept open at all times to ensure the spawning of adults.

In the nineteenth century many small hydro-electric stations were being built adding to the already considerable number of weirs, mill dams, and private water constructions obstructing the passage of ascending adults. Fish ladders were either absent or notably inefficient despite the regulations laid down previously. The once-flourishing salmon fishery was never to recover, not through a lack of protective laws but due to the inability of government to enforce them. In 1920-30 Spanish rivers were yielding a fraction of their historical catches, despite progressive action by government to control overexploitation.

The first salmon hatcheries were being built at this time to restock the rivers and some scientific studies were undertaken by a group of enthusiasts, notably the Marquis of Marzales who made contact with Calderwood and Hutton in Britain and Roule in France. The existence of male precocious parr in the Spanish rivers was well known at this time and there must have been some knowledge of the age and length composition of the salmon populations (Camino, 1940). Salmon fishing was managed independently by two government ministries, one regulating the estuaries, the other regulating the rivers. As in France, divided administration did little to preserve the salmon stocks (Netboy, 1968).

Netting in the estuaries started in mid-February, often for the entire week and operated day and night. Salmon catches in this period amounted to some 4,000 fish per year in the province of Santander and slightly more in Galicia and Asturias. Total salmon catches probably amounted to some 20,000 fish per year (Camino, 1940; MacCrimmon and Gots, 1979). These quite high catch levels, however, may have resulted from increasingly more effective fishing effort rather than from improvements in abundance.

According to Camino (1940) who refers to a survey carried out by W.M. Gallichan (1904) of most salmon rivers in Spain at the turn of the century, the best ones were capable of producing between 50,000 and 90,000 salmon per year. Industrial pollution was still virtually unknown, productivity was high and there were extensive excellent spawning grounds in most rivers.

As a result of the growing concern among anglers and administrators alike, a royal decree in 1927 introduced some limitations on the operation of netting stations, explicitly forbidding the netting of salmon from 0600 hrs on Friday to 0600 hrs on Monday. Banning of all salmon netting (coastal, estuarine and riverine), however, had to wait another 16 years when the catches reached their lowest levels ever, following the Spanish Civil War (Netboy, 1980).

Salmon populations suffered badly during the civil war (1936-1939) when they provided relatively easy food for the population. Fishing regulations could no longer be implemented and there was little or no control resulting in the looting of rivers by a variety of means including the use of explosives, bleach and other poisons, and all kinds of nets.

Reasons for the steady decline were found in the gross overexploitation and poaching of the rivers, the lack of bailiffs and realistic fines, the pollution of some rivers (Miera, Bidasoa, Besaya), and the existence of countless obstructions to the movement of ascending adults (Camino, 1940).

In February 1942 the Government finally banned all salmon netting stations and in this way dedicated the fishery entirely to anglers.

Management and data collection until 1944 had been dealt with by the Forestry Districts with a lack of resources and limited manpower. Data from this period are very unreliable and hard to find except for the accounts provided by Marzales and a handful of other enthusiasts (Gallichan, 1904; Camino, 1940).

Official salmon statistics were collected from 1949 onwards by the present system of logging each fish caught. All legally caught salmon are given a circular tag and issued with a certificate of capture or guia. This includes data on the weight, fork length, girth, lure or bait employed, name and address of the fisherman, site of capture, and sometimes sex of fish based on external characters (Garcia de Leániz, Hawkins, Hay and Martinez, 1987).

In the early 1960's coinciding with the boom in the Spanish tourist industry, a National Association of Bailiffs was created within the Civil Service resulting in the acquisition of much better fishery statistics. The previous Service became the National Service for Freshwater Fisheries and Game (1960-1972) and considerable funds were invested in the creation of fish ladders, bridges, fishing chalets, and facilities for the stripping and maintenance of broodstock and the hatching of salmon ova.

It was in this period that the division of the salmon beats into restricted areas (cotos) and free zones (zonas libres) was introduced and some entire rivers were designated National Salmon Reserves like the rivers Narcea and Cares in Asturias, Deva in Cantabria and Eo in Galicia. However, this was also a time for the construction of artificial reservoirs, large hydro-electric stations and major projects to regulate the flow of rivers all of which subjected the rivers to still more stress. At present, over 70 per cent of the salmon catches are taken from restricted beats or cotos.

Salmon management was transferred in 1972 to the Institute for the Conservation of Wildlife (ICONA) which continued operating until 1984, when the administration of the fishery was handed over to the regional governments and the various Services of Mountains, Game and Wildlife were created.

Today the Spanish salmon rivers are administered independently by the five northern regional governments of País Vasco (province of Guipuzcoa), Navarra, Cantabria, Asturias, and Galicia. Fishing regulations are basically similar and have been summarised elsewhere (Garcia de Leániz et al., 1987).

10.2 THE SALMON RIVERS

Although the number of salmon rivers in Spain was considerable at one time and included some long ones like the river Duero (Netboy, 1968), salmon are now found only in about 20 rivers (Figure 10.1) of which only six, the river Sella, Cares and Narcea in Asturias and Pas, Asón and Deva in Cantabria, account for over 70 per cent of the total rod and line catch (Table 10.1)

Most of these salmon rivers are short, usually less than 50 km, and have small catchment areas. They are spate rivers, quick to rise and fall and prone to droughts during the summer months when flow rates can decline dramatically.

Much of the land in these northern provinces is over 500 m high with peaks up to 2,500 m. Gradients can be quite steep and run-off and erosion pose important problems in periods of heavy rain, particularly in areas where deforestation has taken place. The pH of these rivers is generally not far from neutral becoming slightly alkaline during the summer. Serious industrial pollution is generally not a problem and the best salmon rivers tend to score high in water quality assessments.

Water temperatures differ markedly from those in more northern latitudes and only fall below 7°C for one or two months during the year. The growing season for juvenile salmon (number of days in which water temperature is above or equal to 7°C; Symons, 1979) is considerable and averages over 330 days per

Table 10.1: Spanish salmon statistics 1949-1985

showing numbers of salmon caught by rod and line by river and district

	Galicia									Asturias					Cantabria				Navarra
	Eo	Landro	Lerez	Mandeo	Masma	Miño	Sor	Tambre	Ulla	Canero	Cares	Narcea	Navia	Sella	Asón	Deva	Nansa	Pas	Bidasoa
1949	187	1	2	0	1	0	1	10	51	0	600	666	200	715	200	101	48	10	29
1950	220	1	2	2	2	0	1	15	60	0	711	750	288	822	320	112	60	15	30
1951	221	2	2	1	5	2	2	19	78	0	815	782	300	900	558	210	78	22	35
1952	242	3	3	1	7	0	0	23	100	0	900	800	321	915	621	298	90	29	38
1953	258	4	3	1	9	0	1	30	128	0	965	810	360	1,052	798	320	95	32	40
1954	544	5	22	20	39	0	31	31	443	0	1,478	788	576	2,871	1,321	298	146	69	224
1955	465	3	15	58	34	0	27	164	339	0	1,376	622	725	1,068	702	61	88	39	45
1956	433	3	10	30	39	0	21	138	367	0	1,701	1,108	1,169	1,236	1,008	291	60	110	111
1957	191	3	14	10	23	0	28	177	281	0	855	630	811	978	801	147	51	127	75
1958	191	1	16	16	24	0	15	109	152	0	842	353	755	992	732	140	28	82	65
1959	312	0	21	8	37	0	58	109	96	20	1,335	1,298	701	2,781	1,541	361	129	136	94
1960	390	0	26	5	16	0	13	90	99	0	1,066	785	374	1,381	917	154	51	109	180
1961	56	0	6	2	30	5	21	17	115	7	677	389	280	984	552	55	44	79	96
1962	185	3	16	1	39	38	38	64	140	37	623	592	175	1,151	559	158	68	135	87
1963	361	4	19	3	51	29	29	90	269	28	677	555	276	1,705	476	189	30	111	61
1964	285	6	84	3	82	4	38	89	261	29	1,683	564	359	1,027	741	292	25	79	226
1965	375	1	116	2	77	14	11	45	295	115	1,365	860	787	1,534	1,106	289	49	166	158
1966	348	14	23	8	69	30	46	58	97	76	1,063	1,197	764	1,162	1,343	255	36	305	324
1967	703	17	30	2	105	39	19	24	357	104	1,119	1,928	387	1,194	773	281	18	226	146
1968	366	10	65	15	82	45	18	11	160	70	1,114	860	189	2,078	747	285	47	294	110
1969	409	15	79	15	82	48	42	155	435	195	1,871	1,349	286	2,041	881	742	55	466	135

Table 10.1 (Cont'd)

	Galicia									Asturias					Cantabria				Navarra
	Eo	Landro	Lerez	Mandeo	Masma	Miño	Sor	Tambre	Ulla	Canero	Cares	Narcea	Navia	Sella	Asón	Deva	Nansa	Pas	Bidasoa
1970	505	8	158	3	96	108	46	31	510	209	1,071	1,195	137	1,853	781	334	36	619	74
1971	233	1	54	3	11	61	7	7	74	93	461	279	18	567	422	133	16	211	47
1972	174	2	84	7	66	113	11	39	291	185	1,366	626	10	2,230	603	278	75	699	155
1973	166	7	33	5	18	52	1	7	197	100	781	424	2	1,176	461	166	37	420	108
1974	144	7	94	3	34	181	6	0	112	104	670	183	0	443	250	136	45	301	31
1975	619	6	78	0	122	301	0	4	189	206	463	818	0	630	192	275	74	522	102
1976	622	5	81	2	127	296	6	0	263	260	321	916	2	306	38	121	16	132	91
1977	316	14	23	0	44	74	3	2	88	194	646	455	6	486	113	152	42	473	81
1978	753	12	52	2	44	329	3	0	296	553	1,000	687	22	676	58	308	53	604	41
1979	514	10	3	6	54	30	0	2	24	593	831	793	6	1,211	52	405	52	404	37
1980	1,222	32	39	12	70	142	0	3	543	741	922	1,975	18	1,196	182	467	41	568	30
1981	835	7	16	15	27	113	0	14	256	541	682	756	10	678	54	94	65	214	17
1982	187	2	3	0	19	43	0	0	104	291	233	198	9	382	3	69	30	185	0
1983	489	10	0	0	12	36	0	0	104	336	632	851	19	823	70	143	41	428	35
1984	343	15	11	10	35	106	6	1	180	161	348	956	38	551	94	103	25	186	16
1985	389	12	3	11	41	63	2	2	175	166	312	201	10	384	262	57	27	150	8

year or 90 per cent of the time, compared to 18-41 per cent in some rivers in Norway (Jensen and Johnsen, 1986) and around 50 per cent for some Scottish rivers.

Spanish salmon rivers can also be narrow and deep, their bottom profile being v-shaped in those that run from the higher mountains through narrow gorges of bed-rock. The numerous deep pools formed between waterfalls may offer ideal habitats for resident trout but there may be a shortage of salmon fry habitat in some areas.

Some spawning ground has been lost through water schemes and power plants; in the salmon rivers of Asturias almost 30 per cent of the total river length has been lost in just 36 years (see Chapter 11). There are, however, numerous small streams and tributaries which could be colonised by salmon if fish passes were created or repaired.

10.3   THE SALMON STOCKS

The Spanish rod and line salmon catches, 1949-1985

Figure 10.2 shows the Spanish rod and line catch for the period 1949-1985, with the contribution to the total catch by the three major salmon regions of Asturias (60.2 per cent), Cantabria (20.6 per cent) and Galicia (15.6 per cent). Navarra has been excluded as its only salmon river, the Bidasoa, provides less than 2 per cent of the total catch (Table 10.1).

A series of marked peaks and troughs in the total catch are evident every 10-11 years, beginning in 1949, but it is probably unwise to draw far-reaching conclusions about the reasons for these variations without data on salmon abundance.

In general, Spanish salmon catch statistics may reflect mean salmon abundance in particular periods (Miranda, personal communication) but great care must be exercised when looking at individual rivers, as fishing effort is largely unknown and dependent on highly seasonal factors - weather, holidays, etc. The construction of dams and water abstraction schemes exerts effects upon the catches, and the incidence of poaching is variable and dependent on river topography, weather conditions and vigilance by bailiffs.

The abrupt decline in salmon catches in 1971 (Figure 10.2) is thought to be related to the first high incidence of infectious disease in Spain, when hundreds of fish were found dead along the banks of most salmon rivers. Important losses of adult salmon at around this time were also reported in the British Isles and in France (Mills, 1971; Prouzet, 1984).

Ulcerative dermal necrosis (UDN) was held responsible for the deaths but no proper diagnosis has been carried out to date in Spain although plans were recently made in the region of

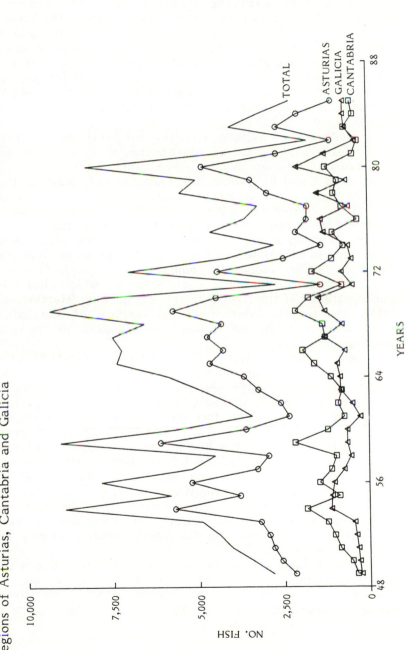

Figure 10.2: Spanish rod and line salmon catches, 1949-1985, showing contribution by the three major regions of Asturias, Cantabria and Galicia

Cantabria to send samples abroad. Patches of skin fungus had been analysed and both Saprolegnia, a fungus, and Aeromonas spp., a bacterial disease, were identified, both these conditions being associated with the syndrome of UDN (Roberts and Shepherd, 1986). Losses were particularly high in the river Sella in Asturias and the rivers Pas and Asón in Cantabria but a pronounced drop in the salmon catches is also apparent elsewhere (Table 10.1, Figure 10.2). In recent times the numbers of diseased fish is thought to have decreased considerably. Strict comparisons are impossible as there are no overall counts of dead salmon and estimates largely come from the information provided by the bailiffs.

Declines in salmon catches in particular rivers can sometimes be related to the construction of dams and to sources of industrial pollution. Salmon were caught in the river Canero (also called Esva) in Galicia, for example, only after a dam had been damaged by a series of heavy spates in the late 1960s. Salmon disappeared completely in the rivers Miera and Besaya in Cantabria as a result of industrial pollution at the turn of the century. Declines in captures in the rivers Navia, Ulla, Nansa and Miño can also be related to hydro-electric projects (MacCrimmon and Gots, 1979; Garcia de Leániz et al., 1987), (Table 10.1). Although opposition to major water schemes, which alter the flow and topography of the rivers, is certainly growing (Martinez-Conde, 1984), the general interest in the welfare of the salmon is understandably slight and usually subordinated to the pace of industrial progress.

Total salmon catches in the period 1949-1985 averaged 5,224 salmon per year which, compared with over 20,000 fish caught annually in the decade 1920-1930 (Camino, 1940), represents a fourfold decrease over 25 years. Comparisons of yearly catches between the three major salmon regions of Asturias, Cantabria and Galicia yield various degrees of correlation. Salmon catches in Asturias are highly correlated with those in Cantabria ($r=0.88$, $p < 0.001$) while comparisons between both Asturias and Galicia and between Cantabria and Galicia show a poor degree of correlation ($r=0.44$, $p < 0.01$ and $r=0.31$, NS, respectively). Correlation of yearly catches between regions, therefore, seems to depend largely on geographical proximity with rivers in Asturias and Cantabria yielding closely related captures (Figure 10.2).

Spanish salmon catches show no clear correlation with total Scottish catches. It must be stressed that grilse are largely under-represented in the catch sample from these rivers, so changes in abundance of this age class will have little or no effect on the reported captures.

Spawning and time of entry to freshwater

There are no fish counters or fences in Spanish rivers nor are there reliable ways of counting adults ascending salmon rivers outside the fishing season. Both spawning time and time of entry into freshwater must come from historical accounts and information provided by the bailiffs. In places where broodstock are caught and held for stripping, some information on the time of entry is available.

Salmon enter Spanish rivers throughout the year (Camino, 1940; Notario, 1971) although the majority of them are probably spring fish entering the rivers in March-July with peaks in catches generally occurring in May and June. Fish are known to enter the rivers during the winter (Camino, 1940; Notario, 1971) and are called 'invernizos' (meaning from the winter). These are usually large multi-sea-winter fish which are still represented in the rod and line catch in March, at the beginning of the fishing season.

Spring fish are sometimes named according to their month of entry and are generally 2-sea-winter fish in the rivers of Cantabria and probably elsewhere. The average size of fish entering in June and July is normally smaller. By the end of the fishing season, and occasionally earlier, catches tend to drop considerably and small numbers of grilse (55-65 cm) may be represented in varying numbers depending on the year.

When netting was still permitted in Spanish rivers prior to 1942, the smallest fish were normally caught in August and the largest in March and April (Camino, 1940). Although the mean weight of salmon does not seem to have changed substantially and is still about 5 kg in the rivers of Cantabria (Table 10.2), large fish (sometimes reaching up to 18 kg; Camino, 1940; MacCrimmon and Gots, 1979) were often caught at the beginning of the netting season in mid February.

Spawning takes place usually from December to February but it can extend over November to March (Camino, 1940). Differences in water temperature exist between rivers due to snow melt and underground springs so there is some variation in the time of hatching of salmon ova which in Spanish rivers is probably between 35 and 45 days based on a development rate of 410-450 degree days observed in a French river of similar temperatures (Prouzet and Gaignon, 1982). Although the construction of dams and small weirs without appropriate fish ladders has greatly reduced the available area for spawning in some rivers, there are still extensive areas in the upper reaches which could be used for the stocking of eggs and fry in large numbers.

Table 10.2: Wet weight (kg) of salmon caught in four different rivers Mean ± 95% CL, sample size in parentheses

		River		
Year	Pas	Deva	Asón	Nansa
1983	5.24±0.14 (428)	5.04±0.32 (143)	4.56±0.23 (70)	4.92±0.33 (41)
1984	5.33±0.20 (186)	4.99±0.23 (103)	4.97±0.27 (95)	5.29±0.74 (25)
1985	5.12±0.22 (150)	4.66±0.36 (57)	4.62±0.11 (262)	5.75±0.70 (27)
1986	4.71±0.16 (250)	5.00±0.21 (93)	5.33±0.25 (146)	4.57±0.53 (49)
N	1,014	396	573	142

Differences in the length of adults returning to different rivers

Mean fork lengths in the period 1983-86 for each of the Cantabrian rivers (Pas, Asón, Deva and Nansa) were analysed for significance by ANOVA tables with unequal sample size (Zar, 1984).

Differences in mean length between rivers were found to be statistically significant (Table 10.3), indicating that the sample

Table 10.3: Yearly differences in mean fork length (cm) between rivers. Mean ± 95% CL, sample size in parentheses

		River				F
Year	Pas	Deva	Asón	Nansa	N	Value
1983	83.44±0.59 (428)	80.84±0.70 (143)	78.30±1.09 (70)	79.46±1.62 (41)	682	23.62*
1984	83.64±0.94 (186)	80.33±1.01 (103)	80.80±1.26 (95)	81.92±3.20 (25)	409	8.04*
1985	83.40±1.02 (148)	80.24±1.84 (57)	79.37±0.54 (260)	83.52±3.13 (27)	492	18.71*
1986	78.96±0.94 (249)	81.55±0.92 (93)	80.51±1.21 (146)	74.74±2.93 (47)	535	10.40*
N	1,011	396	571	140	2118	
F value	28.91*	1.23 (NS)	3.91**	8.44*		

Table 10.3 (Cont'd)

* $p < 0.001$
** $p < 0.01$

catches from these rivers did not belong to the same population. Inspection of the means in fork length and length frequency distributions (Figures 10.3 and 10.4) showed that salmon from the river Pas normally have a greater fork length and heavier weight due to the higher incidence of early run fish, except in 1986 when some numbers of grilse entered this river in June.

Means of fork length for each river were also found to differ significantly from year to year except in the river Deva (Table 10.3). Yearly variations in mean fork length can usually be attributed to the varying contribution of grilse and small two sea-winter salmon to the sample catch.

Although early run fish can sometimes be sexed based on external characters (Shearer, 1972), some bailiffs tend to overestimate the number of males based on the protrusion at the tip of the lower jaw, frequently mistaken for an incipient kype. Sex ratios will not, therefore, be given.

Seasonal changes in the length of returning fish

Table 10.4 shows the seasonal differences in mean fork length in the four major salmon rivers of the region of Cantabria. Data from different rivers were not pooled as significant differences between rivers were found to exist. In all four rivers, mean fork length decreased significantly with season ($p < 0.001$) although individual monthly differences were not always significant.

Inspection of the monthly length frequency distributions for each river reveals the successive removal of larger fish and the entry of smaller individuals as the season progresses. Peak catches occur mostly in May and June except in the river Pas where they occur in April and May. No significant correlation was found between size of monthly catches and monthly total rainfall for this period although peak catches tended to occur after periods of some rain.

Small fish (possibly grilse) are largely absent from the rod-and-line catch (Figures 10.3 and 10.4) except in the Pas and to a lesser extent in the Nansa where they were present in some numbers at the end of the fishing season in 1986.

Figure 10.3: Length frequency distribution of rod-and-line catch, rivers Pas and Ason, 1983-1986

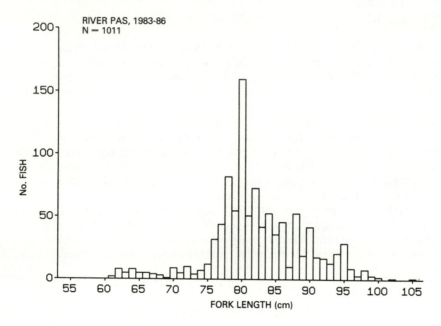

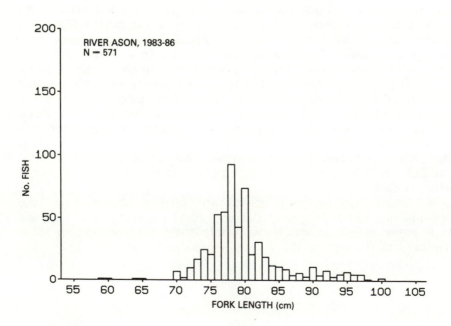

Figure 10.4: Length frequency distribution of rod-and-line catch, rivers Deva and Nansa, 1983-1986

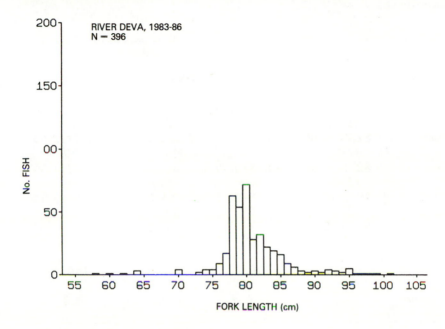

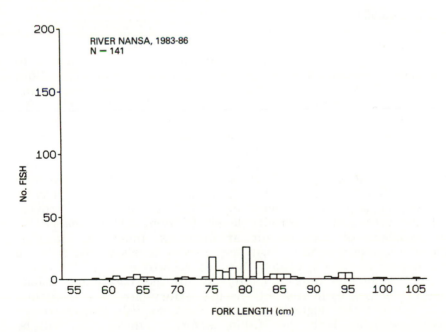

Table 10.4: Monthly differences in mean fork length (cm) 1983-1986 pooled data. Mean ± 95% CL, sample size in parentheses

	River			
Month	Pas	Deva	Asón	Nansa
March	87.50±1.14 (153)	86.93±4.06 (15)	85.31±2.24 (38)	90.17±6.06 (12)
April	82.27±0.64 (284)	83.10±1.51 (60)	81.18±1.09 (148)	82.27±2.63 (30)
May	82.20±0.59 (350)	80.98±0.55 (187)	78.60±0.57 (257)	80.33±1.69 (36)
June	79.04±1.15 (205)	78.98±0.78 (122)	78.85±0.79 (124)	74.69±2.05 (62)
July	81.74±2.38 (19)	76.92±4.94 (12)	78.00±2.24 (4)	-
n	1,011	396	571	140
F value	37.83*	17.02*	16.62*	18.82*

* $p < 0.001$

Distribution of salmon catches throughout the day

In Spain salmon may be caught from one hour before sunrise to one hour after sunset in accordance with the fishing regulations (García de Leániz et al., 1987). There is little information on the daily activity rhythms of wild salmon although continuous tracking studies of adult fish in some Scottish rivers seem to suggest that peaks in activity occur around sunrise and sunset with lows around noon (Hawkins and Smith, 1986).

Adult salmon normally cease to feed when they enter freshwater (Jones, 1959). Anglers, however, catch these fish with both artificial lures and live bait and, at least for juvenile salmon, angling success with equal fishing effort can provide reliable estimates of activity levels (Gibson, 1973). Changes in the numbers of fish caught at different times of the day, therefore, may depend on both the activity levels of the fish and the degree of fishing effort. Detailed records exist in Cantabria since 1983 on the time of capture of all fish during the fishing season though estimates of fishing effort are not available.

Only three anglers have the right to fish the restricted beats and it is likely that fishing effort at these sites will be unevenly distributed throughout the day. Fishing in the free

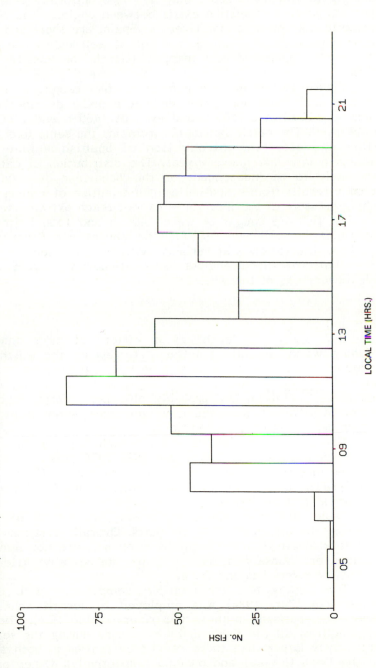

Figure 10.5: Distribution of salmon rod-and-line catches in the free zones throughout the day. Pooled data, rivers Pas, Asón, Deva and Nansa 1983-1986

zones, on the other hand, is done on a half an hour basis if there is more than one angler seeking the same beat at the same time. Fishing effort in these zones may be more uniformly distributed, especially when competition exists between anglers. The rivers in Cantabria, like most salmon rivers in Spain, are short and salmon are mostly caught in a limited number of well-known pools which are clearly signposted and mapped (Garcia de Leániz et al., 1987).

Figure 10.5 shows the number of fish caught in the free zones at each time during the day. A bimodal distribution with modes at 1100 h and 1800 h and lows at 1400 h and 1500 h can be observed. The catch distribution remains the same (clock time) before and after the introduction of Spanish summer time, starting in late March every year. The distribution of catches is thus likely to be independent of the behaviour of salmon and probably results from a non-uniform distribution of fishing effort. Fishing pressure in the free zones cannot reach extreme values as very few fish are caught between 1400 h and 1500 h (the local lunchtime) and only one fish may be caught per fisherman per day. This observation is at variance with the assertions of Netboy (1974) that professional anglers fish continuously '...from dawn to dusk ...every day of the week'.

Lures and baits

Salmon in Spain are caught by a variety of lures and baits, including worms (usually Lumbricus terrestris, the earthworm), cooked or uncooked prawns, mixed worm and prawn, freshly killed minnows (Phoxinus), wet and dry flies, and spoons and devons. Anglers in Cantabria, and probably elsewhere, normally carry a number of different baits and lures and may switch from one to another depending on the conditions.

The number of fish caught by each bait or lure in the Cantabrian rod-and-line fishery for 1983-1986 is shown in Table 10.5. The majority of the catch is taken by prawns (40 per cent), mixed prawns and worm (21.6 per cent), spoons (14.7 per cent) and worms (11.1 per cent).

Table 10.5 also shows the mean fork length of salmon caught by the different lures and baits. Overall variations in the mean fork length of fish caught by each bait are not significant in the rivers Nansa and Asón but are statistically different ($p < 0.05$) in the rivers Pas and Deva.

The Turkey test for multiple comparisons with unequal sample size (Zar, 1984) was employed to identify significant pairwise comparisons in these two rivers (Table 10.6). The mean fork length of all the fish caught by worms during the season is significantly larger than those caught by prawns in both the Pas and the Deva ($p < 0.01$ and $p < 0.05$ respectively). Other pairwise

Table 10.5: Differences in mean fork length of salmon caught by various baits and lures. Pooled data 1983-1986. Mean $\pm$ 95% CL, sample size in parentheses

	Pas	Deva	Asón	Nansa	n	%
Worm	84.39±1.23	82.67±2.04	78.78±1.12	–	230	11.1
	(105)	(52)	(73)			
Mixed	82.64±0.82	80.81±0.54	79.35±3.25	81.61±3.05	448	21.6
	(226)	(182)	(17)	(23)		
Spoon	82.43±1.78	81.44±5.41	80.30±0.87	81.05±3.53	305	14.7
	(76)	(9)	(182)	(38)		
Devon	87.67±16.17	80.50±2.49	75.0[a]	–	16	0.8
	(3)	(12)	(1)			
Minnow	82.17±1.46	–	79.90±1.95	–	121	5.8
	(81)		(40)			
Fly	81.58±1.42	80.60±1.72	81.08±2.74	76.86±9.33	126	6.0
	(85)	(10)	(24)	(7)		
Prawn	81.73±0.70	79.93±0.98	79.49±0.74	77.51±1.77	832	40.0
	(418)	(128)	(216)	(70)		
N	994	393	552	138	2,077	100
F value	2.64**	2.29*	1.12 (NS)	2.29 (NS)		

Note: a, excluded from calculations.

* $p < 0.05$
** $p < 0.01$

Table 10.6: Tukey test for multiple comparisons showing q values. All other comparisons not significant

	River	
Comparison	Pas	Deva
Worm vs Prawn	4.99**	4.73*

comparisons are not different.

      Comparisons of mean fork lengths of salmon caught by mixed bait (only one fish was caught by worm) and those caught by prawns in the river Nansa were also significantly different (t = 2.33, p<0.05).

Differences in the mean fork length of fish caught by various baits and lures may result from one or both of the following factors: (a) differences in lure selectivity, and (b) seasonal trends in the use of particular baits.

It is clear from Table 10.7 that catch by bait is highly dependent on season in all four rivers (p < 0.001). Salmon are caught with worms largely at the start of the fishing season (March) and with prawns mostly at the end (June, July). Since larger fish appear in the catch before smaller fish irrespective of the bait being employed (Table 10.4), a spurious correlation may occur between bait or lure and fish size.

Table 10.7: Contingency tables for hypotheses testing. 1983-1986 pooled data. Ho= bait/lure used is independent of month

River	Contingency table	n	d.f.	Chi square
Pas	6 x 5	998	20	88.7***
Asón	6 x 4	554	15	67.2***
Deva	3 x 5	363	8	32.0***
Nansa	3 x 4	130	6	32.6***

*** p < 0.001

Reject Ho for all rivers.

The seasonal use of baits and lures (worm first, prawn later) may be related to a variety of factors including water flow conditions, temperature, angler's choice, and availability and cost of live bait. Whether the effectiveness of different baits bears any relation to the behaviour and size of the fish at the beginning and end of the fishing season (e.g. large multi-sea-winter salmon vs. small 1-2 sea-winter fish) can only be ascertained by controlled field experiments and fishing effort data.

Age structure

Table 10.8 shows the age at smolting of adult salmon caught by rod and line in the Asón in April-June 1986. Age data from a limited number of smolts caught by electrofishing in April 1986 is also shown for comparison.

Although the combined sample size (n=113) is very small, the results indicate that most of the smolts in the Asón (94.7 per

Table 10.8: Age at smolting from adult rod and line catch data and smolt sampling (in brackets). R. Asón, April-June 1986, n=95 (18)

Smolt age	n	%
S1	35(11)	40.7
S2	54(7)	54.0
S3	6(0)	5.3

cent) emigrate to sea after one or two years of residence in freshwater.

These results are similar to those found in rivers of Asturias where most smolts are S1 (55-85 per cent) or S2 (15-45 per cent) (Martin Ventura personal communication). In the river Elorn in France, the proportion of smolts migrating after one and two years of residence in freshwater is 47 and 53 per cent respectively (Prouzet and Gaignon, 1985). In the river Eo in Galicia most smolts are S1, in the Pas in Cantabria they are S2 (Camino, 1940; Netboy, 1968). The smolt run takes place from March to May although in some rivers, like the Pas, it may extend to June.

Most of the scales examined did not show the clear freshwater winter bands normally found in salmon from more northern latitudes suggesting that growth in these rivers is almost continuous. In fact, the scale growth patterns observed are similar to those often seen in hatchery reared fish (Anon, 1984a).

Spanish smolts are large, normally averaging 15-17 cm in fork length (Camino, 1940). Larger smolts, sometimes reaching 20-25 cm, are occasionally reported by bailiffs.

Qualitative estimates of bottom fauna were carried out in the rivers Asón and Pas during April 1986. Large numbers of insects including mayflies (predominantly Ephemerella, Ecdyonurus, Rhithrogena and Baetis), and flies were found along with some freshwater snails and Echinogammarus, an amphipod normally associated with productive warm habitats. Productivity in these rivers is known to be quite high (Camino, 1940), although detailed data are not available.

Water temperatures are not available for all rivers, but these tend to vary from 6-10 °C in winter to 15-21°C during the summer (Anon, 1984b). These temperatures lie within the optimum range for both egg fertilisation and incubation and juvenile salmon growth (MacCrimmon and Gots, 1979).

Mature male parr occur in many of the rivers of Northern Europe and North America. Some of the juveniles caught in the

Figure 10.6: Length frequency distribution of juveniles caught by electrofishing, river Asón, April 1986

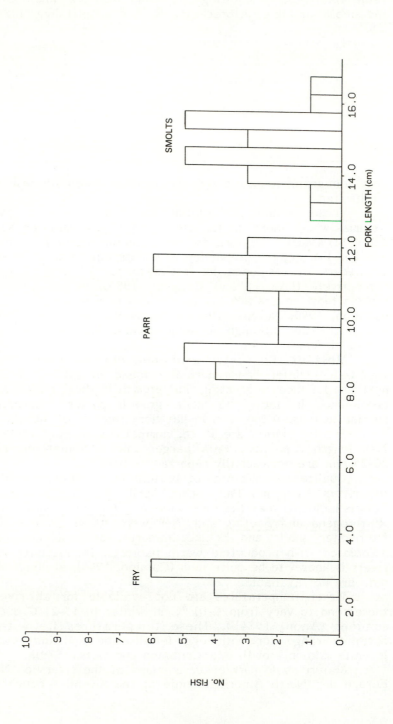

Asón in April 1986 (Figure 10.6) were also mature males. Mature female parr, on the other hand, are extremely rare among anadromous salmon (Mills, 1971), probably due to the difficulty of attaining enough weight in freshwater to produce a sufficiently large number of eggs.

The existence of mature female parr has been reported by some bailiffs in Spain but it has not been properly documented and there are no published accounts. Recently, a large mature female parr measuring 22.5 cm and weighing 150 g was caught in the river Elorn in France; this fish produced 256 viable ova which could be made to hatch (Prouzet, 1981).

The majority of the adult scales examined in the Asón in 1986 were from 2SW fish (86 per cent), although 3SW (13 per cent) and 4SW salmon (1 per cent) were also present (Table 10.9).

Table 10.9: Sea age of rod and line sample catch. River Asón, April-June 1986 (n=97)

Sea age	n	%
2 SW	83	85.6
3 SW	13	13.4
4 SW	1	1.0
Prev. spawners	2	2.1

Most salmon from rivers in Galicia and Asturias are also 2SW fish (Netboy, 1968; Martin Ventura, Chapter 11) but grilse are represented in some numbers (5-15 per cent). The absence of 1SW in the scales examined probably results from the small sample size and the early closing of the fishing season this year. Grilse averaging 2.2 kg in weight enter the river Elorn in France from May to January (Prouzet and Gaignon, 1985).

A relatively high proportion of the adult scales examined from the river Asón (19 per cent) showed a large degree of erosion, this being particularly noticeable from fish caught in June (50 per cent), when rainfall and flow rates normally drop considerably (Table 10.10). This may indicate that salmon could be held in the lower beats of the river during the dry periods before being able to move upstream in the wetter months.

There is very little information concerning the migration of Spanish salmon as no tagging programmes have been carried out in recent years (Miranda, personal communication). An adult salmon tagged on September 1969 in Disko Bay off west Greenland, however, was subsequently recaptured in October 1970

Table 10.10: Mean air temperatures and total rainfall in the city of Santander, province of Cantabria. Standard deviations in parentheses

Mean air temperature (°C)						
Period	March	April	May	June	July	August
1983-86	10.4(0.7)	12.3(1.9)	13.6(1.0)	17.3(0.7)	20.0(0.9)	19.7(0.1)
1926-86	11.2	12.1	14.1	16.9	19.0	19.4

Total rainfall (mm)						
Period	March	April	May	June	July	August
1983-86	122.2(49.2)	103.0(40.8)	119.8(98.2)	38.3(12.9)	50.4(35.9)	121.7(140.6)
1926-86	88.1	95.1	92.6	65.7	56.5	82.5

in the Asón (Netboy, 1974). An additional tagged fish from the same station was caught in the Sella in May 1972, suggesting that Spanish salmon may reach the same distant feeding grounds as fish from some British, Irish and French rivers. The great majority of recaptures of wild smolts tagged prior to 1972 in French rivers was found in Greenland (Swain, 1980), again suggesting that salmon from southern Europe may share the same feeding grounds as those from northern Europe.

Stocking effort

Stocking of salmon eyed ova and fry into Spanish rivers was already being carried out at the turn of the century (Camino, 1940; MacCrimmon and Gots, 1979) although more systematic stocking started only in 1949.

Total stocking effort in recent times is estimated as 200,000-300,000 eyed ova and between 90,000 and 120,000 fry (4-5 cm) annually, coming partially from the stripping of native broodstock and from the purchase of salmon ova from Scotland, Iceland and Scandinavian countries (Miranda, personal communication).

In the region of Cantabria detailed records exist of all the stocking programmes since 1972. Average annual stocking of eyed ova and fry is 180,000 and 9,500 respectively. Eyed ova are all imported from Scotland and Iceland and planted in Vibert boxes between January and March. Stocking densities tend to be very high and planting sites are not always the most appropriate probably resulting in higher than normal mortalities.

Fish were released only in the rivers Asón and Pas in the early 1970s. Planted fish ranged from 6 to 10 cm in fork length and came from local hatcheries or, in the Pas, from the rescue of parr (10 cm) caught in isolated pools during particularly dry summers.

Hatchery-reared fry have been planted out normally in November while rescued parr were transferred from the dry zones in July and August.

Other fish planted in the salmon rivers of this province include large numbers of brown trout (Salmo trutta) of local and foreign origin. Rainbow trout (Salmo gairdneri) have been introduced to the river Deva. Black bass (Micropterus) has also been introduced in a large artificial reservoir draining into the Mediterranean.

Commercial rearing in Spain of both Atlantic and Pacific salmon (mostly Oncorhynchus kisutch, Coho salmon) has increased in recent years but the market is still dominated by the production of rainbow trout at a level of 17,000 metric tons per year (Miranda, personal communication).

The effect of stocking programmes upon the subsequent recruitment of adults is often subject to debate. When the fish being stocked, either from eyed eggs or fry, are not the same as the native population, problems may arise. The introduced fish may not have the most appropriate characteristics and, ideally, some form of marking should be employed to assess the specific contribution of these individuals to the total catch. When this is not possible, recruitment models may be needed to estimate the effect of stocking.

Variable numbers of salmon ova and fry have been planted irregularly in the river Asón since 1973. A simple model to examine the effects of stocking has been constructed based on the following assumptions:

(1) The survival rates from planted eyed ova and fry to smolt are constant from year to year. These survival rates were estimated as 0.05 and 0.2 respectively which are slightly more conservative than those reported by Symons (1979) for low survival conditions.

(2) Five per cent of the smolts from stocked ova leaving the streams return as adults and 25 per cent of these returning adults are taken by the rod and line fishery operating from March to June-July every year.

(3) The contribution of each age class to the total catch is constant for the period studied (1976-1986) and may be summarised as follows:

Total age	Percentage contribution
3 years	33
4 years	50
5 years	15
6 years	2

These are total age estimates obtained from adult age data in 1986.

(4) The contribution to the fishery of second generation adults resulting from the stocking of fish is negligible for this short period.

A null hypothesis formulated as Ho = salmon catches in the period 1976 to 1986, bear no relation to stocking effort in previous years as defined above. Actual values of survival rates and recruitment are irrelevant to the model provided they hold more or less constant for this period. Small violations of the assumptions can also be tolerated.

Figure 10.7 shows the relation between the estimated recruitment from stocking and the actual salmon catch for the period 1976-1986. The correlation coefficient is 0.71 ($p < 0.05$) for a sample size of 11 years. A simple linear regression indicated that half the variance in salmon catches in this period

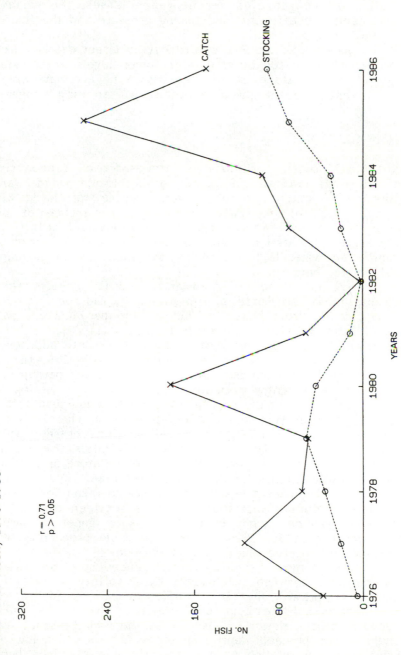

Figure 10.7: Relation between rod and line catches and estimated recruitment from stocking programmes in the river Asón, 1976-1986

r = 0.71
p > 0.05

CATCH

STOCKING

No. FISH

320

240

160

80

0

1976   1978   1980   1982   1984   1986

YEARS

could be explained by the estimated effect of stocking in previous years. Factors of greater importance in controlling yearly catches of salmon in these rivers possibly include the relative strength of different year classes, the variation in catchability coefficients and fishing pressure and the fluctuations in natural populations.

Rivers like the Pas with relatively larger catches are likely to be less affected by similar or lower levels of stocking, but since no age data exist for these rivers no attempts were made to investigate the possible effect of stocking programmes.

CONCLUSIONS

(1) Yearly variations in catches for all salmon rivers in Spain from 1949 to 1985 show no discernible trend. Particular declines in catches in some rivers can be correlated with the construction of impassable dams, the occurrence of diseases (possibly UDN) and the growth of industrial pollution. The resources allocated to salmon data collection have varied greatly and many other factors including the incidence of poaching may affect the catches.

(2) In all four Cantabrian rivers (Pas, Asón, Deva and Nansa) mean fork length of fish caught by rod and line decreased significantly from March to July in the period 1983-1986. Age data from the river Asón suggests that this is due to the early entry of large multi-sea-winter fish and the late addition to the catch of increasingly smaller salmon, mostly 2SW fish. Grilse (1SW) are seldom represented in the total catch perhaps because their time of entry normally lies outside the fishing season.

(3) Statistically significant variations in mean fork length occurred from year to year in all but the river Deva. These differences result mostly from the varying contribution to the catch of small (70-77 cm) 2SW individuals. No significant correlation exists between either air temperature or total rainfall and monthly catches for the period 1983-1986.

(4) Significant yearly differences in mean fork length and length frequency distributions exist between the four rivers studied, salmon caught in the Pas being generally larger and heavier except for 1986, where a sizeable proportion of grilse (60-68 cm) entered this river in May-June.

(5) Although significant differences in mean fork length were found between fish caught by a variety of baits (notably worm vs. prawn), the use of different baits is highly seasonal, some baits and lures being used mostly at the start of the fishing season, others almost exclusively at the end. Seasonal trends in bait choice probably depend on many factors which may or may not bear any relation to lure selectivity and salmon behaviour.

(6) Salmon in the four rivers studied may be caught from one hour before sunrise to one hour after sunset in accordance with the fishing regulations. Catch peaks, however, occur at mid morning and mid afternoon with lows around local lunchtime for both restricted beats (with a limit of three anglers per day) and free zones (one angler per beat on a half an hour basis). It is tentatively concluded that fishing effort at different times of the day in these rivers is unevenly distributed and far from the level expected to result from intense competition between anglers in the free zones (legal saturation level).

(7) At least for one river, the river Asón, it can be shown that stocking effort bears a direct relation to salmon recruitment for the last ten years. In fact, half the variation in yearly catches for this period can be explained by stocking alone. Stocking in these rivers, however, is done with limited resources and this is more likely to have an effect only in rivers yielding moderate to low catches.

REFERENCES

Anon. (1984a) Atlantic salmon scale reading. Report of the Atlantic Salmon Scale Reading Workshop. ICES, Aberdeen, 23-28 April, 17 pp.

Anon. (1984b) Análisis de calidad de aguas 1982-1983. Ministerio de Obras Públicas y Urbanismo. Dirección General de Obras Hidráulicas. Madrid, pp. 1-52

Camino, E.G. (1940) El Salmón, Fuente de Riqueza. Publicaciones de la Dirección General de Turismo, Madrid, 74 pp.

Dunfield, R.W. (1985) The Atlantic salmon in the history of North America. Canadian Special Publication of Fisheries and Aquatic Sciences, 80, 181 pp.

Escagedo Salmøn, M. (1927) Privilegios, escrituras y bulas en el pergamino de la insignie y real Iglesia colegial de Santillana. Tomo II, p. 435. cited in Camino (1940), pp. 37-8

Gallichan, W.M. (1904) Fishing and travel in Spain. F.E. Robinson, London, 227 pp.

Garcia de Leániz, C., Hawkins, A.D., Hay, D. and Martinez, J.J. (1987) The Atlantic salmon in Spain. The Atlantic Salmon Trust

Gibson, R.J. (1973) Interactions of juvenile Atlantic salmon (Salmo salar L.) and brook trout (Salvenillus fontinalis (Mitchell)). The International Atlantic Salmon Foundation. Special Publication Series, 4 (1), 181-202

Hawkins, A.D. and Smith, G.W. (1986) Radio tracking observations on Atlantic salmon ascending the Aberdeenshire Dee. DAFS, Scottish Fisheries Research Report, no. 36, 24 pp.

Jensen, A.J. and Johnsen, B.O. (1986) Different adaptation strategies of Atlantic salmon (Salmo salar) populations to extreme climates with special reference to some cold Norwegian rivers. Canadian Journal of Fisheries and Aquatic Sciences, 43, 980-4

Jones, J.W. (1959) The Atlantic salmon. Collins, London, 192 pp.

Jusué Mendicouague, (1953) Las Regalías Salmoneras. Centro de Estudios Montañeses, Santander, 606 pp.

MacCrimmon, H.R. and Gots, B.L. (1979) World distribution of Atlantic salmon, Salmo salar. Journal of the Fisheries Research Board of Canada, 36, 422-57

Martín Ventura, J.A. (1985) (personal communication) Consejería de Agricultura y Pesca. Sección de Conservacion de la Naturaleza, Calle Uría 10-1, Oviedo 33071, Spain

Martinez Conde, E. (1984) El embalse del Pas, un crimen ecológico. El País, 12 July, p. 24

Mills, D.H. (1971) Salmon and trout: A resource, its ecology, conservation and management. Oliver and Boyd. Edinburgh, 351 pp.

Miranda, J.L. (1986) (personal communication) Instituto para la Conservación de la Naturaleza (Icona). Subdirección General de Recursos Renovables. Sección de Pesca Continental. Gran Vía de San Francisco No. 35. Madrid 28005, Spain

Netboy, A. (1968) The Atlantic salmon. A vanishing species? Faber and Faber. London, 475 pp.

Netboy, A. (1974) The Salmon, their fight for survival. Andre Deutsch, London. 304 pp.

Netboy, A. (1980) Salmon, the world's most harrassed fish. Houghton Mifflin Company, Boston, 613 pp.

Notario, R. (1971) Entrada del salmón en los rios y comportamiento de esta especie piscícola en el agua dulce. Montes. Enero-Febrero, pp. 39-40

Prouzet, P. (1981) Observation d'une femelle de tacon de saumon atlantique (Salmo salar L.) parvenue a maturite sexuelle en riviere. Bulletin Français de Pisciculture, 282, 16-19

Prouzet, P. (1984) Caractéristiques du stock de saumon atlantique (Salmo salar L.) capturé à la ligne sur l'Aulne (rivière de Bretagne-Nord) durant la période 1973-1981. Revue des Travaux de l'Institut de Pêcheries maritime, 46, 285-98

Prouzet, P. and Gaignon, J.L. (1982) Fécondité des saumons atlantiques adultes capturés sur le bassin versant de l'Elorn (rivière de Bretagne Nord) et caractéristiques de leurs pontes. Bulletin Français de Pisciculture, 285, 233-43

Prouzet, P. and Gaignon, J.L. (1985) Caractéristiques du stock de saumon atlantique d'un hiver de mer (Salmo salar L.) capturé sur l'Elorn de 1974 a 1984. Revue des Travaux de l'Institut de Pêches maritime, 47, 167-78

Roberts, R.J. and Shepherd, C.J. (1986) Handbook of trout and salmon diseases, 2nd edn, Fishing News Books Ltd, Surrey, 222 pp.

Shearer, W.M. (1972) A study of the Atlantic salmon population in the North Esk, 1961-1970. MSc. thesis, University of Edinburgh

Swain, A. (1980) Tagging of salmon smolts in European rivers with special reference to recaptures off West Greenland in 1972 and earlier years. Rapport et Procès Verbaux de la Réunion. Conseil International pour l'Exploration de la Mer, 176, 93-113

Symons, P.E.K. (1979) Estimated escapements of Atlantic salmon (Salmo salar) for maximum smolt production in rivers of different productivity. Journal of the Fisheries Research Board of Canada, 36, 132-40

Zar, J.H. (1984) Biostatistical Analysis. 2nd edn, Prentice-Hall, Englewood Cliffs, NJ, 718 pp.

Chapter Eleven

THE ATLANTIC SALMON IN ASTURIAS, SPAIN: ANALYSIS OF
CATCHES, 1985-86. INVENTORY OF JUVENILE DENSITIES

Juan Antonio Martin Ventura
Consejeria de Agricultura y Pesca del Principado de Asturias,
Servicio de Conservacion de la Naturaleza, Uria 10, 1', 33071
Oviedo, Spain

INTRODUCTION

First evidence for the presence of salmon in our waters is
provided by the fossils of the Grotte de la Riera in the Sella
basin. Their age is estimated from strata at between 21,000 and
17,000 to 10,750 years. These coincide with the climatic period
temperate - cold and humid (Strauss, Clarke and Ortea, 1980). In
the same way one finds the vertebrae of salmon in the Solutreen
and Magdalenien levels of the Altamira Grotto (Cantabria) dating
back more than 10,000 years (Jusué Mendicouague, 1953). The
author found some similarities between the Magdalenian culture
and the Lapps and Eskimos in trying to correlate them with the
prehistory of the salmon in the north of Spain.

The first Spanish text which refers to salmon appears in
'Liker Iodiciorum ou Forum Ludicum' (654) translated later by the
ancient Spanish as 'Fuero Juzgo'.

In 1258, King Afonso X, the wise, granted some dispositions
to protect salmon smolts (which he called 'corgones').

At the beginning of 1949 there existed in Spain poor data
on the catches of Atlantic salmon recorded in the 25 salmon
rivers then existing (Ventura, 1986).

This species has never been the subject of a scientific
study in Spain until 1980, excepting the works of the Marquis de
Marzales, those of Camino (1949), Elegido (1958, 1960), Notario
(1971), Dalda and Serantes (1974) and De Miguel (1976). As a
result, the existing basic bibliography is consequently very poor.

In this work we try to resolve and define the morphological
characteristics of Asturias salmon, those occurring the furthest
south. The data analysed refer to the catches over the period
1949-1986 (length and weight) and more precisely to the captures
in 1985 and 1986, during the fishing season which is traditionally
between the beginning of March and the middle of July (135

days). The general characteristics of the salmon rivers of Asturias are given in Table 11.1.

Table 11.1: General characteristics of the salmon rivers of Asturias

River	Catchment (km²)	Accessible length (km)		Average flow (m³/s)
		(1950)	(1986)	
Eo	819	60	40	19.6
Navia	2,572	50	10	62.7
Esva	467	5	33	10.6
Narcea	1,850	70	43	43.2
Sella	1,246	70	50	35.0
Cares	1,120	60	45	22.4
		310	221	

One notices a big reduction in the length accessible to salmon in recent years. The numerous tributaries, not recorded here and now inaccessible, provide excellent natural nursery areas.

Table 11.2 gives an analysis of the catches on the rivers Eo, Navia, Esva, Narcea, Sella and Cares for the years 1985 and 1986.

Table 11.2: Analysis of the catches during the summer, 1985 and 1986

	1985	1986
Eo	286	667
Navia	10	No fishing
Esva	166	803
Narcea	201	537
Sella	384	2,737
Cares	312	679
Total	1,359	5,428[a]

Note: a, Also includes five salmon in other fisheries in the rivers Porcia and Bedon.

211

This estimate of 5,428 salmon in 1986 perhaps appears a big increase when first seen, but it should be mentioned that a total of 1,889 salmon were taken chiefly in the rivers Sella and Esva with symptoms of fungus on the head and fins.

On the other hand, this catch estimate represents a higher average than those recorded in the years 1950 (3,993), 1960 (4,260), 1970 (3,032) and even for the recent year 1980 (3,314).

ANALYSIS OF CATCHES

The river Eo, the furthest west in Asturias, always has the largest catches at the beginning of the season. By the end of March they can reach 9.5 per cent of the total catch compared with 0.12 per cent for the Sella and 2.48 per cent for the Esva. This difference can frequently continue until mid-June. The other rivers have nearly the same percentages (Figures 11.1 and 11.2).

At the beginning of the season one can notice a difference in certain rivers in the weight of catch of large salmon in the two years combined (Figure 11.3).

The percentage distribution of the average annual catch by weight is given in Figure 11.4. The histogram depicted shows a normal distribution corresponding to the large spring salmon of three sea-winters and generally more than 7 kg. From Figure 11.5 it can be seen that, as the season advances (March to July) the catch of large salmon declines. Taking their place are those of 5 kg (April, May and June) and already by mid-June one gets small salmon of 3 kg, grilse, which will constitute the largest part of the catch in July.

The percentage distribution of the average annual catch by length is given in Figure 11.6. The size interval chosen is 4 cm, from 50-114 cm, and the resulting average shows that the 75-79 cm class is the best represented (35 per cent) followed by the 80-84 cm class (30 per cent), then the 70-74 cm class with 15 per cent of the catch. The monthly progression shows, as one might expect, a similar pattern as that for weight. The average length of fish is about 90-94 cm in March, then it is fairly stable at between 75 to 87.9 cm in the period May to June and falls to 60-64 cm and 65-69 cm in June and July (Figure 11.7).

AGE

The study of age classes has been based on scale samples collected in 1982, 1983 and 1984, also on samples from the river Esva in 1985. A total of 4,500 scale samples from the different rivers were analysed following internationally accepted criteria (Bagliniere, 1985).

Figure 11.1: Monthly distribution (%) of salmon catch, 1985

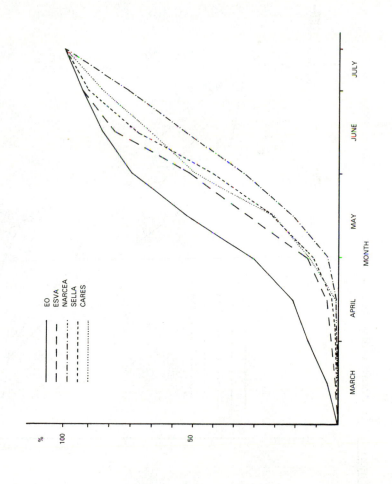

214

Figure 11.2: Monthly distribution (%) of salmon catch, 1986

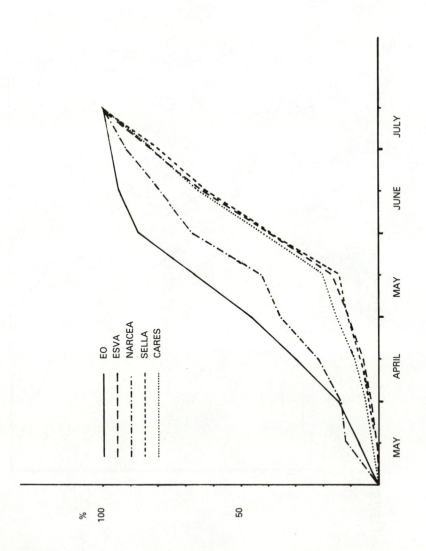

Figure 11.3: Monthly change in average weight (kg) (1985-1986)

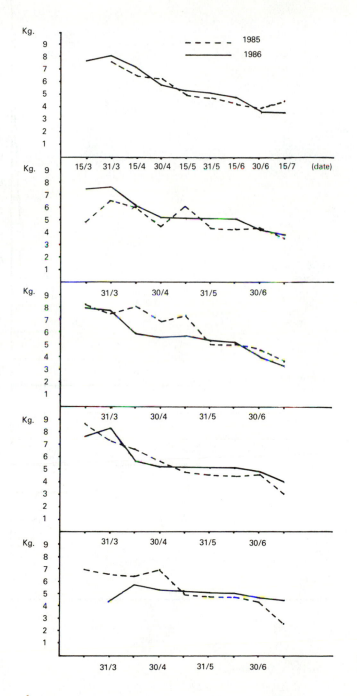

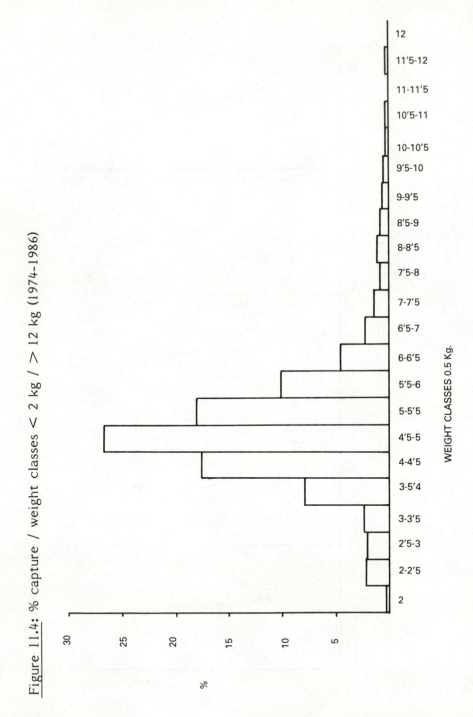

Figure 11.4: % capture / weight classes < 2 kg / > 12 kg (1974–1986)

Figure 11.5: Monthly change (March-July) of catches grouped by intervals of 0.5 kg

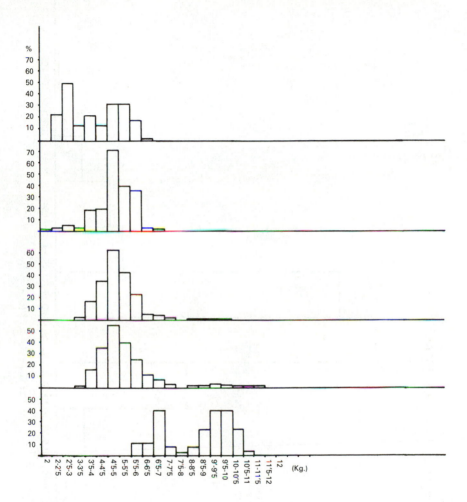

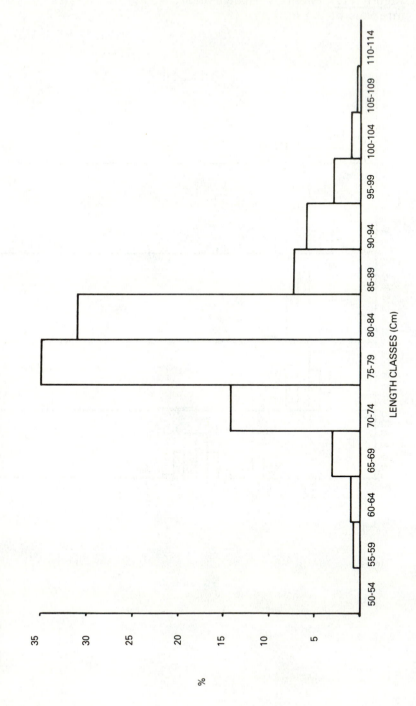

Figure 11.6: % Catch / length classes (1974–1986)

Figure 11.7: Monthly change (March-July) of catches grouped by length classes

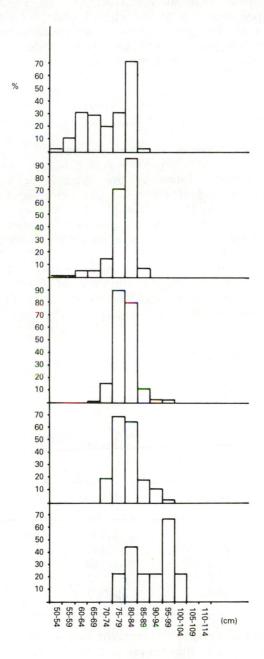

From the analysis (Figure 11.8) the following facts were deduced. More than 65 per cent of the smolts migrated at the end of their first year in the river (14 to 15 months). The remaining 35 per cent left at the end of their second year in the river. More than 75 per cent of the adult salmon captured returned after two years in the sea. The salmon sampled during the fishing season in Asturias were represented in the following categories:

(1)   Large spring salmon of 3 sea-years.
(2)   Small spring salmon of 2 sea-years.
(3)   Grilse of 1+ sea-years.
(4)   Small summer salmon of 2+ sea-years.

The percentages found by the Marquis de Marzales were similar for the freshwater stage (52 per cent for 1+ and 48 per cent for 2+ smolts). In the marine phase three sea-winter salmon were more abundant (36 per cent), also two sea-winter salmon (60 per cent) whereas grilse were less abundant than today (3 per cent against 6-7 per cent).

The present tendency is for an increase in the number of grilse and a decrease in the number of large spring fish.

All the salmon have migrated earlier (see Figure 11.8), 3 per cent of the samples belonged to the age class (1+3).

SALMON DISEASE

During 1986 a total of 1,889 salmon were taken from the rivers with symptoms of disease (Table 11.3).

Table 11.3: Incidence of diseased salmon

River	Total catch	Diseased fish	%
Sella	2,737	1,646	60.13
Esva	807	191	23.60
Cares	686	49	7.14
Narcea	537	3	0.55
Eo	667	0	0

The general symptoms encountered were fungus on the head, the abdomen and the dorsal and caudal fins. The moribund fish swam slowly at the surface in areas of still water. It was noticed that low temperatures of 2-3°C accelerated the develop-

Figure 11.8: Age classes

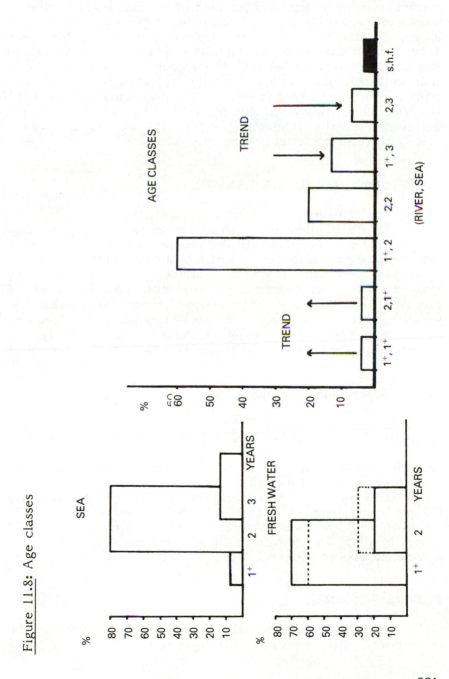

ment of the symptoms. The symptoms are already visible when the fish enter freshwater.

According to an ichthyopathologic analysis (Marques, personal communication) 100 per cent of the fish analysed showed lesions of the liver and in culture produced Aeromonas hydrophila in every case. Similarly, fatty degeneration of the liver was detected. Trials with antibodies showed a resistance to amphicycline, tetracycline and chloramphenicol (used generally in fish culture), and a high sensitivity to gentamycin and trimethopin sulphametoxazol. Studies are continuing in 1987 towards immunoglobulins (in relation to a possible general deficiency in the immunity system leading to a greater susceptibility to bacterial and mycotic diseases).

INVENTORY ON JUVENILE DENSITIES

During the summers of 1984-85 and 1986 a study was made of the spatial distribution of young salmon in freshwater, and also on the presence of accompanying species and the relationship with the physical characteristics of the sampled sites.

Each year 20 stations were studied in the summer. The average densities recorded for young salmon were 18.75/100 m^2 and only 0+ and 1+ age classes were found. The presence of salmon at the sampling sites is closely related to the speed of the current, the optimum being 55 and 85 cm/s. Similarly, the influence of the type of substrate on the density, the most important being stones with an average diameter greater than 20 cm. The average length of fish sampled in August was 84 mm for 0+ and 167 mm for 1+ (Figure 11.9).

SPANISH IMPORTS OF SALMON

According to the statistics of the Director General of External Commerce, in 1983, 1984 and 1985, there was imported a total of 1,166, 1,258 and 1,672 tons of salmon respectively, the majority (90 per cent) fresh. The chief supplier is Norway with 1,370 tonnes in 1985, followed at some distance by Denmark with 108 tonnes in 1985. The remaining 10 per cent consisted of salmon frozen and smoked (Figure 11.10).

FUTURE ACTION

(1)   A knowledge of the salmon population dynamics in the river.

Figure 11.9: Length range of young salmon

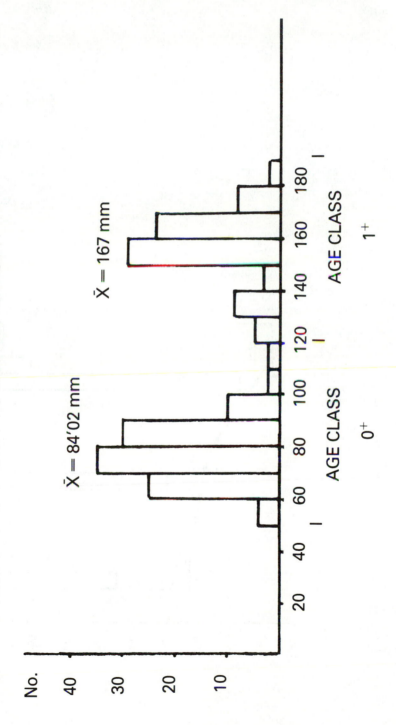

223

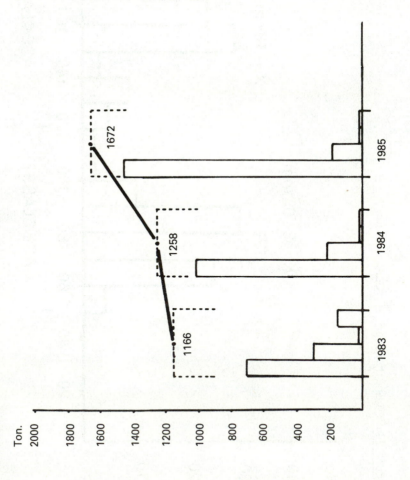

Figure 11.10: Spanish salmon imports (1983–1985)

(2)   Provide for the negotiation of natural and artificial barriers by spawners and smolts during their downstream migration.
(3)   The creation of salmon hatcheries to:
      (a) compensate for the loss of spawning ground
      (b) increase the survival rate of alevins.
(4)   To provide better administration and legislation to protect the salmon rivers against all aggression.

In Asturias one has been in the process of carrying out these objectives for fourteen years in three stages. The last of these stages intends to restore the salmon in the river Naton (4,866 km^2), the greatest salmon river at the end of the nineteenth century and today polluted by the products of coal-mines, industries and domestic wastes. In 1986 a fish trap was constructed with holding ponds and counter on the river Sella (Cano), which has already been used this year, with an expenditure of 20 million pesetas (1 million French francs or £107,527).

CONCLUSIONS

The balance sheet of the last 40 years of salmon catches shows clearly the decline of the Atlantic salmon in Asturias.
      One can pin-point the principal causes:

(1)   Pollution of waters: (a) chemical; (b) physical (barrages).
(2)   The drastic reduction in the available area accessible to salmon.
(3)   Disease.
(4)   Overfishing in (a) sea; (b) river.

If one considers a few of these causes one finds that many of these problems can be solved with a concerted effort. The Council of Agriculture and Fisheries in the principality of Asturias began in 1986 a programme of work with the primary objectives of:

(1)   Restoring the abundance of a diminished species and at the same time those other species present in our rivers.
(2)   Restoring the whole river ecosystem.
(3)   Directing towards the traditionally depressed economic zones, the riches which it will be imagined that the restoration of Atlantic salmon can provide.

The Atlantic Salmon in Asturias, Spain

SUMMARY

Over the years 1985-86 a study has continued on the salmon angling catches from the rivers Eo, Navina, Esva, Narcea, Sella and Cares in the Principality of Asturias. During the fishing season (March to July) the capture rate is greatest in May and June with more than 35 per cent of the total annual catch being taken in each of these months. It is on the river Eo, the furthest west, that the catches are most important at the start of the season.

The average weight of fish is 8 kg in March (large spring salmon that have spent three winters in the sea), then the weight remains stable at 5 kg in May and June (spring salmon of two sea-winters), and at the end of the season the weight drops to 3.5 to 4 kg in July with the arrival of the summer salmon or grilse which have only spent a year at sea. Between 35 and 60 per cent of the catch weighs between 4 and 5 kg and 50 to 80 per cent of the fish are between 75 to 85 cm long. The samples collected on the river Esva in 1985 are considered to be characteristic of catches in Asturias.

About 75.3 per cent of the parr migrated as smolts at the end of their first year in freshwater (14-15 months), the remainder went to sea at the end of two years. Of the adults caught, 71.7 per cent had spent two years at sea; 60.2 per cent of the catch belonged to the age class 1+ 2.

Similarly, during the summers of 1985 and 1986 a study was made of the spatial distribution and also the presence of other species and the relationship with the physical characteristics of the sampling zone.

A study was made of 20 different stations each summer for three years and an average density of 18.75/100 m^2 juvenile salmon was obtained. Only age classes 0+ and 1+ were represented. The average length in the second fortnight of August of age class 0+ was 84 mm and of age class 1+ was 167 mm.

The presence of juvenile salmon was associated with the speed of the current and the greatest numbers were found between 55-85 cm/s. There was also a correlation between the presence of juveniles and the sizes of the stony substrate (av. diam. 20 cm).

REFERENCES AND FURTHER READING

Bagliniere, J.L. (1979) Les principales populations des poissons sur une rivière a salmonidés de Bretagne-Sud, Le Scorff. Cybium 3eme Série, 1979(7), 53-74

Bagliniere, J.L. and Arribe, D. (1985) Microrepartition des populations de truite commune (Salmo trutta L.) de juveniles de saumon atlantique (Salmo salar L.) et des autres espèces présentes dans la partie haute du Scorff (Bretagne). Hydrobiologia, 120, 229-39

Bagliniere, J.L. and Porcher, J.P. (1980) Principales caractéristiques des adultes de saumon atlantique (Salmo salar L.) capturés par pêche à la ligne dans trois fleuves cotiers du Massif Armoricain: Le Scorff, La Sée et La Selune. Bulletin Francais Pisciculture, 279, 65-75

Camino, E.G. (1949) La riqueza piscícola de los ríos del Norte de Espana. Pub. Div. Gen. Turismo, 84 pp.

Dalda, J. and Serantes, R. (1974) El salmón atlántico, Salmo salar L. en aguas ibéricas: Estudio ecológico y biométrico. Trabajos Compostelanos de Biología 1974

De Miguel, A. (1976) Aeromonas salmonicida en los ríos de Santander. IEA Institución Cultural de Cantabria Vol II 1976 (Abr. 1977)

Elegido, M. (1958) La pesca fluvial en Lugo: Apreciaciones estadisticas. Dir. Gen. de Montes. Caza y Pesca Fluvial, 66 pp.

Elegido, M. (1960) Métodos escalimétricos aplicados al salmón. Revue Montes, 1960, 90

Jusué Mendicouague, P. (1953) Las regalías salmoneras. Cent. de Estudios Montañeses, Santander 1953

Larios y Sanchez de Piña, Pablo (1930) Rios salmoneros de Asturias. Consejeria Superiere de Pesca y Caza, 1930. Dip. Prov. de Asturias

Notario, R. (1971) Entrada del salmón en los ríos y comportamiento de esta especie piscícola en agua dulce. Revue Montes, 157, 39-40 (En-Feb)

Prouzet, P. and Jezequel, M. (1982) Caractéristiques des populations de saumon atlantique (Salmo salar L.) capturés à la ligne sur L'Elorn (rivière de Bretagne Nord) durant la période 1974-1981. Bulletin Français Pisciculture, 289, 94-111

Strauss, G., Clark, G.A., Altuna et Ortea, J.A. (1980) Subsistencia en el Norte de España durante la última glaciación. Invest. y Ciencia Ed. española. junio.

Ventura, J.A.M. (1986) Situation du saumon en Espagne. Coloque Franco-Quebecquoise sur la restauration de rivières à saumon Proceedings of Symposium. Bergerac, 1985

Chapter Twelve

EXPLOITATION OF SALMON IN IRELAND

T.K. Whitaker
Salmon Research Trust of Ireland, Farran Laboratory,
Co. Mayo

INTRODUCTION

Although over 80 per cent of Irish salmon are caught in coastal waters, salmon is not treated statistically as a sea fish. Salmon accounts for about 10 per cent of the total value of fish landed at Irish ports and ranks in this respect after mackerel, herring, cod and crustaceans. Tonne for tonne, however, salmon has been by far the most valuable fish caught in Irish waters, followed by sole, turbot and brill.

THE IRISH SALMON CATCH

The Irish catch in 1985 represented about one-quarter of the total catch of wild Atlantic salmon. By weight the Irish catch consists predominantly (over 80 per cent) of grilse or one sea-winter fish.

Table 12.1 shows the quantity of salmon caught by all methods in Ireland in the years 1975 to 1985.

The figures show a precipitous decline to a low point in 1981, when the total catch was one-third of the 1975 figure. There has been a spasmodic recovery since 1981 but never enough to regain the 1975 total, admittedly unusually high, the average for the previous five years having been 1,850.

To the total Irish catch, the Republic has contributed an average of 91 per cent, Northern Ireland an average of 9 per cent.

At the Salmon Research Trust in Co. Mayo, where the conditions enable a complete count to be made of the movement of salmon to and from freshwater, it has been found that the numbers available for spawning have been adequate for self-sustaining populations in only three of the past twelve years and half of what was required in six of those twelve years.

Exploitation of Salmon in Ireland

Table 12.1: Irish salmon catch (tonnes)

Year	Republic of Ireland	Northern Ireland	Total
1975	2,216	164	2,380
1976	1,561	113	1,674
1977	1,372	110	1,482
1978	1,230	148	1,378
1979	1,097	99	1,196
1980	947	122	1,069
1981	685	101	786
1982	993	132	1,125
1983	1,656	187	1,843
1984	887	78	965
1985	1,588	97	1,685

Note: figures from Department of Fisheries and Forestry (Republic of Ireland) and Department of Agriculture (Northern Ireland), the Foyle catch being divided evenly between the two areas.

Table 12.2 gives details (a minimum of two survivors per female grilse is required for population stability).

Table 12.2: Survival ratios to spawning grilse

Brood-year class	Survival to spawing grilse (4 years later) per grilse female
1970	1.48 - 2.06
1971	2.83 - 3.29[a]
1972	0.97 - 1.39
1973	0.80 - 0.89
1974	1.10 - 1.11
1975	2.03 - 2.37[a]
1976	1.60 - 1.80
1977	0.93 - 1.02
1978	1.77 - 1.96
1979	1.16 - 1.28
1980	0.83 - 0.91
1981	2.2 - 2.5 [a]

Note: a, years of adequate spawning escapement.

The general evidence is of a decline in salmon stocks for which excessive exploitation at sea is the main cause, aggravated at times by a drop in the natural survival rate of salmon in the sea.

CONTRIBUTION OF REARED SMOLTS TO SALMON CATCH

The adverse trend in stocks has occurred despite the contribution which is being made in recent years by reared smolts.

It has been assessed that in the Republic of Ireland in 1980 about 7 per cent of the salmon caught in drift nets came from reared smolts. A similar estimate for 1985 indicates an average contribution from reared smolts of the order of 3 per cent, with the proportion as high as 12 per cent in one important area (Galway/Limerick). There was a moderate increase (from 360,000 to 420,000) in the number of reared smolts released in 1985 as compared with 1980. This represents about one-third of the total production of reared smolts, the remaining two-thirds being used for sea farming.

The reared smolts released to the sea are a contribution by bodies such as the Salmon Research Trust, the Electricity Supply Board and the Office of Public Works to the replenishment of salmon stocks. The benefits of this public-spirited form of sea-ranching accrues mainly to the coastal and estuarine fishermen, free of charge to them. Every 100,000 smolts so released would have cost about £50,000 to rear and of the 10,000 grilse expected to return as a result to Irish coastal waters, 8,000 with a market value of £80,000 are likely to be taken by drift nets.

Irish legislation does not permit the construction of traps in tidal or freshwater other than those already licensed before 1923.

Reared smolt production in Northern Ireland is confined to the River Bush and was less than 24,000 in 1985. These smolts are released as a contribution to sea-ranching, but the return of adult fish to the river is much below the economic level.

METHODS OF EXPLOITATION

The proportion of the total catch in the Republic of Ireland which is taken by drift nets has risen from a 20 per cent average in the 1950s, to an average of 33 per cent for the 1960s, 69 per cent for the 1970s and a peak of 85 per cent in 1985. Draft nets, which used to take the biggest proportion of the catch, have dwindled to under 10 per cent and rod and line from the 16 per cent of the 1950s to a mere 3 per cent today.

The number of commercial drift net licences in the Republic has increased from 318 in 1960 to 817 in 1970 and 827 in 1985 but, because of illegal fishing, these figures understate the intensity of drift-netting.

In Northern Ireland, the bag net has traditionally been the principal instument of capture of salmon, accounting on average for half the total catch.

VALUE OF IRISH SALMON CATCH

The estimated value of the total Irish salmon catch rose between 1975 and 1985 from about £3.3m sterling to over twice that value. The salmon of 1985 realised more than three times the value per tonne of their 1975 counterparts, thus keeping pace with inflation. Because of the reduced catch, however, the gross value, allowing for inflation, showed a real decline.

Despite buoyancy of demand, the price of salmon is coming increasingly under the influence of the expanding supply of farmed fish. Conservation of wild fish stocks may thus be indirectly aided.

Studies by the Economic and Social Research Institute and by Mr C.P. Mills of the Salmon Research Trust show that, from a national standpoint, the most valuable form of exploitation of salmon is by rod and line. Visiting anglers make a contribution to national income which is a high multiple of the value of a salmon caught by other means.

A policy which allows exploitation by drift nets to increase disproportionately, while exploitation by rod and line decreases, appears to be at variance with the national economic interest.

CONSERVATION REQUIREMENTS

Salmon fishing is regulated in a variety of ways in the hope of conserving stocks. The main provisions are:

(1)   Fishing is banned outside 12 miles from coastal baselines.
(2)   Annual close seasons are prescribed for fishing by drift nets, draft nets and rod and line.
(3)   Commercial fishing is banned at week-ends (from 0600 h on Saturday to 0600 h on following Monday).
(4)   The number of licences to fish commercially by the various means - drift nets, draft nets and rod and line - is limited by law.
(5)   The maximum length and depth of drift net are regulated by law.
(6)   Monofilament netting of salmon is banned.

(7)   The minimum mesh size of a commercial salmon net is prescribed by law.
(8)   There is a limit on the size of boat allowed for salmon fishing.
(9)   A commercial net must be moving at all times.
(10)   Netting in freshwater is prohibited.
(11)   All who deal in salmon must be in possession of a salmon dealer's licence.

ENFORCEMENT PROCEDURES

There is a high incidence of breach of the regulations just mentioned. The most serious stock-depleting effects are caused by fishing at prohibited times and seasons, the use of monofilament nets of illegal length, depth and mesh size, the use of fixed nets, unlicensed fishing by drift nets at sea and poaching in estuaries, rivers and lakes.

An indication of illegal out-of-season netting of salmon is given by the proportion of fresh-run net-marked fish passing through the traps at the Salmon Research Trust in the later autumn months. This proportion was of the order of 12 per cent in August-September 1985, not much below the ratio observed at the peak of the legal season (about 15 per cent for June-July).

Enforcement of the regulations in the Republic of Ireland rests with the protection staff of fishery boards, supported by the police, the air corps and the navy. Evasion of the law is widespread and not only protection staff but even police and naval personnel meet at times with obstruction and violence. Two naval patrol vessels (mine sweeper class) were specifically assigned to fishery protection in the Republic but the vessels are old, have been out of commission for long periods, and need to be replaced. Larger naval vessels have been made available from time to time for protection purposes. As yet only one fishery board (North-West Region) has its own sea-going patrol boat but another is due to go into service in the Shannon region.

Despite inadequacies in staff numbers and equipment, considerable results are achieved by the protection service in confiscation of illegal fishing nets and gear and the prosecution of offenders. For instance, nearly 170 miles of illegal netting were confiscated at sea in the two years 1983-1984. The situation, however, remains far from satisfactory.

In Northern Ireland, the regulations and protective arrangements are similar. As in the Republic, protection and enforcement measures are not successful to the degree required to ensure maintenance of a high level of stocks.

The cost of even partly effective protection is high in relation to the purely commercial value of the Irish salmon catch.

POLLUTION AND ARTERIAL DRAINAGE

For salmon, the principal pollution danger is the destruction of spawning streams and their populations by noxious effluent such as slurry, peat silt, seepage from silage, waste from factories and creameries and water contaminated by mining spoil. The law forbids such discharges but they occasionally occur and detection of the offender is not always easy nor are prosecutions certain to result in fines of a sufficiently deterrent size. Unduly liberal granting of planning permission for houses on lakeshores and riverbanks also poses pollution problems.

Arterial drainage, by channelising the natural flow has, over a protracted period, impaired the spawning potential of a number of major rivers such as the Boyne. In time, and with remedial measures, there has been a reasonable degree of recovery, as in the River Moy.

SALMON FARMING

The exploitation of salmon as they return from the sea to freshwater to spawn is, of course, no longer the only method employed. In recent years there has been a marked development of sea-farming, based on the artificial feeding in sea cages of reared smolts and the culling for sale of most of the maturing salmon.

In the Republic of Ireland, the number of producers has risen from three, with an output of 35 tonnes in 1981 to 10 with an output of 500-700 tonnes in 1985 and an estimated 1,200-1,500 tonnes in 1986. This is still small by comparison with the Scottish output of 5,000 tonnes and the Norwegian output of 25,000 tonnes but Irish production is expected to reach 10,000 tonnes per annum by 1990 and to increase further thereafter. Large investors have entered the field and joint ventures with Norwegian interests are being undertaken. Marketing is being co-ordinated.

Salmon farming has not so far developed in Northern Ireland.

Some 70 per cent of Irish salmon exports are of farmed salmon.

Research in support of salmon farming, and aquaculture generally, needs to be extended and co-ordinated.

SUMMARY

(1)   The total Irish salmon catch has fallen over the ten years 1976-1985 to an annual average almost 30 per cent below that for the previous five years. The spawning escapement

to freshwater has been insufficient in many recent years to ensure population renewal. This depletion of salmon stocks is occurring despite a contribution from reared smolts. The principal cause is excessive exploitation of salmon in coastal waters.

(2)   In the Republic of Ireland, drift nets now account for 85 per cent of the salmon caught, as against 20 per cent in the 1950s. The catch by rod and line, of far greater potential value to the national economy, has contracted from 16 per cent to a mere 3 per cent.

(3)   Illegal fishing is of serious dimensions and difficult to control with present resources and methods. Pollution, and channelisation of river flow by arterial drainage, are significant but lesser evils.

(4)   Salmon farming is rapidly increasing but production is still small compared with that of Norway and Scotland.

(5)   Research in support of salmon farming needs to be extended and co-ordinated.

Chapter Thirteen

CATCH RECORDS - FACTS OR MYTHS?

Alex T. Bielak[1] and Geoff Power[2]

1   The Atlantic Salmon Federation, Suite 1030, 1435 St
     Alexandre, Montreal, Quebec H3A 2G4, Canada
2   Dept of Biology, University of Waterloo, Waterloo, Ontario
     N2L 3G1, Canada

What is a fisherman without hope?
What can he learn without experience?
Who will believe him without the record?

> Henry van Dyke
> Senior Member
> Sainte Marguerite Salmon Club
> 1 April 1924

INTRODUCTION

The ideal barometer of Atlantic salmon management success
would be complete and accurate records of how many, where and
what sizes of salmon were caught in the past, the types of gear
used and the effort expended. While such records would prove
most helpful, they still would not be easy to interpret due to
varying response times to management decisions, and the
confounding effects of climate and other environmental
influences on salmon stock density and behaviour. The scarcity of
good records and the difficulty of interpreting them provides a
real challenge, stimulating both the historian and the detective in
us to sort myth from reality, trying to provide a reliable
background against which present day results can be judged.
      Atlantic salmon catch records come in many forms and
disguises; all have inherent value if used in the right context. In
the broadest sense, anything preserved relating to a catch
constitutes a catch record. Most commonly these range from
anecdotal accounts found in various writings, through private
fishing logs, to the more formal statistical reports published by
government agencies. They can be more or less precise and span
just a few years or over a century. Bielak (1984) defined
long-term records as those over 50 years. There is some

confusion in the terminology used in the literature - for instance Niemela, Niemela and McComas (1985) refer to an eleven-year record for the Teno river, Finland, as long term. We would suggest modification and subsequent general adoption of Pirie's (1986) definitions of short (1-3+ yr), medium (10-30 yr) and long (over 30 yr) records to short, 1-9 yr, medium, 10-50 yr and long, over 50 yr.

At a time when the Atlantic salmon resource is generally considered as depleted throughout most of its range, there is an ever-increasing interest in assessing the present status of stocks in an historical perspective. The apparent past plentitude of the salmon provides an impetus for its protection and restoration. Historical records are used to bolster arguments regarding the potential yields from existing, improved and new habitats, or even to caution against certain management actions. Such increased interest in historical data is demonstrated by a number of recent general publications on the subject (e.g. Dunfield, 1985; Taylor, 1985; Jenkins and Shearer, 1986) as well as some controversy over the general, and specific utility of such catch records as indicators of stock abundance.

Such publications have also underlined the general paucity and inconsistencies in the data available to researchers, as well as the problems encountered in examining historical data. One of the conclusions presented in a synthesis paper at the recent conference on Age of Maturity in Salmonids (Porter, Healey and O'Conell, 1986) was that there was 'a serious lack of time series of information on characteristics of individual wild stocks', and the recommendation that 'government agencies should compile existing data sets in a common format such that these data are comparable... and readily accessible.'

CATCH RECORDS

At the most basic level perhaps the earliest catch records are the paleolithic images of salmon, found in the caves of the Perigord in France. Similarly, early accounts by travellers and historians chronicle the importance of salmon runs as well as distribution and local abundance. For a comprehensive view of the salmon resource in early North American history it would be difficult to find a more extensive review of archival material than that presented by Dunfield (1985).

Among the more reliable indicators are records kept for the purposes of taxation and valuation of fisheries - the earliest of which may well be the Domesday book. However, given human nature, these values might be minimised or exaggerated depending on the case. Similar problems pertain to records of commercially caught salmon to the point where the government publication reporting Norwegian catch statistics makes specific reference to this being a reason for the unreliability or lack of some of the data (Anon., 1984; Hansen, 1986). Shearer (1986) maintains that Scottish records before 1850 are better than those of later years, because after that time people began to realise the value of the fishery. Dunfield (1985) reports that accounts of abundance of salmon were over-inflated to lure European settlers to North America. In recent years, quotas imposed on Quebec North Shore commercial fishermen have been based on previously declared catches. It is popularly thought that in many, if not most, instances the quotas are far less than actual previous catches.

Shipping records and bills of lading also provide valuable insights into the magnitude and size of individual and total salmon catches (e.g. Power, 1976; Dunfield, 1985; Taylor, 1985; Shearer, 1986; Bielak and Power, in press), although such records almost always require the application of various correction factors to account for gutting of fish and preservation and transportation procedures.

With the development of angling both in Europe and North America, the competitive spirit and curiosity of anglers and eventually the regulatory activities of government agencies led to the establishment of personal and club logs of catches. Even today anglers returning to a fishing lodge can expect to be met by a welcoming committee interested in the real weight of the catch! These records are among the most extensive and complete data series available to researchers, although recent Canadian management measures (e.g. the institution of catch and release of large salmon) probably diminish the value of data collected in the last few years.

As the need for research, stock assessment and regulation of both angling and commercial catches grew, so did the volume of information amassed by government agencies. These data are generally the basis of long-term catch curves such as those presented by Lindroth (1965) for the Baltic, Taylor (1985) for Newfoundland, Hansen (1986) for Norway and Anon. (1986) for eastern Canada.

Catch curves for Newfoundland, Canada and Norway (Figures 13.1-13.3) illustrate the general nature of such records. Catches are summed totals from many fisheries exploiting stocks over large areas. It is interesting to note the extent to which records from adjacent areas track together (Figures 13.1 and 13.2) and whether salmon fisheries on both sides of the Atlantic (Figures 13.2 and 13.3) show similar trends. Whatever the conclusion, the field is wide open for speculation to explain either differences or similarities in the patterns of landings. When the records are plotted as smoothed running averages to remove some of the extreme annual variations in catches, and

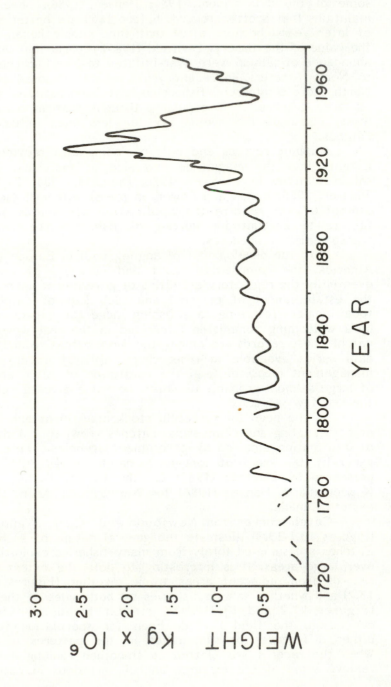

Figure 13.1: Trends in the salmon landings for Newfoundland and Labrador. Free hand smoothed curve modified after Taylor (1985) and brought up-to-date using recent Canadian statistics

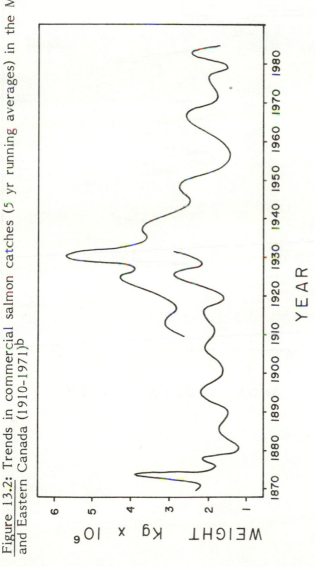

Figure 13.2: Trends in commercial salmon catches (5 yr running averages) in the Maritimes (1870-1930)[a] and Eastern Canada (1910-1971)[b]

Source: a, Huntsman (1931); b, Anon. (1986).

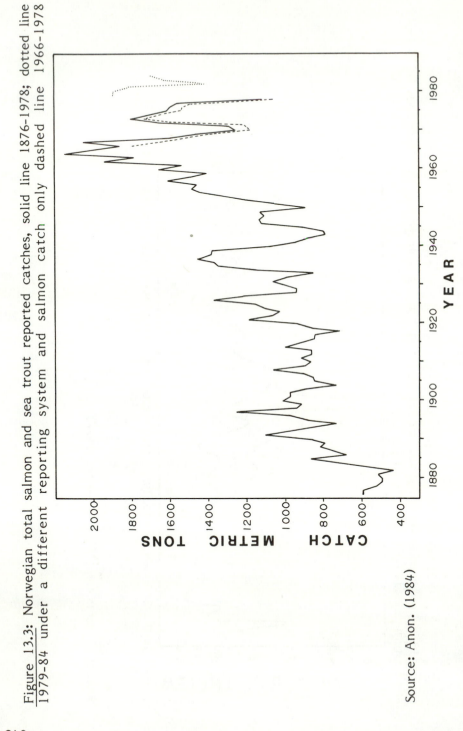

Figure 13.3: Norwegian total salmon and sea trout reported catches, solid line 1876-1978; dotted line 1979-84 under a different reporting system and salmon catch only dashed line 1966-1978

Source: Anon. (1984)

240

emphasise trends as in Figures 13.1 and 13.2, regular short-term (10-12 year) and longer-term cycles become evident. Again these are ripe for speculation. An analysis by Huntsman (1931), influenced by contemporary work on animal population dynamics, related the short-term cycles to sunspot periodicity and such discussions still appear in the literature (e.g. Pirie, 1986). The longer-term cycles are now felt to be related to variations in the marine environment influencing sea temperatures (Scarnecchia, 1984); altering migration (Martin and Mitchell, 1985); modifying ocean circulation and shifting feeding areas (Dunbar and Thomson, 1979; Dunbar, 1985; Reddin and Shearer, 1987). It is interesting to note that the Baltic catches differ markedly from the Norwegian, which would imply marine rather than freshwater conditions are important in determining harvest.

A criticism of pooled data is that much gets lost and hidden in the summation. Berg (1964) provides many long-term angling catch records for individual north Norwegian rivers. These show the usual year to year variability, and a general trend towards increased catches in the post World-War II period. This is ascribed to the results of fish ladder construction putting into production much new water, but increased leisure and affluence may also have added to the angling returns. Mills (1986) graphed rod catches from a beat on the lower Tweed from 1862, and the only obvious fluctuations in numbers landed were short-term in nature, whereas his analysis of the commercial catch in the same river, by the Berwick Salmon Fisheries Company, revealed marked medium and long-term fluctuations in landings over a similar period.

It is examination of the full gamut of historical records that eventually leads us to question whether any credence can be attached to them, whether they represent 'myths or facts'. As will become apparent, they, like the curator's egg, are 'good in parts'.

CATCH RECORDS - SOME OF THE ISSUES

A controversy in the French literature over the historical abundance of salmon exemplifies some of the issues raised in dealing with historical accounts.

On the one hand Thibault (often publishing with Rainelli) essentially argues for a scientific, rather an historical, approach to dealing with catch records (e.g. Thibault, 1981, in press, personal communication; Thibault and Rainelli, 1980, 1981, 1982; Rainelli and Thibault, 1985). These authors maintain that actual knowledge of French Atlantic salmon stocks truly resides in the results of scientific studies made over the past 15 years, although the data are marred by incomplete sampling of runs.

They state that there are no <u>documented</u> cases of over-exploitation of French salmon stocks, and that the claims of 'a fabulous super-abundance' of salmon during a 'golden age' are exaggerations. A particular point of contention relates to alleged over-estimates, by Vibert, of historical salmon production of the rivers of Brittany. Thibault does, however, admit to a general decline in French stocks, and attributes this largely to damming of rivers and some instances of pollution.

Thibault and Rainelli (1982) bemoan the lack of good long-term capture data which can be fully related to spawning stocks, particularly with respect to natural fluctuations and general ecology of the species. They conclude that comparison of catches has little meaning under such circumstances, and also that because of lack of data, it is not possible to say whether characteristics of stocks have changed over the last 30 years. Notwithstanding the above, Thibault (1981) has drawn conclusions from historical data examined in a comparative fashion, for rivers in the Finistere region (1951-1980) and Cornwall and Devon (1952-1979).

On the other hand Vibert (1980 a,b, 1982 a,b and personal communication) believes salmon were much more abundant in the past, in Brittany in particular and France in general, than at present. He bases this conclusion on a concept of climatic potentiality, supported by extrapolations of selected data from stocks believed to be producing maximally, and reference to the historic abundance of salmon in North American river systems such as the Restigouche. Notwithstanding the quite different approaches taken by certain scientists to the question of former abundance of salmon in France, it is generally agreed that the resource is now seriously depleted. Whether the more optimistic estimate of Vibert, or the conservative results anticipated by Thibault regarding the future of France's Atlantic salmon resource are realised is unimportant. Salmon stocks in France were once more abundant than at present and investment in restoration should pay large dividends.

Throughout his publications Thibault raises a number of important questions regarding the interpretation of historical data. These concerns are also reflected in many of the papers recently published on the status of Atlantic salmon stocks in Scotland (Jenkins and Shearer, 1986). In this publication, Mills (1986) states that an increasing proportion of Scottish salmon runs are outside the legal netting season, a fact which will be of importance in the value of statistics obtained in later years. He concludes that management decisions are often based on casual observations, limited data and speculation. Dunkley (1986) adds that declining catches may reflect changes in availability of fish to the nets rather than absolute numbers.

Year-to-year changes in fishing effort will also affect

reliability of statistics (Laird and Needham, 1986; Hansen, 1986; Shearer, 1986) as will the accuracy and constancy of data collection (Browne, 1986). In the case of the 'official' Norwegian data, Hansen (1986) concludes that one may safely assume that the statistics give a picture of the tendency within the fisheries, and he draws some general conclusions from data trends between 1876 and 1980. In contrast, Browne (1986) reports that export figures for Irish salmon exceeded, at times, published catch figures, and states that although it is apparent that major changes in the stock of Irish salmon are probably continuing to occur, the data are inadequate for monitoring fluctuations in the stock.

Perhaps the most significant papers in this overview of Scottish salmon stocks are those by Shearer (1986) and Lakhani (1986) and the conference summary by McIntyre (1986). Shearer (1986) evaluates the data available for assessment of Scottish salmon stocks as well as the limitations thereof. He refers to the fragmentary nature of the national catch data between the mid-1850s and before 1952 (confirmed if we look at the data presented and interpreted by George (1984)), and the usefulness of railway/steamship returns in comparing catches between years during that period. He refers to the growing reluctance to disclose information as values of fisheries increased and hostility between rods and nets grew. Fish counts at North of Scotland hydro-electric facilities are mentioned as are the data series pertaining to smolt production.

Shearer (1986) also deals extensively with biases and limitations in the data: in particular problems of grilse error, whereby large grilse are mis-classified as salmon; biases arising due to limited season sampling in rivers where the salmon run throughout the year; changes in effort and gear; and the effects of off-shore fisheries established since the 1960s. Although he concludes that the present data are insufficient to draw <u>firm</u> conclusions about the strength of past and present stocks, he does summarise national catches in a series of useful graphs, shows present catches are equivalent to historical ones in some cases, and demonstrates some underlying similarities as well as differences in catch trends between rivers, as well as different components of salmon stocks.

Lakhani (1986) takes a strict statistical approach, and states that statutorily collected (Scottish) salmon statistics cannot be regarded as satisfying the basic requirements of a sampling study, and cannot therefore satisfactorily indicate the status of Atlantic salmon in Scotland. Furthermore, even if all statistical objections were overcome, he maintains that catch statistics are inherently poor stock indicators.

A purely theoretical approach to the problem of catch records, while important for pointing to weaknesses in data collection, is only helpful to the extent that practical improvements in recording can be identified. Unfortunately for fishery biologists, population dynamics is far more advanced and refined in theory than in practice, and there are very few examples of successful application that can demonstrably meet all the theoretical assumptions. The world goes on and the fishery biologist/manager has to do the best he can with what data are available. McIntyre (1986) makes a realistic assessment of the statistical problems, and we concur with his view that, ultimately, management decisions must be made on the facts available, and be based on a sound biological awareness of the species and its ecology.

Notwithstanding the problems in interpretation of catch statistics alluded to, and in contrast to the type of dispersed data examined by George (1984),there are a number of discrete continuous medium or long-term commercial, angling, as well as trap count data which avoid many of the sorts of pitfalls mentioned above, and are susceptible to detailed examination (e.g. Miramichi River, Ruggles and Turner 1973; River Wye, Gee and Milner, 1980; Icelandic Rivers, Scarnecchia 1984; River Dee, Martin and Mitchell 1985; Restigouche River, Chadwick, Brazeau-Carrier and Leger, 1985; Godbout River, Bielak and Power, 1986).

For instance, in the case of the Aberdeen Harbour Board salmon catch records examined by Martin and Mitchell (1985), in their investigation of the influence of sea temperature upon numbers of grilse and MSW salmon caught in the vicinity of the Aberdeenshire Dee, the data proved suitable for study for a number of reasons, particularly because catch-per-unit-effort could be calculated. In the course of their study the authors established that the Aberdeen harbour data generally reflected similar ones from the northeast of Scotland for the period after the statutory collection of data began in 1952.

Chadwick et al. (1985) presented historical catches of Atlantic salmon at four sports fishing camps on the Restigouche River, NB, and concluded that these data could be useful in estimating total annual angling harvest as well as spawning escapement in the river.

ANALYSES OF ANGLING DATA; SOME EXAMPLES

In the course of research on the Quebec North Shore of the Gulf of St Lawrence (Bielak, 1984) we had the opportunity to examine angling records at fishing clubs situated on various rivers. Some of these data have already been published, and the problems involved in locating and treating them previously thoroughly discussed (Riley, Bielak and Power, 1984; Bielak and Power, 1986 and in press). The new data presented here, regarding salmon angled in the St Marguerite river, strongly corroborate previous findings.

Before presenting these data it is useful to revew briefly some of the methodologies more commonly used to interpret historical records such as these. Firstly, and most simply, one can look at whatever raw data are available. For instance, in the case of a study of the potential of rehabilitating the stocks of the Rivière à Mars, at the head of the Saguenay fjord, researchers simply examined angling catches made by the Price Bros. Club in order to establish historical catch composition (Legault, Dunont, Thibault and Boudreault, 1985). A slightly more sophisticated approach involves appraising the data as before and using whatever other 'hard' data might be available to assist in drawing conclusions. In the case of the analyses by Chadwick et al. (1985), long-term records of daily catches from the Restigouche River were categorised as grilse (less than 5 lb), small salmon (5-15 lbs inclusive) and large salmon (greater than 15 lbs), based on weights of aged fish taken at the Department of Fisheries and Oceans Dalhousie trap, in the estuary of the same river, between 1972 and 1980. This approach does not give a precise indication of the percentage of fish arbitrarily assigned to one category which should actually have been placed in another. For example, large or 3SW salmon at Dalhousie weighed 22.62 ± 7.89 lbs (95 per cent CL, n=43) which means that an unknown number of 3SW fish would mistakenly have been classified as 2SW. There is also no separate category pertaining to multiple spawners, presumably because no weight data for previously spawned salmon appear to be available in the publication (Peppar, 1983) used to categorise catches.

The significance of the above probably varies depending on the depth of subsequent analyses attempted. We would suggest that this approach is largely adequate in most instances, where one is simply trying to discern broad trends, or to consolidate similar data series for a single river system, as in the Restigouche study just cited. In any event the limited ancillary data are often, all that is available to researchers without embarking on new studies.

The ideal approach, especially in a situation where one is examining records from more than one river system, is to base the weight/sea-age conversion on simultaneously gathered biometric data from salmon of each study river. This facilitates a more accurate grouping of fish into age-specific weight-classes by the use of weight frequency distributions of fish of known sea-age. The weight-age conversion and classification can then be retroactively applied to fish recorded in club log-books. The conversion of weight to sea-age makes inference possible regarding temporal changes in (1) the proportion of the run made up by each age-class, and (2) the mean weight of the fish of each class. More details of the analyses applied, which are tailored according to the peculiarities of each data-set, may be found in Bielak (1984), Riley et al. (1984) and Bielak and Power (1986).

Most authors appear to favour graphic or simple robust statistical analyses for the treatment of salmon catch data (e.g. Ruggles and Turner, 1973; Ricker, 1981; Chadwick et al., 1985). There is perhaps some merit in this approach since it avoids the problem of statistical overkill, and trying to extract too much from data which are known to have various limitations and shortcomings. This is the option we have adopted; at the same time we have used assumptions (for example in our weight-age conversions) that lead to the most conservative interpretation of any observed trends in the data. At the other end of the spectrum, Dempson, Myers and Reddin (1986) used complex time series analysis procedures in attempting to relate sea surface temperatures to year-to-year variations in stock characteristics in two Canadian rivers. The authors assumed that angling catch data from the Restigouche Angling Club (which show a substantial increase in grilse) and from the Moisie River adequately reflected the size composition of fish returning to these rivers. They failed to find evidence that sea surface temperatures had an effect on the within-population annual variations in the ratios of grilse to MSW salmon. The problem here is that it might be impossible to verify the basic assumption that angling records accurately reflect the size composition of returning fish, and unless this can be done, any analyses are questionable.

Bielak (1984) treated the Moisie records cited by Schiefer (1971) with some circumspection, and did not even attempt to use the limited data found in Weeks (1971). In the case of angling catches recorded by the St Marguerite club for the period 1894-1982, Bielak (1984) found that grilse were only sporadically recorded before 1955. The obvious increase in proportion of grilse (Figure 13.4) from the 1950s on was attributable to more consistent recording of smaller fish, as well as extended fishing seasons in recent years. Because of the unreliability of the data in this respect, grilse were eliminated from an analysis of age-class proportions. The manager of the Restigouche club (Al Carter, personal communication) and long-term fishing guides on the river confirm that the early Restigouche grilse data, used by Dempson et al. (1986), are unreliable for similar reasons, a view also shared by biologists who have recently worked in the region (J.L. Peppar - DFO, E.M.P. Chadwick - DFO, and A. Madden - DNR, personal communication). The evident similarity in reported catches of grilse on both the St Marguerite and Restigouche rivers (Figure 13.4) strongly suggests that in both rivers one need look no further than the sociological explanations for the apparent increase in percentage of grilse, and the

Figure 13.4: Changes in the proportions of grilse in the recorded angling catches of the St Marguerite salmon club and four clubs on the Restigouche River 1894-1984. Values for the Restigouche River from Chadwick et al. (1985)

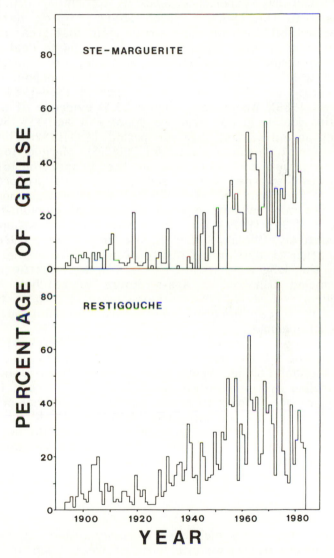

analyses performed by Dempson et al. (1986) were not appropriate.

This must not be construed as a denigration of the value of all historical catch records, but rather as a caveat not to neglect such basics as going back to the source when possible. In fact, where angling records for the Restigouche are concerned,

Chadwick (personal communication) states that 'there is no doubt that sport club data are one of the best measures of stock abundance for (the) Restigouche River' whereas Madden (personal communication) believes 'that the data are priceless'.

Our own experience leads us to similar conclusions with regard to the Quebec North Shore angling data we have collected, although we must stress again that great care must be taken in their analysis, and that it is crucial to deal with as long a time series as possible. For example, we reported that the mean weight of 2SW salmon angled in the Godbout had declined at a rate of 0.01 lb/yr over the period 1864-1983 (Bielak and Power, 1986). We present (Figure 13.5) evidence of a remarkably similar decline in the mean weight of 2SW and 3SW St Marguerite River angled salmon over the period 1894-1982. This change in age-specific mean weight, for the St Marguerite salmon, is accompanied by decreases in the proportion of 3SW and previously spawned salmon and an increased proportion of 2SW fish (Figure 13.6). It is important to note that none of these changes were evident upon examination of medium-term angling records (1963-1982/3 for the St Marguerite and Godbout Rivers respectively (Bielak, 1984; Bielak and Power, 1986)), a clear indication of the limited utility of medium-term data series for detecting long-term trends. Changes in age-specific mean weight of angled fish and in age-structure of catches, in both the Godbout and St Marguerite rivers, were related to the selective effects of commercial fisheries rather than any long-term climatic change.

There are indications of another long-term selective effect on timing of the return migration, perhaps also related to commercial fishing. Where time of river entry is not rigorously controlled by temperature, as it is on the Quebec North Shore, migrants are tending to return outside the normal netting season and thus escape capture and reporting (Dunkley, 1986; Mills, 1986; Shearer, 1986). Such conclusions could not have been reached without careful examination of historical data, and would certainly not have been evident from short-term catch records.

CONCLUSIONS

It is evident from the above discussion, that although the data used in the studies we cite must be treated judiciously, Thurso's (1986) admonition not to get too excited by short-term fluctuations is apposite. Historical catch records should rarely, if ever, be taken simply at face value, nor should undue significance be accorded to short or medium-term records as indicators of absolute or irreversible changes in salmon stocks. If interpreted by investigators with some familiarity with both the animal and

Figure 13.5: Changes in mean weight of angled Atlantic salmon, Godbout (1864-1983) and St Marguerite (1894-1982) rivers, Quebec North Shore. Trend lines are as follows: Godbout, Previously spawned wt (lbs) = 62.88-0.0231 (year), Two-sea-winter wt (lbs) = 35.16-0.0134 (year); St Marguerite, Three-sea-winter wt (lbs) = 42.81-0.0123 (year), Two-sea-winter wt (lbs) = 26.37-0.0079 (year)

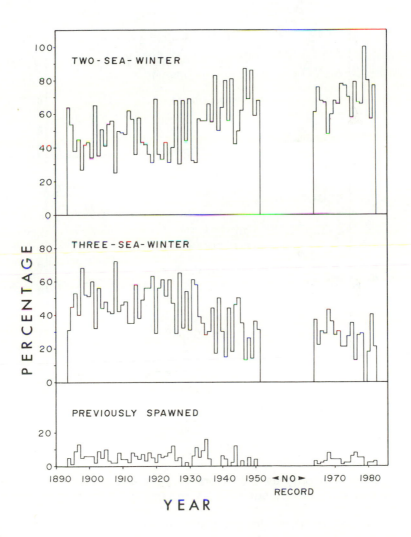

fishery in question, and an understanding of the limitations of the data, we believe historical records can ultimately be of major

Figure 13.6: Estimated proportion of two-sea-winter, three-sea-winter and previously spawned salmon taken by members of the St Marguerite Salmon Club (1894-1982). Grilse, salmon less than or equal to 5.5 lbs, have been eliminated before calculating the percentages

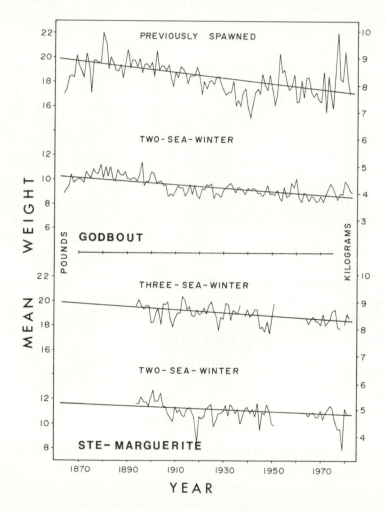

significance, not only as indicators of past and present stock and fishery characteristics, but also as indicators of potential future production.

The answer then, to the question addressed in the title of this paper, 'Catch statistics - facts or myths', is that although considerable work may be required to sift nuggets of fact from a motherlode of myth, the results often enable the scientist and historian to strike it rich.

REFERENCES

Anon. (1984) Salmon and sea trout fisheries 1983. (Norwegian Official Statistics B 491). Oslo - Kongsvinger: Central Bureau of Statistics of Norway. (In Norwegian with English summary)

Anon. (1986) Yearly commercial salmon landings for four of the five eastern Canadian provinces for the period 1910-1983, estimated Canadian share of Greenland Landings (40%), and Recreational Landings since 1965. Dept. of Fisheries and Oceans, Canada, 1 p.

Berg, M. (1964) Nord-Norske Lakseelver, Johan Grundt Tanum Forlag, Oslo, 300 pp.

Bielak, A.T. (1984) Quebec North Shore Atlantic salmon stocks. PhD dissertation, University of Waterloo, Waterloo, Ont. 236 pp.

Bielak, A.T. and Power, G. (1986) Changes in mean weight, sea-age composition, and catch-per-unit-effort of Atlantic salmon (Salmo salar) angled in the Godbout River, Quebec, 1859-1983. Canadian Jounal of Fisheries and Aquatic Sciences, 43, 281-7

Bielak, A.T. and Power, G. (In press) Utilisation des données historiques pour évaluer les stocks de saumon. In Colloque pour la Restauration des Rivières à saumon Bergerac. 28 mai-1 juin 1985

Browne, J. (1986) The data available for analysis on the Irish salmon stock. In D. Jenkins and W.M. Shearer (eds), Abbots Ripton, The Status of the Atlantic Salmon in Scotland, (ITE symposium no. 15). Institute of Terrestrial Ecology, pp. 84-90

Chadwick, E.M.P., Brazeau-Carrier, D. and Leger, C.E. (1985) Historical catches of Atlantic salmon (Salmo salar) at four sport fishing lodges on Restigouche River NB. Canadian Technical Report on Fisheries and Aquatic Sciences. No. 1362. iii+ 27 pp.

Dempson, J.B., Myers, R.A. and Reddin, D.G. (1986) Age at first maturity of Atlantic salmon (Salmo salar) - influences of the marine environment. In D.J. Meerburg (ed.), Salmonid age at maturity. Canadian Special Publication on Fisheries and Aquatic Sciences, 89, pp. 79-89

Dunbar, M.J. (1985) Evidence for a 50-year cycle in animal populations and in climate, with particular reference to the Atlantic Salmon (Salmo salar L.) In Proceedings of the 1985 Northeast Atlantic Salmon workshop. Moncton, N.B. April 1985. Publ. by Atlantic Salmon Federation, pp. 92-101

Dunbar, M.J. and Thomson, D.H. (1979) West Greenland salmon and climatic change. Meddeland Grønland. 202(4), 19 pp.

Dunfield, R.W. (1985) The Atlantic salmon in the history of North America. Canadian Special Publication on Aquatic Sciences, 80, 181 pp.

Dunkley, D.A. (1986) Changes in the timing and biology of salmon runs. In D. Jenkins and W.M. Shearer (eds), The Status of the Atlantic Salmon in Scotland, (ITE symposium no. 15) Institute of Terrestrial Ecology, Abbots Ripton, pp. 20-7

Gee, A.S. and Milner, N.J. (1980) Analysis of 70-year catch statistics for Atlantic salmon (Salmo salar) in the River Wye and implications for management of stocks. Journal of Applied Ecology, 17, 41-57

George, A.F. (1984) Scottish salmon and grilse return-migration variation over 200 years. In A. Holden (ed), Proceeding of the Institute of Fisheries Management 15th Annual Study Course 10-13 Sept. 1984, Stirling University, Scotland, pp. 23-32

Hansen, L.P. (1986) The data on salmon catches available for analysis in Norway. In D. Jenkins and W.M. Shearer (eds), The Status of the Atlantic Salmon in Scotland (ITE symposium no. 15). Institute of Terrestrial Ecology, Abbots Ripton. pp. 79-83

Huntsman, A.G. (1931) The maritime salmon of Canada. Biological Board of Canada Bulletin, 21, 99 pp.

Jenkins, D. and Shearer, W.M. (1986) The Status of the Atlantic Salmon in Scotland. ITE symposium ISSN 0263-8614; No. 15. Institute of Terrestrial Ecology, Abbots Ripton, 127 pp.

Laird, L.M. and Needham, E.A. (1986) Salmon farming and the future of Atlantic salmon. In D. Jenkins and W.M. Shearer (eds), The Status of the Atlantic Salmon in Scotland, (ITE symposium no. 15). Institute of Terrestrial Ecology, Abbots Ripton, pp. 66-72

Lakhani, K.H. (1986) Salmon population studies based upon Scottish catch statistics: statistical considerations. In D. Jenkins and W.M. Shearer (eds), The Status of the Atlantic Salmon in Scotland, (ITE symposium no. 15). Institute of Terrestrial Ecology, Abbots Ripton, pp. 116-20

Legault, M., Dumont, R., Thibault, P. and Boudreault, V. (1985) Plan de Restauration du saumon Atlantique dans la Rivière à Mars. Publ. Association des pêcheurs sportifs de la Rivière à Mars. Ville de la Baie. 134 pp.

Lindroth, A. (1965) The Baltic salmon stock. Mittelungen International Verhandlungen Limnologae. 13, 163-192

Martin, J.H.A. and Mitchell, K.A. (1985) Influence of sea temperatures on the numbers of grilse and multi-sea-winter Atlantic salmon (Salmo salar) caught in the vicinity of the River Dee (Aberdeenshire). Canadian Journal of Fisheries and Aquatic Sciences, 42, 1513-21

McIntyre, A.D. (1986) Summary. In D. Jenkins and W.M. Shearer (eds), The Status of the Atlantic Salmon in Scotland, (ITE symposium no. 15) Institute of Terrestrial Ecology, Abbots Ripton, pp. 121-5

Mills, D.H. (1986) The biology of Scottish salmon. In D. Jenkins and W.M. Shearer (eds) The Status of the Atlantic Salmon in Scotland, (ITE symposium no. 15) Institute of Terrestrial Ecology, Abbots Ripton, pp. 10-19

Niemela, E., Niemela, M. and McComas, L. (1985) Long term catch statistics for Atlantic salmon from the Teno and Naatamo Rivers. Working paper 85/, Working group on North Atlantic Salmon. International Commission for the Exploration of the Sea

Peppar, J.L. (1983) Adult Atlantic salmon (Salmo salar) investigations, Restigouche River system, New Brunswick, 1972-80. Canadian MS Report on Fisheries and Aquatic Sciences, No. 1695. viii+33 pp.

Pirie, J.D. (1986) Is there a basis here for prediction? In D. Jenkins and W.M. Shearer (eds), The Status of the Atlantic Salmon in Scotland, (ITE symposium no. 15). Institute of Terrestrial Ecology, Abbots Ripton, pp. 112-15

Porter, T.R., Healey, M.C. and O'Connell, M.F. (with Baum, E.T., Bielak, A.T. and Côté, Y.) (1986) Implications of varying the sea age at maturity of Atlantic salmon (Salmo salar) on yield to the fisheries, In D.J. Meerburg (ed.) Salmonid age at maturity. Canadian Special Publication on Fisheries and Aquatic Sciences, 89, 110-17

Power, G. (1976) History of the Hudson's Bay Company salmon fisheries in the Ungava Bay region. Polar Record, 18 (113), 151-61

Rainelli, P. and Thibault, M. (1985) La surabondance de consommation de saumon autrefois: Une surabondance veritablement ...fabuleuse. Cahiers de le Nutrition et de Diétetiques, XX, 4, 292-7

Reddin, D.G. and Shearer, W.M. 1987 Migration and distribution of Atlantic salmon (Salmo salar L) in the Northwest Atlantic. Special publication, American Fisheries Society: Common Strategies of anadromous and catadromous fish

Ricker, W.E. (1981) Changes in the average size and average age of Pacific salmon. Canadian Journal of Fisheries and Aquatic Sciences, 38, 1636-56

Riley, S.C., Bielak, A.T. and Power, G. (1984) The Atlantic salmon stock of the Grand Watshishou River, Quebec: a historical perspective. Naturaliste Canadien, 111(3), 219-28

Ruggles, C.P. and Turner, G.E. (1973) Recent changes in stock composition of Atlantic salmon (Salmo salar) in the Miramichi River, New Brunswick. Journal of the Fisheries Research Board of Canada, 39, 779-86

Scarnecchia, D.L. (1984) Climate and oceanic variations affecting yields of Icelandic Stocks of Atlantic salmon (Salmo salar). Canadian Journal of Fisheries and Aquatic Sciences, 41, 917-35

Schiefer, K. (1971) Ecology of Atlantic salmon, with special reference to occurrence and abundance of grilse, in North Shore Gulf of St. Lawrence rivers. PhD dissertation, University of Waterloo, Waterloo, Ont. 129 pp.

Shearer, W.M. (1986) An evaluation of the data available to assess Scottish salmon stocks. In D. Jenkins and W.M. Shearer (eds), The Status of the Atlantic Salmon in Scotland, (ITE symposium no. 15). Institute of Terrestrial Ecology, Abbots Ripton, pp. 91-111

Taylor, V.R. (1985) The early Atlantic salmon fishery in Newfoundland and Labrador. Canadian Special Publication on Fisheries and Aquatic Sciences, 76, 71 pp.

Thibault, M. (1981) Aménagement et gestion des rivières à saumon atlantique du massif armoricain. In Ministère de l'Environnement, 3èmes Assises internationales de l'Environnement, Paris 1980, vol. 4, Société en Environnement, Etudes et recherches, Documentation française, Paris, pp. 103-9

Thibault, M. (In press) La problèmatique du saumon atlantique en France. In Colloque pour la Restauration des Rivières à saumon. Bergerac, 28 mai - 1 juin, 1985

Thibault, M. and Rainelli, P. (1980) L'abondance passée du saumon atlantique: mythe ou réalité? (Essai de synthèse à partir de l'exemple de la Bretagne). Bulletin of the Scientific and Technical Department of Hydrobiology. Institut Nationale Recherches Agronomique, 9, 78 pp.

Thibault, M. and Rainelli, P. (1981) Difficulties in the management of Atlantic salmon stocks in France. Anadromous and Catadromous Fish Committee, International Council for the Exploration of the Sea. CM 1981/M, 17

Thibault, M. and Rainelli, P. (1982) L'exploitation des populations naturelles de saumon atlantique en France de 1950 à 1980. In Colloque sur la production et la commercialisation du poisson d'eau douce. Association Internationale des Entretiens Ecologiques, 30 mars - 1 avril 1982, 22 pp.

Thurso, The Lord (1986) The Management of a rod and line and a commercial fishery. In D. Jenkins and W.M. Shearer (eds), The Status of the Atlantic Salmon in Scotland, (ITE symposium no. 15). Institute of Terrestrial Ecology, Abbots Ripton, pp. 55-9

Vibert, R. (1980a) Primauté des décisions politiques dans l'épuisement ou le développement de la ressource saumon. In Le
saumon en France (numéro spécial). Saumons, 34, 7-13
Vibert, R. (1980b) Perspectives. In Le saumon en France (numéro
spécial). Saumons, 34, 72-6
Vibert, R. (1982a) Potentialités climatiques de capture et de
production en saumon atlantique. Saumons, 39, 7-10
Vibert, R. (1982b) Discordance entre potentialités climatiques de
captures en saumon atlantique et captures avancées pour la
Bretagne au XVIIIe siècle. Saumons, 42, 19-24
Weeks, E. (1971) The Moisie Salmon Club. Barre Publishers, Mass.
240 pp.

Chapter Fourteen

RELATING CATCH RECORDS TO STOCKS

W. M. Shearer
Freshwater Fisheries Laboratory Field Station, 16 River Street,
Montrose, Scotland

14.1   INTRODUCTION

This paper describes the catch data available to assess Scottish
salmon stocks, discusses the limitations of the data and indicates
the differences which can arise when catch records are used to
measure the strength of salmon stocks and to make comparisons
between years.

Although fish counters are presently installed at North of
Scotland Hydro-Electric Board (NSHEB) dams on more than 20
rivers, each one recording the upstream and downstream
movement of salmon, the counts refer to tributaries rather than
complete river systems. Furthermore, the accuracy of resistivity
counters can change as the pre-set trigger level alters as the
conductivity of the water changes, i.e. when the conductivity of
the water decreases, the counter will count smaller fish. The
reverse will occur when the conductivity of the water rises. As
changes in conductivity are unlikely to be constant between years
or for the same duration, any error is unlikely to be constant
between years. False counts or fish being missed may arise from
other electrical and mechanical faults. To take an extreme
example, during the summer and autumn of 1984, the counts
produced by the counter sited at Dunalastair Dam on the River
Tummel, Perthshire were 20 per cent of the number of fish
known to have ascended, these fish having been caught upstream
of the counter (Struthers and Stewart 1984, 1985). Thus
comparisons of the strength of spawning stocks between rivers
and between years based on the counts produced at NSHEB
counter sites could be misleading.

In Scotland, salmon are caught along the coast by bag-nets
(usually on rocky shores) and stake nets (on sandy beaches)
collectively called fixed engines, and by net and coble
(seine-nets) and rod and line in estuaries and in freshwater
(Strange, 1981). In addition to an annual close time which, for the
majority of nets, extends from the end of August to roughly

256

mid-February, there is a weekly close time from noon on Saturday until 0600 hours on the following Monday morning. The rods can normally begin fishing earlier and cease fishing later each year than the corresponding nets. Depending on the Salmon Fishery District, the rod fishing season begins between 11 January and 25 February and ends between 30 September and 30 November. The weekly close time for rods is Sunday.

14.2   AVAILABLE CATCH DATA

National catch statistics

Most countries presently collect catch figures and these catches in 1960-1985 are summarised in the Report of the ICES North Atlantic Salmon Working Group (Anon, 1986). In Scotland catch figures have been available to the Department of Agriculture and Fisheries for Scotland (DAFS) since 1952. Prior to that date there was no statutory obligation to make catch and effort data available to the Department. As a statutory requirement under the Salmon and Freshwater Fisheries (Protection) (Scotland) Act 1951 catches divided between salmon, grilse and sea trout are provided for each month by number, by weight and by method of capture (fixed engine, net and coble and rod and line). In addition, net and coble fisheries are requested to supply the minimum and maximum number of netting crews and persons engaged in netting operations each month and fixed engine fisheries are also asked for the minimum and maximum numbers of traps operated each month. Operators of fixed nets are also requested to give some additional details of gear used. For example, bag, fly and stake net fishermen are asked to supply details of the minimum number of bags or pockets fished in any month. Although at one time a separate form had to be completed for each fishery, operators of more than one fishery in the same Fishery District can now combine their returns on one form.

Brief summaries have been published each year between 1952 and 1981 giving the reported catch for Scotland as a whole, divided both between salmon, grilse and sea trout for each of the gears, fixed engine, net and coble and rod and line. Following requests for the salmon catch statistics to be available in greater detail, a compendium covering the years 1952-1981 was published (Anon, 1983, 1984a) and from 1982, publication of the annual catch figures has been on a regular basis (Anon, 1984b, 1985a, b).

Historical catch records

Many private estates and old established commercial fishing companies have long series of catch records. The information recorded varies widely, but most records give the date, number of salmon, grilse and sea trout caught at each fishing station or on a beat of a river and their weight. No measure of effort is given.

Published catch records.
Another source of information is the published catch records, but they are fragmentary particularly after the mid 1850s and before 1952. The growing secrecy of net fishing proprietors and lessees stemmed from a hostility between the netsmen and proprietors of rod and line fisheries. The monetary value of leased fishings which periodically came up for renewal, imposed further restrictions on both the free exchange of information and its quality. Once again no measure of effort was given.

Nevertheless, catch figures from widely separated fishings are scattered throughout the Annual Reports of the Fishery Board for Scotland and in the reports of the various Salmon Commissions which heard evidence in the nineteenth and twentieth centuries. Prior to 1952, estimates of the total Scottish catch each year were based mainly on the returns provided by the railway and steamship companies of the weight of salmon and trout carried by them.

More recently, George (1982) has examined and commented on the catches taken between 1790 and 1976 by anglers and net fisheries fishing within Fishery Districts stretching the length and breadth of Scotland.

14.3   BIASES IN CATCH DATA

Salmon runs do not necessarily conform to fishing seasons. Fish which arrive on the coast and enter rivers after the end of the fishing season contribute neither to the fixed engine catch nor to the catches taken by net and coble and rod and line in the rivers. Changes in the timing of runs could bias estimates of annual stock abundance based solely on catch data in the fishing season. Similarly changes in the starting and finishing dates of fishing seasons could also bias estimates of annual stock abundance. Furthermore, it is unlikely that the fishing effort, or the rate at which the stock has been exploited has remained constant throughout a fishing season, from one season to another or from one area to another (Shearer, 1986). In general, not only has the number of sites fished decreased but also the length of the season fished at many of these sites has diminished. In addition, the catches at a number of net fisheries are limited by quota and some owners of rod fishings specify not only the number of rods which can be fished at the same time but also the length of each fishing day, the method of fishing and the lures which may be fished.

The basic design of salmon nets has remained remarkably constant since they were first introduced but the material used in their manufacture has changed from natural to man-made fibres. Most salmon cobles have been motorised and at some net and coble fishing sites the nets are hauled by powered winches rather than by man-power. Although it is generally accepted that fixed engines manufactured from synthetic materials are less prone to damage and remain in fishing order longer when storms occur at sea than comparable gear made from natural fibres, differences in the overall catching efficiency resulting from these changes have not been quantified. Thus, it is not clear whether the trends observed in catch data were correlated with the availability of fish or with the efficiency of the effort put into catching them. Nor, with the data available, is it possible to quantify the effect which observed reductions in the number of sites fished and the length of their fishing season have had on total catches. Stations which stop fishing first tend to have the lowest catch rates. Therefore, reductions in total catch due to the closing of fishing stations may not be directly proportional to the number of stations which have closed.

Similarly, a reduction in the number of crews operating a net and coble fishery may not produce a comparable decrease in the exploitation rate since netting efficiency can be enhanced by restricting fishing to those periods which, from past records, have produced maximum catches. These times are likely to coincide with the main influx of fish into freshwater.

No data describing the effort expended to take the annual reported rod catches are available. Even if the numbers of rod days were known, it cannot be assumed that all anglers and tackle are equally efficient.

The fish caught may not be typical of the stock. Differences could occur for various reasons, including gear selectivity or differing levels of exploitation on different components of the stock. Shearer (1984) demonstrated a relationship between both river and sea age and the calendar date when fish belonging to particular cohorts returned to fresh water. The age composition of the catch will be biased towards those age groups which return in greatest proportion during the fishing season.

The rod catch could be particularly sensitive to physical changes in the river system which could include the deposition of gravel in pools, and changes in both the patterns and rates of discharge and temperature. Changes in the sea age composition of the stock could also be important as they would alter the timing of runs and the availability of catchable fish not only to the river system as a whole but also to different regions of the river by differing proportions.

14.4   LIMITATIONS OF CATCH DATA

Annual fluctuations in catches

Fluctuations occur in annual reported catches for a number of reasons. A low catch in a particular year must not be assumed to be evidence of a decline in salmon stocks, though increased catches may usually be related to an increase in the numbers of catchable fish. The spawning escapement (i.e. the proportion of returning salmon that is not caught by nets and rods) cannot be estimated directly from the numbers of fish caught because a large number could enter the river after the end of the fishing season having made no contribution to catches.

Grilse error

In the catch returns submitted to DAFS, lessees and owners of salmon fisheries, with few exceptions, have separated their catches into salmon and grilse on the basis of weight. Fish weighing less than 3.6 kg have been classed as grilse and the rest as salmon when fish in both sea age groups are present. In 1952-1985, the proportions of fish classed as grilse varied because of changes in the growth of grilse in the sea. Furthermore, as grilse generally increase in weight as the season advances, the magnitude of this reporting error does not remain constant throughout the fishing season. In August and September, 'salmon' catches will contain relatively more over-weight grilse than in June and July. Generally, those years when grilse were most abundant were also characterised by above-average proportions of over-sized grilse. In 1850-1950, the relative number of over-weight grilse classed as 'salmon' was probably less than in more recent times because grilse tended to be on average lighter and market requirements were less rigidly tied to weight.

Effort

Although effort data are available for net fisheries, both the units used to record it and the time interval between each record are too imprecise to describe adequately real changes in fishing effort (each month fixed engine and net and coble fisheries respectively record the minimum and maximum number of traps in operation and the number of crews employed).

Unreported catches

Annual recorded catches describe the minimum number of fish caught and landed each year because some legal fisheries fail to report all or part of their catches and illegal fisheries make no return. In either instance, if the number of fish involved were significant, reported catches would underestimate the strength of the total potential spawning stock.

14.5   STUDY AREA AND FISHING METHODS

The North Esk is one of Scotland's major salmon rivers. It rises as three streams in Invermark forest, high in the Cairngorm massif and flows through the glacial valley of Glen Esk and the rich agricultural land of the 'Howe of the Mearns', to enter the North Sea 4.8 km north of Montrose. The Atlantic salmon population of the North Esk has formed the subject of an intensive research study since the early 1960s as described by Shearer (1972).

The only permissible method of net fishing within the estuarial limits of the North Esk is by net and coble (drag, draught seine, or sweep net). Fishing for salmon by net and coble in the North Esk is confined to the lowermost 3 km of the river. There are 13 recognised fishing sites along the river from the Gauge to the sea, but the major fishing effort is concentrated at Morphie Dyke betwen February and May and in the lower reaches at the Flats and Nab between June and August (Figure 14.1). The Morphie

Figure 14.1: Lower reaches of the River North Esk showing sites of net and coble fishing stations

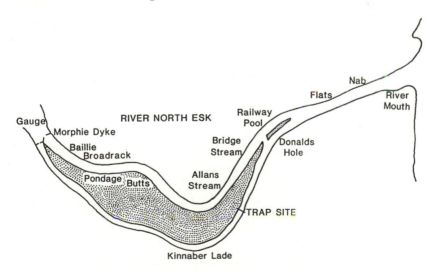

Dyke station is normally fished daily, while the Flats and Nab are similarly fished from early June onwards except when the river is in spate. Fishing at the other stations is less regular and depends upon the day-to-day management of the fishery. Shearer (1972) gives a detailed description of the operation of this fishery. In 1985 fishing upstream of the Flats was limited to angling until 1 April.

14.6 EXPLOITATION OF SALMON IN THE NORTH ESK

Estimates of the exploitation of salmon and grilse by the net and coble fishery in the North Esk were obtained from the recapture of fish tagged at Kinnaber Mill trap during the commercial netting season (Figure 14.2). The trap, which is at Kinnaber Mill Lade, is fully described by Pratten and Shearer (1981).

Figure 14.2: Diagram of Kinnaber Mill fish trap

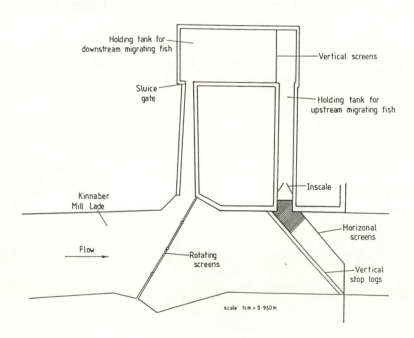

Upstream migrating salmon trapped in the holding tank of the trap were removed by means of a specially designed, heavy-duty, plastic bag (Figure 14.3) the mouth of which was held open by a triangular shaped metal frame. Once a fish had been coaxed into the bag, the frame was removed and the neck of the bag held closed; in this way a quantity of water was retained with the fish as it was lifted from the trap. The fish was then manipulated within the bag until its dorsal fin protruded through a slit in the side. Tagging was achieved through a slit using individually numbered plastic floy tags inserted into the musculature, below the anterior edge of the dorsal fin so that the T-bar anchor rested between the inter-neural rays. Fish were released into the main river through a plastic pipe supplied with a flow of water running from the trap to the main river. This tagging operation was successfully designed to minimise the damage to fish caused by handling. Only fish which showed no external symptoms of disease or damage were tagged.

Figure 14.3: Bag and frame used to remove adult salmon from Kinnaber Lade trap and to hold fish during tagging

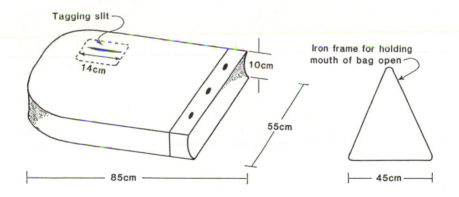

In order to measure tag loss, all fish tagged at Kinnaber Mill trap since 1982 were panjetted at the same time as they were tagged, leaving a small blue spot which was easily recognised during normal sampling of the catch in the fish house. As it was important that the tagged fish were undamaged, they were not measured, sexed or scaled before release.

Net and coble catches at each fishing station along the length of the river, accurately broken down into salmon and grilse, were readily made available by Messrs Joseph Johnston and Sons Ltd.

Once the fish had been tagged and released, the success of this experiment depended upon receiving details of all recaptures. Thus every effort was taken to inform both netsmen and anglers of the possible presence of tags, posters were displayed at appropriate sites, netting stations were regularly visited, and random checks for the presence of tagged fish were made on catches in fish houses. It is considered that the non-reporting of tags recovered was not significant.

Regular sampling of the net and coble catch from the North Esk has resulted in the recovery of only one (0.1 per cent) fish with a blue spot and no tag in situ. In addition, tagged fish were caught after a lapse of several months between tagging and recapture, and a smaller number of fish were recaptured in the year following that in which they had been tagged with the tag still in place; these fish had spawned in the interval. The conclusion is that tag loss is negligible.

The catch figures and recaptures were used to estimate the rate of exploitation by the net and coble fishery on grilse and salmon migrating into the North Esk in 1976-1985. The population passing through the main fishing area during the netting season was estimated using a stratified mark-recapture method (Schaefer, 1951) incorporating a small adjustment for tags taken outside the river. The numbers of fish which ascended the river but were not caught in the commercial fishery during the netting season (the escapement) were then calculated by subtracting the catch at the principal netting site from the estimated number of salmon and grilse available to that fishery. The netting season exploitation rates were then estimated, knowing the escapement of salmon and grilse and the total river catch of the two sea age groups (Table 14.1). The validity of the assumptions which this method makes is fully dealt with by Pratten and Shearer (1981).

The exploitation rates on salmon and grilse varied from year to year with no apparent trend (Figure 14.4). They also fluctuated independently of one another. When the annual rates for salmon and for grilse were each tested independently for significance, 12 and 20 of the 45 possible comparisons of exploitation rate in each of the two sea age groups were significant at the 5 per cent level. Furthermore, instances when the rates in two consecutive years were not significantly different at the 5 per cent level were rare and occurred on grilse only three times, in 1976 and 1977, 1981 and 1982 and 1983 and not at all on salmon. In most years in the time series, therefore, the net and coble fishery in the North Esk was removing significantly different proportions of the available fish.

Figure 14.4: Estimated annual exploitation rate by North Esk net and coble fishery on fish available during fishing season

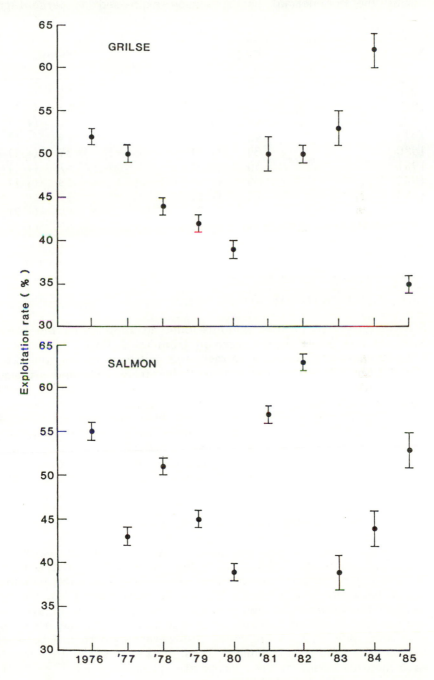

Table 14.1: Exploitation rates on grilse (one sea-winter) and salmon (multi sea-winter) by North Esk net and coble fishery during the commercial fishing season expressed as percentages

Year	Grilse		Salmon	
1976	52	(+ 1)	55	(+ 1)
1977	50	(+ 1)	43	(+ 1)
1978	44	(+ 1)	51	(+ 1)
1979	42	(+ 1)	45	(+ 1)
1980	39	(+ 1)	39	(+ 1)
1981	50	(+ 2)	57	(+ 1)
1982	50	(+ 2)	63	(+ 1)
1983	53	(+ 2)	39	(+ 2)
1984	62	(+ 2)	44	(+ 2)
1985	35	(+ 1)	53	(+ 2)

(+ 95 Confidence limits)

14.7   THE NORTH ESK STOCK

A resistivity fish counter commissioned in 1980 spans the North Esk at Logie, some 5 km upstream from its mouth (Figure 14.5). This counts the number of salmon and grilse ascending the river above Logie. These counts are available from 1981 and although it is possible to allocate them to the appropriate spawning stock, they cannot be split between salmon and grilse.

One of the problems associated with the counters sited at NSHEB dams has been circumvented at Logie by installing another module in a standard NSHEB counter which compensates for changes in the electrical resistance between the electrodes. More recently, a new design of counter has been installed which automatically compensates for changes in conductivity and analyses and responds to the electrical waveforms produced by fish as they cross the electrode array rather then the electrical imbalance which occurs between the electrodes.

Figure 14.6 shows the number of fish which crossed the Logie fish counter combined with the corresponding catch taken by net and coble and rod and line below Logie in each calendar month between 1 December 1980 and 30 November 1985. On the basis of the information obtained from the regular examination of the salmon caught in Kinnaber Mill trap since 1970, the period of migration of each spawning stock was determined to fall between 1 December and 30 November. Thus, it was possible to allocate each monthly count to the appropriate spawning stock.

Figure 14.5: Logie fish counter

Figure 14.5: Upstream counts of salmon at the Logie fish counter site plus catches taken below Logie

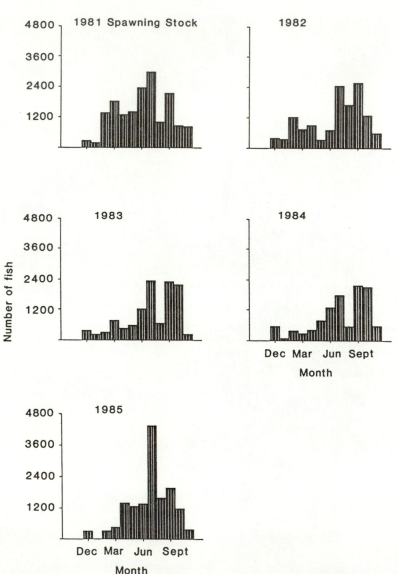

Although a proportion of the fish which crossed the counter after 31 August could have been available to both the net and coble fishery and the rod and line fishery below Logie before 1 September (the commencement of the annual close time for nets), it has been assumed that this number in any one year was not significant. This assumption was based on the results obtained from tagging adult salmon in the lower reaches of the North Esk. They showed that at this time of year fish moved upstream of Morphie Dyke relatively quickly and mainly within two days of tagging. Therefore, it is not unreasonable to assume that the majority of the fish recorded at Logie after 31 August each year were not available to the fisheries below Logie before 1 September.

The fraction of each spawning stock which has been assumed to have entered freshwater after the end of the netting season and therefore not available to contribute to the catch fluctuated widely between years ranging between 23 per cent in 1981 and 44 per cent in 1984 (Table 14.2). Even making an

Table 14.2: Percentage of each annual spawning stock entering the river after the end of the net fishing season (31 August)

Year	Percentage
1981	23
1982	33
1983	41
1984	44
1985	24

allowance for the number of fish which migrated over Logie before the start of each fishing season and therefore not available to the fisheries below Logie when they commenced fishing, the proportion of the stock which was available, particularly to the net and coble fishery, differed markedly between years. One result of this difference was that the annual mean rates of exploitation by the net and coble and the rod and line fisheries on the total spawning stock each year in 1981-1985 were significantly different at the 5 per cent level from the corresponding value in all other years (Figure 14.7). Thus, catch figures during the period for which data were available were not only unreliable indicators of the strength of the spawning stock but also of the magnitude of the difference in stock size between years.

Figure 14.7: Annual exploitation rate by all fishing methods within estuarial limits on North Esk stock

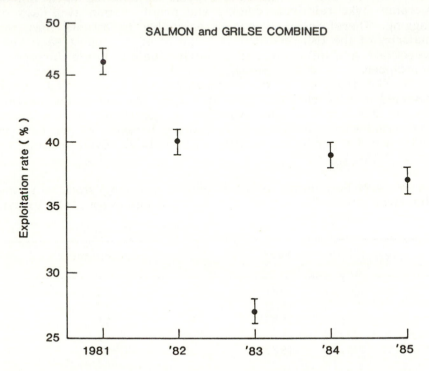

14.8 DISCUSSION

Most of the information examined in this paper came from fisheries which by statute are only allowed to fish for rather less than half the year, although salmon are known to enter freshwater throughout the 12 months. For example, if the timing of runs has changed as a result of a decline in the multi sea-winter component in the stock, the proportion of the total stock available for exploitation during the fishing season could have declined. As a result, a below-average catch would not signify a small total stock as the spawning escapement would have been augmented by those fish which normally would have been caught in the fishing season. Even in the short term, the use of catch figures to compare the strength of spawning stocks between years could give misleading results since the proportion of the total stock entering freshwater before the end of the net fishing season can vary significantly from one year to the other. Such behavioural changes can easily distort the underlying real trends in population abundance.

The lack of reliable effort data describing the effort put

into catching salmon is another difficulty in attempting to relate catches to abundance. Thus, the results of any comparisons between catches by years and by gear and their relationship to the spawning stock must be suspect. There are obvious difficulties in defining 'fishing effort' which cannot be the same quantity for different persons or fisheries. Furthermore, the catch effort by each fishery is not fixed at the outset but is determined voluntarily by each fishery. Thus, reduced fish abundance or the prevailing market conditions may induce a fishery to step up its catching effort or, alternatively, to reduce or abandon fishing.

Despite the deficiencies in the catch statistics collected under the 1951 Act which have been described, these data indicate the size of the annual harvest as well as the breakdown of the harvest by weight and by district and months. Such basic information can be essential for the formulation of policy affecting salmon fisheries.

14.9 SUMMARY

This paper describes the main sources of catch data which are available to assess the state of Atlantic salmon in Scotland and discusses the inadequacy of this material as an indicator of stock size. In addition, recent data from the North Esk describing the proportion of the total stock which migrates into the river after the end of the netting season and the rate at which the stock is exploited after it enters the river are discussed in the context of the relationship between stock and net catch.

Since 1952, it has been a statutory requirement to make catch and effort data available, in confidence, to the Department of Agriculture and Fisheries for Scotland. Many private estates and long-established fishing companies have long series of catch records and only some of these are published. However, no records of the effort required to take these catches are recorded.

The catch figures come from fisheries which are only allowed to operate for rather less than half the year, although salmon are known to return to freshwater throughout the 12 months. The proportion which entered the North Esk after the end of the netting season was not constant between years, varying between 23 and 44 per cent of the total stock in 1981-1985. This can cause difficulties in interpreting the catch data. For the period over which data are available, a below-average catch would not signify a small stock as the spawning escapement would have been augmented by those fish which normally would have been caught during the fishing season.

The lack of reliable data describing the effort put into catching salmon is another major drawback in interpreting catch trends. Exploitation rate data for the North Esk show that it can vary significantly between years. Thus, the results of any comparisons between catches by years and by gear and their relationship to the spawning stock must be suspect.

At best, the present catch data indicate the size of the annual harvest by gear, area and month. If reliable catch effort data without bias were available then catch statistics, adjusted for the fishing effort employed, could provide an index of abundance of salmon stocks in Scottish home waters in a given season.

REFERENCES

Anon. (1983) Scottish salmon catch statistics 1952-1981. Department of Agriculture and Fisheries for Scotland, Edinburgh

Anon. (1984a) Scottish salmon catch statistics 1952-1981. Department of Agriculture and Fisheries for Scotland, Edinburgh

Anon. (1984b) Scottish salmon and sea trout catches, 1982. (Statistical Bulletin Number 1/1984). Department of Agriculture and Fisheries for Scotland, Edinburgh

Anon. (1985a) Scottish salmon and sea trout catches, 1983. (Statistical Bulletin Number 1/1985). Department of Agriculture and Fisheries for Scotland, Edinburgh

Anon. (1985b) Scottish salmon and sea trout catches, 1984. Statistical Bulletin Number 2/1985). Department of Agriculture and Fisheries for Scotland, Edinburgh

Anon. (1986) Report of meeting of the Working Group on North Atlantic Salmon, Copenhagen, 1986, International Council for the Exploration of the Sea, CM 1986/Assess, 17

George, A.F. (1982) Scottish salmon returns – Migration variations C. 1790-1976. M Phil Thesis, The Open University

Pratten, D.J. and Shearer, W.M. (1981) Fishing mortality of North Esk salmon. CM 1981/M:26, International Council for the Exploration of the Sea, Copenhagen

Schaefer, M.B. (1951) Estimation of the size of animal populations by marking experiments. Fishery Bulletin of US Fish and Wildlife Service, 52, 189-203

Shearer, W.M. (1972) A study of the Atlantic salmon population of the North Esk 1961-1970. MSc Thesis, University of Edinburgh

Shearer, W.M. (1984) The relationship between both river and sea age and return to home waters in Atlantic salmon, CM 1984/M:24, International Council for the Exploration of the Sea, Copenhagen

Shearer, W.M. (1986) The exploitation of Atlantic salmon in Scottish home water fisheries in 1952-1983. In D. Jenkins and W.M. Shearer (eds), The Status of the Atlantic salmon in Scotland, (ITE Symposium Number 15). Institute of Terrestrial Ecology, Abbots Ripton, pp. 37-49

Strange, E.S. (1981) An introduction to commercial fishing gear and methods used in Scotland. Information Pamphlet Number 1, Department of Agriculture and Fisheries for Scotland, Aberdeen

Struthers, G. and Stewart, D. (1984) A report on the composition of the adult salmon stock of the upper river Tummel, Scotland, and the evaluation of the accuracy of a closed-channel resistivity counter. International Council for the Exploration of the Sea. CM 1984/M:20

Struthers, G. and Stewart, D. (1985) The composition and migrations of the adult salmon stock in the upper River Tummel, Scotland, in 1984 with observations on the accuracy of resistivity counters at two fish passes, CM 1985/M:14, International Council for the Exploration of the Sea, Copenhagen

Chapter Fifteen

THE USE OF LESLIE MATRICES TO ASSESS THE SALMON POPULATION OF THE RIVER CORRIB

John Browne
Fisheries Research Centre, Castleknock, Dublin 15

15.1   INTRODUCTION

The normal approach to management of Atlantic salmon (Salmo salar L.) is to develop stock-recruitment relationships. Target spawning criteria can be estimated from such relationships and a river system can be managed by allowing specified escapement to achieve the required ova deposition.

There are a number of problems associated with this approach not least of which is that data for stock-recruitment curves are difficult to obtain and require a long time series. Even with specific total escapement and target spawning criteria there is no way of ensuring that equitable distribution of the stock among the tributaries has occurred or that all stock components are adequately represented in the spawning population.

In the one river in Ireland where data are available on ova to smolt survivals in a natural population the correlation between ova deposition and smolt numbers is poor. This may indicate that spawning levels are adequate there most of the time. Allowable harvests cannot be predicted without using a wide range of probabilities to allow for the variable nature of marine survival. Harvest therefore cannot be predicted directly by smolt production.

Matrices seem to offer a tool in the preliminary investigation of populations where some good data are available but data on parts of the life history are lacking. Existing data can be tested in the model and areas where further data are required can be pinpointed. Simulation also provides a method of investigating the effects of proposed management changes.

The use of deterministic models based on matrices has been developed by Leslie (1945, 1948). Lewis (1942) also suggested the use of matrix notation to show the growth of complex populations. Moran (1962), Pielou (1969), Usher (1966), Usher and Williamson (1970) and Williamson (1959, 1967) have described and extended the work of Leslie. The model put forward by Leslie (1945) predicted the age structure of a female population given the age structure of a past time and given age-specific survival and fecundity rates. The model in matrix notation is

$$\underline{A} \ \underline{at} = \underline{at + 1}$$

A is a matrix with fecundities across the top row and survival probabilities across the diagonal which describe the transition probabilities from one time to another. $\underline{at}$ is a column vector representing the population age at time t. $(\underline{at + 1})$ is a column vector similar to $\underline{at}$ representing the age structure at time $(\underline{t + 1})$.

Usher (1971) shows that one of the latent roots of A (denoted by λ) has properties which ensure (1) that the matrix will always determine a meaningful age structure for the population and (2) the age structure will be unique since whatever the size of the matrix used there will be only one biologically meaningful solution.

The intrinsic rate of natural increase (r) is related to λ by $(r = \ln \lambda)$.

This equation has been used frequently in animal population studies (Keyfitz, 1968; Pielou, 1969; Usher, 1971). The equation has been applied to whale populations by Usher (1971) and to fish populations. Horst (1977) applied it to a marine shore fish, the Cunner, (Tautogolabrus adspersus, and it has been used on brook trout (Salvelinus fontinalis). The use of matrices was demonstrated on a population of salmon in the River Boyne but as there were few specific data the exercise was largely theoretical (Browne, 1980). A general description of the matrix model and how it works is given by Blackith and Albrecht (1979).

If a salmon population exists in a river it is either increasing, declining or is stable. If the population has maintained itself at a level for a long time period it can be assumed to be stable or its intrinsic rate of natural increase (or decrease) is low. If average values for a stable population are ascertained for the various life stages, and used in a matrix and run over a large number of generations then the intrinsic rate of natural increase should be stable or low. If the increase is too swift then the population will either expand too quickly or disappear.

In this paper it is proposed to use the data available on the River Corrib to produce average values for a population model for the system using Leslie matrices.

15.2 THE CATCHMENT

In 1979 work began on a population model for the Corrib River in the West of Ireland (Co. Galway). It was proposed to collect information on all life stages of the population and produce a model based on the matrix approach for the system. Some progress has been made but a major aspect of the work, quantifying the spawning escapement, has not been tackled.

The catchment has an area of 3,300 square km (Figure 15.1). There are no salmon in the system in or above Lough Mask, but all the tributaries below Lough Mask have salmon populations. There are two distinct water types typified by low alkalinity, total hardness expressed as $CaCO_3$ ranging from 13.8 to 76.8 mg/l in tributaries to the west of Lough Corrib and high alkalinity, total hardness greater than 348 mg/l in the tributaries to the east of Lough Corrib (Browne and Gallagher, 1981).

The tributaries to the east produce only 1+ and 2+ smolts while the tributaries to the west produce only 2+ and 3+ smolts.

15.3 DATA AVAILABLE

There are four main sources of data: (1) annual population estimates; (2) analyses and estimates of the smolt runs; (3) analyses of the catch data and biological characteristics of the catch; (4) micro-tagging of smolts.

Annual population estimates were carried out on the Corrib from 1979-1984. Fish were caught, marked and returned to the rivers during one day. On the second day the river population now composed of marked and unmarked fish were sampled and the population estimated using the Petersen formula modified by Bailey (1951).

$$N = \frac{T(M+1)}{(R+1)}$$

The variance is given by

$$Var\ N = \frac{M\ (T+1)(T-R)}{(R-1)(R+2)}$$

Where

        N = the estimate of the population
        M = the number marked in the population
        R = the number of recaptures in a sample
        T = the number in an unbiased sample

Figure 15.1: Corrib River system

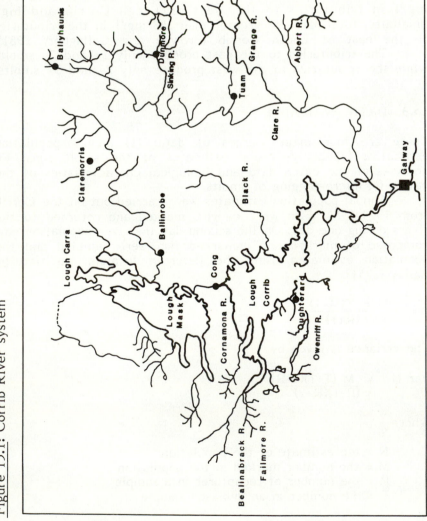

Detailed surveys of each tributary were carried out to determine the area of stream bed suitable for rearing salmon. A number of estimates were made on each tributary with a minimum of two sections being fished, one near the confluence of the tributary and one in the head reaches. From 1982 onwards the emphasis was placed on investigating large areas of individual tributaries to establish their contribution to the salmon stocks in the system. Average numbers per square metre were adopted for each tributary based on these investigations and these are presented in Tables 15.1 and 15.2. The area available, the numbers in each age group present and the survival from one year to the next are given.

It can be seen by inspection that the survival of 0+ to 1+ salmon in the Bealinabrack River (Table 15.1) is too high. In 1983 1+ numbers exceed the 0+ estimate. For the other years survival was high. This can be explained by the fact that there are numerous small streams in the head waters of the system which could not be surveyed. They are now thought to produce large numbers of 0+ fish which migrate downstream because of considerations of space. This is true of the Failmore River also whereas the survival figures for the Owenriff and the Cornamona are lower and fairly constant annually.

Some 0+ fish move from the western tributaries into Lough Corrib because of the high 0+ densities. The survivals given for many western tributaries are therefore lower than actual survivals as some contribution to the smolt run is thought to come from these fish that move into the lake. The densities of 1+ fish in the Bealinabrack and the Failmore rivers are similar to the densities in the Owenriff and Cornamona although the 0+ densities are much lower. Table 15.3 shows the average number by tributary for each group of salmon. The second figure for the Bealinabrack is an estimate of the 0+ production based on the number of 1+ found using a survival of 0.26. This figure is closer to the east side tributary survivals where the total numbers per square metre are lower. The Bealinabrack averages 1.38 salmon/m^2 while the Grange and Sinking average 1.04 and 2.43/m^2 respectively. Survival figures of 0.26 for under yearling densities of 0.009/m^2 to a low of 0.08 at high densities of 2.56/m^2 are given by Symons (1979) for hatchery-reared fish. The survival of wild fish is higher than reared fish so that a figure of 0.26 survival is thought to be realistic. These are probably minimum figures for the Bealinabrack and Failmore given the average survivals in the Owenmore and Cornamona rivers of 0.09 and 0.13.

The figures for survival in tributaries on the east side are relatively constant except in 1983. This was a particularly wet year and survival throughout the system was extremely high.

Table 15.1: Juvenile population estimates on tributaries to the west of Lough Corrib 1979-1984. 1. River Owenriff, area available 91,000 m²; 2. River Cornamona, area available 32,000 m²; 3. River Bealinabrack, area available 96,500 m²; 4. River Failmore, area available 29,900 m²

River	Year	0+ $No./m^2$	0+ No. in river	1+ $No./m^2$	1+ No. in river	1+ Survival	2+ $No./m^2$	2+ No. in river	2+ Survival
1	1979	1.7	154,700	0.2	18,200	-	-	-	-
	1980	1.4	127,400	0.4	36,400	0.23	0.10	9,100	0.50
	1981	4.5	409,500	0.1	9,100	0.07	0.02	1,820	0.05
	1982	3.4	309,400	0.3	27,300	0.06	0.02	1,820	0.20
	1983	3.5	318,500	0.3	27,300	0.08	0.01	910	0.03
	1984	1.0	91,000	0.1	9,100	0.03	0.01	9,100	0.33
	Average	2.58	235,083	0.23	21,233	0.09	0.05	4,550	0.22
2	1979	3.3	105,600	0.3	9,600	-	-	-	-
	1980	2.5	80,000	0.1	3,200	0.03	0.04	1,280	0.50
	1981	4.0	128,000	0.2	6,400	0.08	0.05	1,600	0.10
	1982	3.4	108,800	0.3	9,600	0.08	0.02	640	0.16
	1983	6.9	220,800	1.1	35,200	0.32	0.05	1,600	0.05
	1984	5.3	169,600	0.9	28,800	0.13	0.06	1,920	
	Average	4.2	135,466	0.48	15,466	0.13	0.04	1,408	0.20

Table 15.1 (Cont'd)

River year	0+		1+			2+		
	No./m²	No. in river	No./m²	No. in river	Survival	No./m²	No. in river	Survival
3 1980	0.3	28,950	0.1	9,650	–	0.02	1,930	–
1981	0.6	57,900	0.3	28,950	1.0	0.01	965	0.10
1982	0.6	57,900	0.4	38,600	0.66	0.02	1,930	0.06
1983	1.4	135,100	1.5	144,750	2.5	0.03	2,895	0.07
1984	1.2	115,800	0.5	48,250	0.36	0.02	1,930	0.01
Average	0.82	79,130	0.56	54,040		0.02	1,930	0.06
4 1981	0.7	20,930	0.3	8,970		0.01	299	0.17
1982	0.2	5,980	0.1	2,990	0.14	0.01	299	0.20
1983	1.4	41,860	0.2	5,980	1.00	0.02	598	0.20
1984	0.8	23,920	0.4	11,960	0.28	0.04	1,196	
Average	0.77	23,172	0.25	7,475			598	

Table 15.2: Juvenile population estimates on tributaries to the east of Lough Corrib 1979-1984. 1. River Sinking, area available 28,700 m^2; 2. River Grange, area available 15,000 m^2; 3. River Abbert, area available 82,800 m^2; 4. River Black, area available 10,000 m^2

River year	0+		1+		Survival
	No./m^2	No. in river	No./m^2	No. in river	
1 1979	1.4	40,180	0.4	11,480	
1980	2.8	80,360	0.2	5,740	0.14
1981	1.2	34,440	0.3	8,610	0.11
1982	2.6	74,620	0.2	5,740	0.17
1983	3.4	94,710	1.1	31,570	0.42
1984	0.9	25,830	0.3	25,830	0.27
Average	2.0	58,357	0.43	14,828	0.22
2 1979	0.7	10,500	0.1	1,500	
1980	0.8	12,000	0.2	3,000	0.28
1981	0.4	6,000	0.2	3,000	0.25
1982	0.3	4,500	0.1	1,500	0.25
1983	2.1	31,500	0.1	1,500	0.33
1984	1.0	15,000	0.3	4,500	0.14
Average	0.88	13,250	0.16	2,500	0.25
3 1979	0.7	57,960	0.2	16,560	
1980	0.1	8,280	0.03	2,484	
1981	1.7	140,760	0.2	16,560	
1982	0.4	33,120	0.1	8,280	0.06
1983	2.4	198,720	0.2	16,560	0.50
1984	0.9	74,520	0.4	33,120	0.16
Average	1.22[a]	101,016	0.2	18,216	
4 1981	1.0	10,000	0.04	400	
1982	2.4	24,000	0.05	500	
Average	1.7	17,000	0.045	450	

Note: a, 1980 omitted because of poor weather conditions and floods

Table 15.3: Average number of tributaries of each age group of salmon produced on the east and west sides of Lough Corrib

	River	Area	(0+)	(1+)	S	(2+)	S
West side	Owenriff	91,000	235,083	21,233	0.09	4,550	–
	Cornamona	32,000	135,466	15,466	0.11	1,480	0.07
	Bealinabrack	96,500	79,130	54,040	0.68	1,930	0.12
	Bealinabrack		207,846		0.26		–
	Failmore	29,900	23,173	7,475	0.32	598	–
Total/Average		249,400	746,395	98,214	0.13	8,558	–
East side	Sinking	28,700	58,357	14,828	0.22		
	Grange	15,000	13,250	2,500	0.25		
	Albert	82,800	101,026	18,216	0.18		
	Black	10,000	17,000	450	0.03		
Total/Average		881,700	189,623	35,994	0.19		
Totals		1,131,100	936,018	134,208		8,558	

Estimates of the smolt run were made annually using two methods. The first was a normal tag and recapture method using numbered tags and the second was data from micro-tagging.

On the River Corrib at Galway there is a smolt trap which samples a proportion of the run. The efficiency of this trap changes depending on the number of gates open on the flood control weir just upstream of the trap. Estimates of the run were obtained by tagging smolts at the trap with numbered Floy tags on a daily basis, transporting them upstream approximately 1 mile and releasing them to mix with the normal populations. Table 15.4 shows the releases and recaptures in the various years. The numbers migrating were estimated using the method of Schaefer (1951). Some smolts migrated after the main run and an estimate of the number of these was made using the numbers recorded in the trap and average efficiencies for the trap (Table 15.5).

Table 15.4: Numbers of salmon tagged and recovered in the River Corrib 1980-1984

Year	No. of fish tagged	No. of fish recovered	% recovery
1980	732	125	17.1
1981	1,059	234	22.1
1982	-	-	-
1983	411	41	9.9
1984	680	174	25.6

Table 15.5: Estimates of the smolt run in the River Corrib 1980-1984 by the Schaefer method

Year	No. smolts estimated by the Schaefer model period	No. smolts estimated migrating outside the tagging	Totals
1980	109,137	10,000	119,137
1981	98,678	22,000	120,678
1982	-	-	
1983	94,877	15,000	109,877
1984	75,901	10,000	85,901

The Use of Leslie Matrices

During the run daily samples of scales and measurements were taken to estimate the proportion of the three age-classes occurring in the run (Table 15.6).

Table 15.6: Biological characteristics of Corrib smolts

Smolts migrating in year	(1+)	(2+)	(3+)
1980	28.3	62.7	9.0
1981	29.7	61.3	9.0
1982	29.2	63.7	7.1
1983	35.4	57.3	7.3
1984	39.4	54.8	5.8
Average	32.4	59.96	7.64
Mean lengths	13.6 cm	14.8 cm	15.9 cm
SD	0.87	1.37	1.47

Wild smolts are micro-tagged annually at the smolt trap at Galway. The method and tag have been described by Jefferts, Bergman and Fiscus (1963). Sufficient smolts are tagged to produce a catch of tagged 1 sea-winter salmon in the riverine trap in the next year and a catch of tagged 2 sea-winter salmon in the subsequent year. Trapping effort was constant throughout the period. The catch at the trap over the appropriate two years can be looked on as a sample of fish taken from the stock. The smolts are randomly tagged over the entire smolt run. This type of sampling is discussed in a review by Schaefer (1951) and the number of smolts which migrated can be estimated using the formula suggested by Pearson (1982).

$$S = \frac{St.\ Nc}{Nt}$$

where

S = the number of smolts migrating
St = the number of smolts tagged
Nt = the number of returning adults caught with tags
Nc = the number of returning adults caught at the trap without tags.

Var of N

$$N = \frac{St.\ Nc\ (St-Nc)\ (Nt + Nc)}{Nt}$$

The results are shown in Table 15.7.

The two estimates are similar except in 1984 when the Schaefer method gave a low figure of 85,901. There was a period in 1984 in the early part of the run when the trap was not operating and the numbers estimated to have migrated outside the sampling period are thought to have been substantial. If 1984 is excluded the average number migrating is approximately 116,000 smolts according to both estimates.

The data for adult catches for the 31 years 1954 to 1984 is presented in Figure 15.2. The mean for 2 sea-winter salmon is 491 fish and for 1 sea-winter is 3,888. Since 1974 these means have not been exceeded except in the case of 2 sea-winter fish in 1975. This apparent decline, however, is not obvious when the catch is presented in terms of effort. Figure 15.2 also shows the catch in number of fish per day fished, graphed in log form. Since the 1980s the catch has been over the mean value or close to it in three of the five years in respect of 1 sea-winter fish and four of the five years with respect to 2 sea-winter fish. Also up to the mid seventies when water conditions were unsuitable for the trap a draft or circling net was used downstream and these figures are included in the catch. It is not clear how much effort this added to the earlier catches but certainly the effort was greater.

The catch has been relatively stable over a long period and so by implication is the stock.

A sample of the catch has been analysed and broken into age-classes (Table 15.8). The most important group is the 2-year-old smolt group which from average values produces 73 per cent of the 1 sea-winter catch and 53 per cent of the 2 sea-winter catch. The numbers of 1 sea-winter fish produced from 1-year-old smolts is low. This is because the 1-year-old smolts on average are smaller and because a high proportion of returning 2 sea-winter fish are produced from the 1-year-old smolt group.

The adults in the catch at Galway have a mean weight of 2.0 kg for 1 sea-winter fish and 5.2 kg for 2 sea-winter fish. Data available from ova counts at the trap and from hatchery data given by McCarthy (in press) suggest that the average fecundity is 4,300 ova per 1 sea-winter fish and 7,000 ova for 2 sea-winter fish. External examination of fish at the riverine trap suggests that the male to female ratio is close to 1:1.

Table 15.7: Number of smolts migrating using data from micro-tagging

Smolt year	Tagged	Sampled Grilse	Tags recovered	Sampled salmon	Tags recovered	Total fish sampled without tags	Total tags	Population estimate
(Y)	(M)	(Y + 1)	(Y + 1)	(Y + 2)	(Y + 2)	(C)	(R)	
1980	9,293	1,561	117	344	34	1,905	151	117,240
1981	2,259	3,037	34	270	21	3,307	55	135,828
1982	2,551	3,335	74	436	17	3,771	91	105,712
1983	4,092	1,712	64	243	13	1,955	77	103,894
1984	6,870	2,388	139	351	-	-	-	118,025
							Average	116,139

Table 15.8: The proportion of each age-class of smolts represented in the appropriate 1 sea-winter and 2 sea-winter adult run

Smolts migrating	% returning 1 sea-winter (Y + 1)			% returning 2 sea-winter (Y + 2)		
Year	1+	2+	3+	1+	2+	3+
1980	17.7	67.5	14.8	58.5	39.1	2.4
1981	23.5	63.9	9.0	30.5	67.6	1.9
1982	10.4	77.7	11.5	47.5	49.5	3.0
1983	17.6	73.3	8.5	38.0	57.0	5.0
1984	12.8	82.4	4.8	-	-	-
Average	16.4	72.9	9.72	43.6	53.3	3.1

Wild smolts are micro-tagged annually at the trap at Galway. This tagging programme is designed to study the contribution of these fish to the various drift net fisheries and the riverine catch. It also provides relative annual figures for marine survival. The estimated numbers of tags recovered in marine and river fisheries are shown in Table 15.9 together with the percentage of marine recoveries in each sea area and the estimated percentage mortality due to nets at sea and riverine fisheries. The figures for the mortality due to all fisheries include a figure for tag mortality and an assumed figure for non-reporting of catches. Non-catch fishing mortality includes non-reported catches and all source of fishing mortalities where the fish are not landed. As the figure for non-catch fishing mortality is purely an estimate based on experience of the fishery these figures must be regarded as approximations of the actual values.

Although the trap fishery is relatively stable in its catching effort, from Table 15.9 we can see that relative figures for marine survival are very variable. The tag returns at this trap in conjunction with the estimated numbers caught outside in the drift net fisheries is a relative measure of marine survival. The variation here is a factor of just over 2 from 1.3 to 2.9 per cent, the estimated fishing mortality increasing at sea as the riverine fishing mortality increases. The overall fishing mortalities are estimated at between 10 and 29 per cent.

Table 15.9: The estimated recovery of microtags (in parentheses) and the percentage recoveries of microtags in various marine fisheries

	1981		1982		1983		1984		1985
	1 SW	2 SW	1 SW	2 SW	1 SW	2 SW	1 SW	2 SW	1 SW
Donegal	(24) 7	(13) 57	(45) 50		(53) 47	(4)	(13) 57	(2)	(7) 4
Mayo	(44) 13	(8) 35	(29) 32	(7)	(22) 20		(5) 20	(2)	(26) 17
Galway/Limerick	(266) 78	(2) 9	(9) 10		(32) 29		(5) 20		(79) 50
Kerry	(2) 1		(6) 7		(2) 2		(2) 8		(40) 25
West Cork	(3) 1								(3) 2
S. Coast									(2) 1
Others	(3) 1		(1) 1		(3) 3				
River	(117)	(62)	(34)	(17)	(74)	(21)	(64)	(17)	(139)
No. of fish tagged	9,293		2,259		2,551		4,092		6,870
% tags in nets estimated	3.6	0.25	3.9	0.31	4.4	0.16	1.8	0.10	6.2
% tags in river catch	1.3	0.68	1.5	0.75	2.9	0.82	1.6	0.41	2.0
Total % fishing mortality	4.9	0.93	5.4	1.06	7.3	0.98	3.4	0.51	8.2
Estimated total mortality including NCFMa (%)	14		19		23		10		29

Note: a, NCFM = Non-catch fishing mortality which includes all sources of mortality including non-reported catches.

Figure 15.2: Salmon catches in the Corrib River 1954-1984

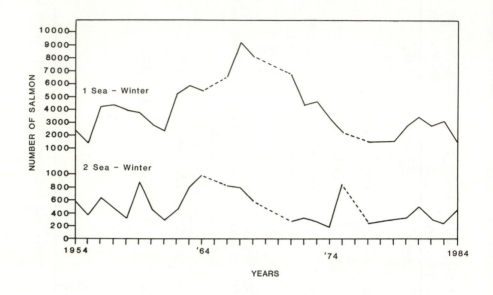

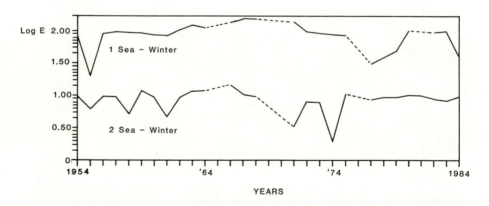

15.4: RESULTS

How realistic are the estimated numbers of migrating smolts in the light of the estimated numbers being produced by the various tributaries as determined by the population inventories?

On average 116,000 smolts go to sea. This can be broken down into the various age-class components using the proportions in Table 15.6. The results are shown in Table 15.10. The average numbers of 1+, 2+ and 3+ smolts can therefore be estimated. The flow chart in Figure 15.3 shows these figures in brackets and estimated figures are also shown in brackets. On the east side

The Use of Leslie Matrices

Table 15.10: The proportion of each smolt age group in 1 sea-winter and 2 sea-winter catches

Smolt migration year	1 sea-winter Year + 1			2 sea-winter Year + 2		
	1 + smolt	2 + smolt	3 + smolt	1 + smolt	2 + smolt	3 + smolt
1980	17.7	67.5	14.8	58.5	39.1	2.4
1981	23.5	63.9	9.0	30.5	67.6	1.9
1982	10.4	77.7	11.5	47.5	49.5	3.0
1983	17.6	73.3	8.5	38.0	57.0	5.0
1984	12.8	82.4	4.8	-	-	-
Average	16.4	72.9	9.7	43.6	53.3	3.0

189,623 0+ fish give rise to 37,120 1+ smolts and have a further survival of 0.19 to September parr producing 35,994 parr. The fate of these parr is unknown but some survive to produce the 69,600 2+ smolts. On the west side 746,395 0+ fish produce 98,214 parr at a survival of 0.13. There is a further survival of these 0+ fish through migration into the lake. Smolt scale readings in 1985 suggest that 26 per cent of the 2+ smolts produced have scale patterns consistent with lake growth and 20 per cent of 3+ smolts have a similar pattern.

The survival of 1+ parr to 2-year-old smolt can only be ascertained by bulking the east side and west side tributaries together. They have a survival of 0.38 to produce 51,504 2+ smolts which is supplemented by 18,096 2+ smolts of lake origin to make up the total of 69,600 2+ smolts. The survival from the west side will be poorer than the survival from the east side as the average east side 1+ parr are 12.1 cm while the average on the west side is 7.6 cm. There is a further survival of 8,558 2+ parr from the 98,214 1+ parr. These survive at a rate of 0.87 to produce 7,424 3+ smolts which are supplemented by 1,856 3+ smolts of lake origin. The only transition considered to be unlikely is the survival of 0.87 from 2+ parr in September to 3+ smolts in May. This appears rather high and may mean that the 3+ smolt component is over-estimated. The transition from one life stage to another is called survival but as in some cases there is more than one resultant group they are not true survivals. Nor can they be compared with figures from other systems as they are particular to the conditions in the Corrib. Given direct survival from 0+ west side parr to 1+ parr at 0.13 cannot be

Figure 15.3: Life history of Corrib salmon

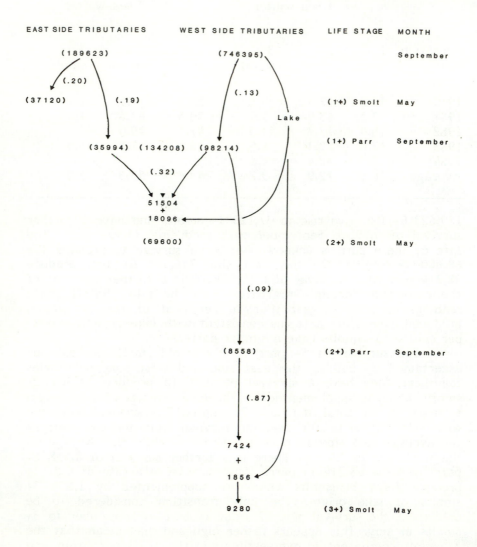

EAST SIDE TRIBUTARIES	WEST SIDE TRIBUTARIES	LIFE STAGE	MONTH
(189623)	(746395)		September
(.20)			
(37120) (.19)	(.13)	(1+) Smolt	May
	Lake		
(35994) (134208) (98214)		(1+) Parr	September
(.32)			
51504 + 18096			
(69600)		(2+) Smolt	May
	(.09)		
	(8558)	(2+) Parr	September
	(.87)		
	7424 + 1856		
	9280	(3+) Smolt	May

compared as there is an indirect survival to 2+ smolts through the lake system. Egglishaw and Shackley (1973) give figures for 0+ fry planted in May and surviving until the following September ranging between 10.0 and 51.1 per cent. Their overall figure for survival was 11.2 per cent. The survival of 0+ to 1+ parr on the east side at 0.19 falls well within the range found by Egglishaw but of course there is a further survival of 0.20 to 1+ smolts. The overall survival of 0.39 on the east side given the lower densities of 0+ fish seems reasonable. Overall the figures appear to offer a reasonable survival pattern and would suggest that the estimate of smolt production and the proportion by age are reasonable based on the information obtained from population surveys.

We have estimated the number of 2+ smolts produced at 69,600 but we do not know their origin. Using the overall survival figure of 0.38 between 1+ parr in September and 2+ smolts produced we can obtain an estimate of survival on the east side of 13,677 2+ smolts. It is then possible to give an estimate of the numbers of smolts produced on both sides.

How many ova are required to produce these 0+ fish? There are no direct survival figures for the Corrib. There is information from two rivers in Ireland where complete counts of ascending adults and migrating smolts are possible: the River Bush in Northern Ireland and the Burrishoole River in the West of Ireland. The River Bush is the most similar to the tributaries to the east of Lough Corrib and is estimated to have minimum survival figures of 0.014 from ova to 2-year-old smolt (Kennedy, personal communication) while the Burrishoole River tributaries to the west has average survival values from ova to 2-year-old smolt of 0.0052 (Anon., 1984). Using these values as average figures for the east and west side we can estimate the number of ova laid down as 50,797 east side smolts require 3,628,357 ova; 65,203 west side smolts require 12,540,600 ova, or approximately 16,000,000 ova.

Using the fecundity of 4,300 per 1 sea-winter salmon we find that 3,726 female 1 sea-winter equivalents would be required to produce this ova deposition. That is a total spawning stock of 7,452 1 sea-winter equivalents.

An analysis of the catch and effort data suggests that 10 per cent of the catch is 2 sea-winter and on this basis the probable make up of the stock is 3,200 1 sea-winter female salmon producing 13,760,000 ova, 320 2 sea-winter female salmon producing 2,240,000 ova which gives a total of 16,000,000 ova. This suggests a stock of 6,400 1 sea-winter spawners and 640 2 sea-winter spawners.

The average proportion of adult 1 sea-winter fish returning (Table 15.11) can be used to apportion the 6,400 1 sea-winter and 640 2 sea-winter fish. The survival flow chart for the population is now as shown in Table 15.12.

Table 15.11: The proportion and number of each age class of smolt in the run

Year	No. est. migrating	% 1+ smolts	No. 1+ smolts	% 2+ smolts	No. 2+ smolts	% 3+ smolts	No. 3+ smolts
1980	121,594	28.3	34,411	62.7	76,239	9.0	10,943
1981	148,121	29.7	43,992	61.3	90,798	9.0	13,331
1982	102,556	29.2	29,946	63.7	65,328	7.1	7,281
1983	104,438	35.4	36,971	57.3	59,843	7.3	7,624
1984	102,400	39.4	40,346	54.8	56,115	5.8	5,939
Average	115,822	32	37,063	60	69,493	8	9,266

Adults

1 sea-winter		16.4		73.0		9.7	
2 sea-winter		43.6		53.3		3.1	

Table 15.12: The number of each life stage estimated

16,000,000	Ova
936,018	0 + parr
37,120	1 + Smolts
69,600	2 + Smolts
9,280	3 + Smolts
1,088	(1+) 1 sea-winter
4,672	(2+) 1 sea-winter
640	(3+) 1 sea-winter
281	(1+) 2 sea-winter
339	(2+) 2 sea-winter
19	(3+) 2 sea-winter

Average fecundity	1 sea-winter 4,300
	2 sea-winter 7,000

The recovery rates (Table 15.9) range from 15 to 30 per cent or a mean of 19 per cent from smolt to 1 sea-winter fish caught and approximately 1.1 per cent for 2 sea-winter fish caught.

The figure of 19 per cent implies a figure of 22,040 fish caught added to the riverine spawning stock of 6,400 or a figure of 28,440 1 sea-winter fish in home waters. The number of fish caught by nets is 22,040 minus the average riverine catch which based on an effort of 40 days fishing, the recent norm is 3,340 including angling. The number of fish caught by nets is then 22,040 - 3,340 = 18,700. The exploitation by nets of fish returning to home waters is 66 per cent and by trap and angling 12 per cent. (The traps and angling actually exploit 34 per cent of fish entering the river.)

The figure of 1.1 per cent for 2 sea-winter fish implies that 1,277 fish were caught, added to the spawning stock of 640 = 1,917. The average riverine catch based on a recent effort of 55 days fishing is 456. The number of fish caught by nets is 821 or 43 per cent exploitation and 24 per cent by trap and angling.

The 1 sea-winter fish caught in home waters, 18,700, can be roughly apportioned to the various sea fisheries using the average values for the proportion caught shown in Table 15.8. The numbers are as follows. Donegal 5,236, Mayo 3,927, Galway 7,106, Kerry 1,683, West Cork < 187 and South Coast < 187.

15.5   MATRICES

A number of matrices can be run using the data outlined. A ten by ten matrix was used for convenience as it was available. The population has been broken down into convenient sub-groups. The population could be modelled on a large matrix which would incorporate all the life stages.

The matrix shown in Table 15.13 was used to investigate the survival of ova to 1, 2 and 3 year old smolts and the number of each age group of 1 sea-winter and 2 sea-winter spawners. The egg input was taken at half the fecundity as the males contribute no ova.

From the data obtained (Table 15.12) we can produce the following sector of the population.

Life stages		Numbers t		Numbers t + 10
Ova		16,000,000		16,093,830
Smolts 1 year		37,120		38,217
Smolts 2 year		69,600		70,001
Smolts 3 year		9,280		9,313
1 sea-winter/1 year smolt	X	1,080	=	1,062
1 sea-winter/2 year smolt		4,672		4,659
1 sea-winter/3 year smolt		640		633
2 sea-winter/1 year smolt		281		284
2 sea-winter/2 year smolt		339		339
2 sea-winter/3 year smolt		19		18

Table 15.13: Matrix

Ova	Smolt 1	Smolt 2	Smolt 3	1 sea 1	1 sea 2	1 sea 3	2 sea 1	2 sea 2	2 sea 3	Vector
0	0	0	0	2,100	2,160	2,160	3,550	3,550	3,550	1.6E + 07
0.00238	0	0	0	0	0	0	0	0	0	37,120
0.00436	0	0	0	0	0	0	0	0	0	69,600
0.00058	0	0	0	0	0	0	0	0	0	9,280
0	0.028	0	0	0	0	0	0	0	0	1,080
0	0	0.067	0.0685	0	0	0	0	0	0	4,672
0	0.0075	0	0	0	0	0	0	0	0	640
0	0	0.0048	0	0	0	0	0	0	0	281
0	0	0	0.002	0	0	0	0	0	0	339
0	0	0	0	0	0	0	0	0	0	19

The survival and fecundities associated with these life stages are shown in Table 15.13. When this matrix was run over 10 generations the vector above resulted.

The total population is increasing at a rate of 1.00006 and the intrinsic rate of natural increase (r) = 0.0001. The population does not settle down to a single rate of natural increase even when run over a very large number of cycles or generations. This is probably because the survivals do not flow from one life stage to the next as they would in a normal animal population. For instance the 2-year-old smolts do not arise from the 1-year-old smolts but both could arise from the same ova deposition. It may also arise because we are dealing with two independent populations within the same matrix, a 1 sea-winter and a 2 sea-winter population. Although they have a common ova stage they appear to be acting independently and providing two pulses to the population.

Each of the survivals across the matrix was tested in turn for sensitivity by increasing the value by 1 per cent and looking at the resultant average intrinsic rate of natural increase. The results were as follows:

Transition	Average value of (r)
ova to 1 year smolt	0.206
ova to 2 year smolt	0.321
ova to 3 year smolt	0.316
1 year smolt to 1 sea-winter spawner	0.023
2 year smolt to 1 sea-winter spawner	0.030
3 year smolt to 1 sea-winter spawner	0.004
1 year smolt to 2 sea-winter spawner	0.027
2 year smolt to 2 sea-winter spawner	0.048
3 year smolt to 2 sea-winter spawner	0.007

By this analysis it appears that the survivals from ova to 2 and 3 year-old smolts are most significant with respect to the matrix. The most significant transition from smolt to grilse are the two-year-old and this is true of the transition into 2 sea-winter spawners also. The transition from 3-year-old smolt to 2 sea-winter spawners has very little influence on the population matrix. The areas that where more precise information is required are on the survivals of ova to smolts and particularly the survivals from ova to 2-year-old and 3-year-old smolts.

Increasing all of the ova inputs by 1 per cent produces an average intrinsic rate of natural increase of 0.009 so that the fecundity estimates are not critical.

Two further matrices were run, one of which allowed the investigation of the contribution of each age class of smolts to the 1 sea-winter and 2 sea-winter fish in home waters and the other allowed the investigation of the catches at sea and the resultant increases if various fisheries were curtailed. The 3+ smolts in the case of 1 and 2 sea-winter fish did not produce a great fluctuation. The main contribution both to 1 sea-winter and 2 sea-winter fish came from the 1-year-old and 2-year-old smolts. In this case the survival of ova to fry was highlighted as a sensitive transition. Further data on this survival will have to be obtained. The effect of each of the fisheries on the population was investigated by reducing the catch in each fishery in turn. This provided a measure of the size of reduction necessary in the catches to result in significant changes in the population structure.

A large matrix incorporating all the available life stages would be useful in studying management implication and planning research in a system. Matrices would be useful in salmon management particularly as a preliminary tool in coming to terms with the very complex life history of the normal salmon population. It would provide a test of available data, a method of deciding where the major research effort should be placed and offers an opportunity for judging management policies by simulation.

It is possible to introduce variables into the matrix to represent the range of variation recorded by experimentation. The method, however, would never be predictive in the sense that the run for the following year can be assessed

It could provide assessments based on different survivals to smolts and different marine survivals so that tables could be produced for a system which would predict runs based on estimates of smolt numbers and marine survival. Unfortunately marine survival can only be measured during the fishing year at present. There is the possibility however that if we knew more about the contribution of various riverine stocks to the high seas fisheries marine survival could be evaluated there.

15.6 SUMMARY

The results of annual population estimates in the River Corrib are used to judge the usefulness of estimates of the smolt run obtained by tagging methods. Estimates of the average survival for each life stage are obtained or inferred from the literature. A population model using Leslie matrices was used to test the data. The matrices are considered very useful in the preliminary investigation of a river system.

REFERENCES

Anon. (1984) Salmon Research Trust Annual Report, XXIX

Bailey, M.T.J. (1951) On estimating the size of mobile populations from recapture data. Biometrika, 38, 293-306

Blackith, R.E. and Albrecht, F.O. (1979) Locust Plagues: The interplay of endogenous and exogenous control. Acrida, 8, 83-94

Browne, J. (1980) Salmonid population dynamics and the use of Leslie matrices. Msc Thesis Trinity College, Dublin

Browne, J. and Gallagher, P. (1981) Population estimates of juvenile salmonids in the Corrib system 1981. Fishery Leaflet, 115, 1-6

Egglishaw, H.J. and Shackley, P.E. (1973) An experiment on faster growth of salmon Salmo salar (L.) on a Scottish stream. Journal of Fish Biology, 5, 197-204

Horst, T.J. (1977) Use of the Leslie matrix for assessing environmental impact, an example for a fish population. Transactions of the American Fisheries Society, 106 (3) 253-7

Jefferts, K.B., Bergman, P.K. and Fiscus, H.F. (1963) A coded wire identification system for macro-organisms. Nature, London, 198(487), 460-2

Keyfitz, N. (1968) Introduction to the Mathematics of Population. Addison-Wesley, Reading, MA

Leslie, P.H. (1945) On the use of matrices in certain population mathematics. Biometrika, 33, 183-212

Leslie (1948) Some further notes on the use of matrices in population mathematics. Biometrika, 35, 213-45

Lewis, E.G (1942) On the generation and growth of a population. Sankhya, 6, 93-6

Moran, R.A.P. (1962) The statistical processes of evolutionary theory. Oxford University Press, London

Pearson, K. (1982) On a method of ascertaining limits to the actual number of marked members of a population of given size from a sample. Biometrika, 20, 149-74

Pielou, E.C. (1969) An introduction to mathematical ecology. Wiley-Interscience, New York

Schaefer, M.B. (1951) Estimation of the size of animal populations by marking experiments. US Fish and Wildlife Service, Fisheries Bulletin 52, 189-203

Symons, P.E.K. (1979) Estimted escapement of Atlantic salmon Salmo salar for maximum smolt production in rivers of different productivity. Journal of Fisheries Research Board of Canada 36, 132-40

Usher, M.B. (1966) A matrix approach to the management of renewable resources, with special reference to selection forests. Journal of Applied Ecology, 335-67

Usher, M.B. (1971) Developments in the Leslie matrix model. In J.N.R. Jeffers (ed.), Mathematical models in ecology. Blackwell Scientific, London

Usher, M.D. and Williamson, M.H. (1970) A deterministic matrix model for handling the birth, death and migration processes of spatially distributed populations. Biometrics, 26, 1-12

Williamson, M.H. (1959) Some observations of the use of matrices in population theory. Bulletin of Mathematical Biophysics, 21, 261-3

Williamson, M.H. (1967) Introducing students to the concepts of population dynamics, the teaching of Ecology. J.M. Lambert (ed.) Symposium of the British Ecological Society 169-76

Chapter Sixteen

RELATIONSHIP BETWEEN ATLANTIC SALMON SMOLTS AND
ADULTS IN CANADIAN RIVERS

E. Michael P. Chadwick

Department of Fisheries and Oceans Science Branch, Gulf
Fisheries Centre, PO Box 5030, Moncton, NB E1C 9B6, Canada

INTRODUCTION

Predicting abundance and stock structure of Atlantic salmon
(Salmo salar L.) adults can best be achieved by examining these
characteristics in smolts. In this paper, previously published
information and new data illustrate four aspects of the predictive
relationship between smolts and adults. First, counts of smolts in
a particular river are the best means for predicting returns of
adults and are necessary to measure sea survival. Second, because
there is variation in sea survival, environmental factors must also
be considered when predicting returns of adults. Third, low sea
survival of smolt migrations can be offset by spreading egg
depositions over several years through multiple spawning. Finally,
the ability to determine sea age of adults by examining smolts
will greatly improve predictive relationships between smolts and
adults.

PREDICTING RETURNS OF ADULTS

The principle of predicting returns of adults from smolts has been
described by Chadwick (1985). Basically, smolts are counted in a
particular river and used to predict the returns of adults to that
river in the following and subsequent years. Smolt counts in one
index river may also be used to predict returns to a group of
rivers in a geographically homogeneous area. Because adult
salmon may be captured in several distant commercial fisheries,
home water commercial fisheries and in freshwater sport
fisheries, it is most useful to predict total returns or the sum of
returns to all fisheries plus spawning escapement.

There are several significant relationships between smolts
and returning adults, although there are not many suitable data
sets. Smolts and adults have been carefully counted on four rivers

Table 16.1: Counts of smolts, 1 SW salmon and 2 SW salmon on Little Codroy River, Newfoundland

Year (i)	Smolts year i	1 SW in year i+1 salmon		2 SW in year i +2 salmon	
		Numbers	% survival	Numbers	% survival
1954	12,030	98	0.8	35	0.4
1955	11,154	79	0.7	43	0.5
1956	14,401	118	0.8	56	0.5
1957	8,607	80	0.9	43	0.5
1958	8,874	53	0.6	22	0.4
1959	11,473	45	0.4	23	0.5
1960	7,165	28	0.4	24	0.9
1961	7,523	41	0.5	33	0.8
1962	7,579	117	1.5	-	-
Mean			0.7		0.6

Source: Murray (1968b)

Table 16.2: Counts of smolts and 1 SW salmon on Western Arm Brook, Newfoundland

Year (i)	Smolts year i	1 SW salmon year i + 1	
		Numbers	% survival
1971	5,734	415	7.2
1972	11,906	827	7.0
1973	8,484	526	6.2
1974	12,055	640	5.3
1975	9,733	552	5.7
1976	6,359	376	5.9
1977	9,640	317	3.3
1978	13,071	1,576	12.1
1979	9,400	470	5.0
1980	15,675	471	3.0
1981	13,981	467	3.3
1982	12,477	1,146	9.2
1983	10,552	238	2.3
1984	20,653	164	0.8
Mean			5.6

Table 16.3: Counts of smolts, 1 SW salmon and 2 SW salmon on Sand Hill River, Labrador

Year (i)	Smolts year i	1 SW in year i +1 salmon		2 SW in year i + 2 salmon	
		Numbers	% survival	Numbers	% survival
1969	54,600	3,560	6.5	267	0.5
1970	50,494	3,487	6.9	169	0.3
1971	55,000	1,877	3.4	489	0.9
1972	37,007	4,525	12.2	-	-
Mean			7.3		0.6

Source: Pratt, Hare and Murphy (1974)

Table 16.4: Counts of smolts and 1 SW salmon on Big Salmon River, NB

Year (i)	Smolts year i	1 SW salmon year i+1	
		Numbers	% survival
1966	18,370	779	4.2
1967	7,250	1,210	16.7
1968	16,640	172	1.0
1969	9,638	426	4.4
1970	19,766	476	2.4
1971	18,384	995	5.4
Mean			5.7

Source: Jessop (1986)

in Atlantic Canada: Little Codroy (Table 16.1), Western Arm Brook (Table 16.2), Sand Hill (Table 16.3) and Big Salmon (Table 16.4). First, returns of adults to neighbouring Grand Codroy and Robinson's rivers (Figure 16.1) could be predicted from Little Codroy smolts (Figure 16.2). These returns were estimated from catches of 1SW salmon and 2SW salmon in the sport fishery, where about 25 per cent of the sport catch was 2SW or older salmon. It had been shown (Chadwick, 1982c) that sport catch was correlated with river escapement on many Newfoundland rivers.

Figure 16.1: Location of place names mentioned in text

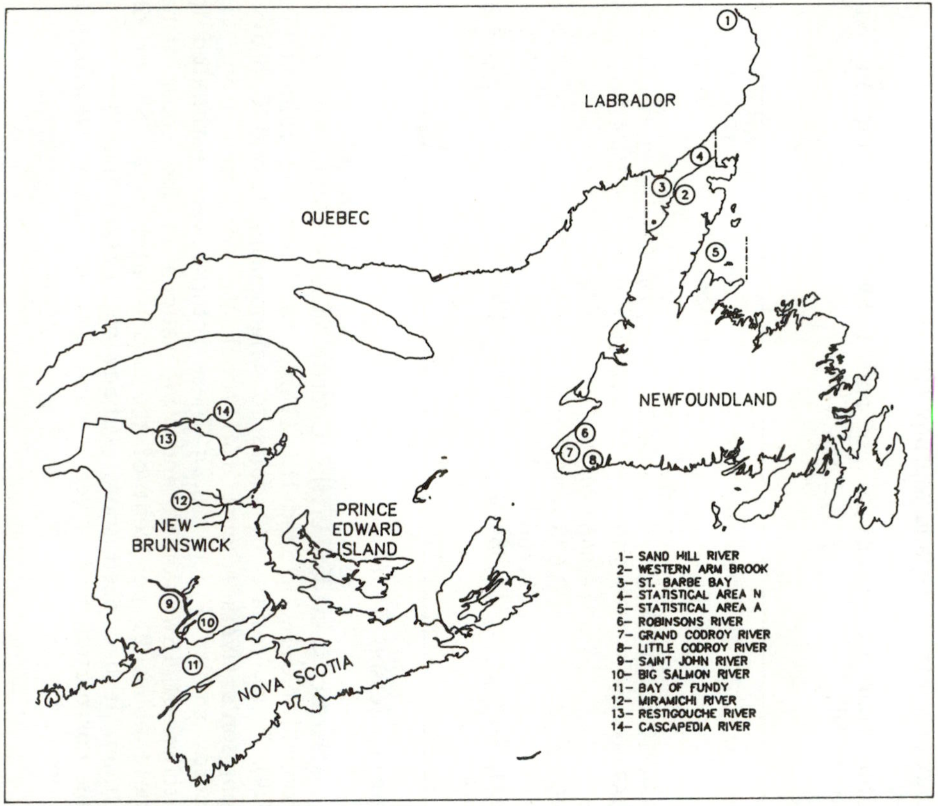

LABRADOR

QUEBEC

NEWFOUNDLAND

PRINCE
EDWARD
ISLAND

NEW
BRUNSWICK

NOVA SCOTIA

1— SAND HILL RIVER
2— WESTERN ARM BROOK
3— ST. BARBE BAY
4— STATISTICAL AREA N
5— STATISTICAL AREA A
6— ROBINSONS RIVER
7— GRAND CODROY RIVER
8— LITTLE CODROY RIVER
9— SAINT JOHN RIVER
10— BIG SALMON RIVER
11— BAY OF FUNDY
12— MIRAMICHI RIVER
13— RESTIGOUCHE RIVER
14— CASCAPEDIA RIVER

Figure 16.2: Relationship between smolts counted on Little Codroy River in year i and sport catch of grilse (year i+1) and salmon (year i+2) in two neighbouring rivers. Year of smolt migration is indicated

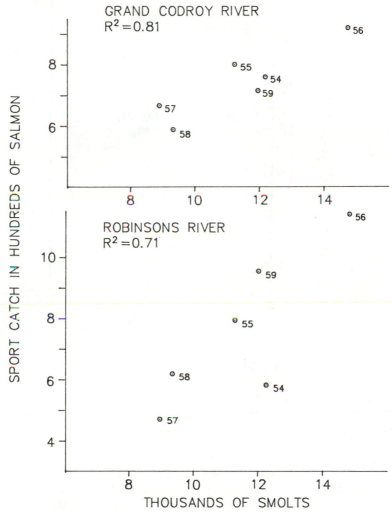

Source: Chadwick (1982c)

Second, until 1985, Western Arm Brook smolts could be used to predict total returns, to Statistical Area N (Figure 16.3). This relationship was based on three assumptions: that Western Arm Brook produced 10 per cent of the smolts in Statistical Area N; local stocks were harvested in the commercial fisheries; and the interception rate in distant fisheries (those outside Statistical Area N) was constant. Although this relationship was weak and

Figure 16.3: Correlation between numbers of smolts counted in Western Arm Brook and returns of grilse (1 SW salmon) to home waters in the following year

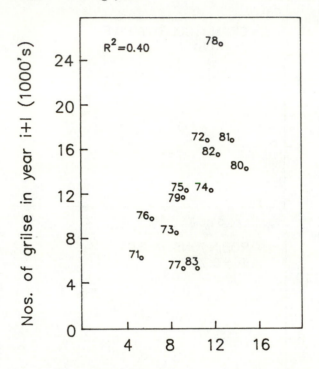

Nos. of smolts in year i (1000's)

Source: Chadwick (1984)

not significant if 1985 returns were included, it was better than when commercial catches were not included in adult returns (Table 16.2).

There were not any significant relationships between smolts and 2SW salmon returning two years later. This lack of a correlation may have been partly because of the handling of smolts on Little Codroy River and because of the greater exploitation of 2SW salmon; > 50 per cent of 2SW salmon are harvested in distant commercial fisheries (Pippy, 1982).

While it is not yet possible to predict returns of 2SW salmon from counts of smolts, there were a number of significant relationships between counts or catches of 1SW salmon and those of 2SW salmon in the following year. In Figure 16.4, it is clear that counts or catches of 1SW salmon could be used to predict counts or catches of 2SW salmon in the following year on

Figure 16.4: Correlations between numbers of grilse (1SW salmon) and numbers of large salmon (2SW salmon and older) in the following year for (a) Cascapédia River, (b) Restigouche River, (c) Saint John River and (d) Miramichi River

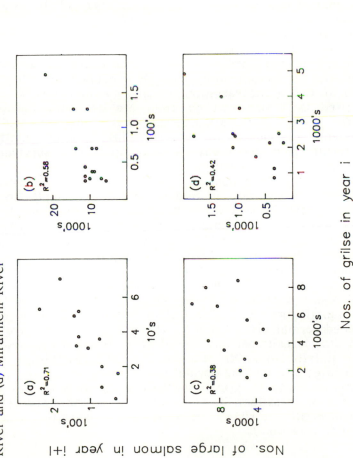

Nos. of grilse in year i

Nos. of large salmon in year i+1

Sources: (a) and (d) Chadwick et al. (1986); (b) Randall et al. (1985b); (c) Marshall (1984

Miramichi (Randall, Chadwick and Schofield, 1985a), Saint John (Marshall, 1984), Cascapédia (Chadwick, Randall and Léger, 1986), and Restigouche (Randall, Chadwick and Pickard, 1985b) rivers. Thus, it appears that if sufficiently long time series of smolt and adult data were available, it would be possible to predict returns of 2SW salmon from counts of smolts.

VARIATION IN SEA SURVIVAL

The simplest way of measuring sea survival is from counts of smolts and adults in the same river. Any measure of sea survival must include natural losses, for example predation or fish which stray from their natal rivers, and fishing losses. It is difficult to separate these two sources of mortality and therefore in this paper I have combined them and defined sea survival as the proportion of smolts which return to their natal river as adults. It follows that if sea survival is constant, then smolts would be useful for predicting the return of adults.

As we have seen in Western Arm Brook, our best data set (Figure 16.3), only 40 per cent of the variation in returns of adults could be explained by smolts. Thus about 60 per cent of the variation in adult returns may be explained by other ways. Because of this great variation, mean survival rates, which are summarised in Tables 16.1 to 16.4, are not very meaningful.

The pattern of variation in sea survival suggests that most of the noise introduced into relationships between smolts and adults results from the great variation in marine survival. One convenient way of comparing variation in sea survival among different rivers is to standardize annual values as a percentage of the highest survival rate in a data set. For example, in Table 16.2, the highest sea survival in Western Arm Brook was 12.1 per cent for the 1978 smolt migration; the lowest sea survival, the 1984 smolt migration, was 0.8 per cent or 0.1 (10 per cent) of the highest. Therefore the 1978 smolt migration receives a value of 10, and the 1984 smolt migration a value of 1. The distribution of survival rates for Western Arm Brook is skewed to the left, that is, there were more lower survival rates in proportion to higher ones (Figure 16.5). This distribution could be changed into a balanced normal distribution by transforming the survival rates into natural logarithms; and therefore it is usually called a log normal distribution. All other rivers with estimates of sea survival rates also had log normal distributions.

Another way of quantifying this variation is to look at the coefficient of variation, the standard deviation as a proportion of the mean. In all five rivers the coefficient of variation was > 40 per cent (Figure 16.5).

Figure 16.5: Relative variation in marine and freshwater survival rates for Atlantic salmon stocks on several Canadian rivers. See text for details of methods

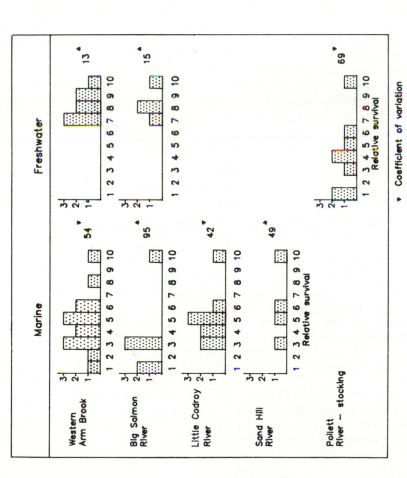

Variation in sea survival is much greater than freshwater survival. Egg to smolt survival rates on two wild rivers, Western Arm Brook and Big Salmon River, were fairly constant; the coefficients of variation were < 15 per cent compared to over 50 per cent for sea survival. Egg depositions greater than one million eggs were not included on Western Arm Brook. These values were omitted because density-dependent mortality is known to occur at high egg depositions and the purpose of this paper is to examine the variation in survival caused by environmental factors (Figure 16.5)

These results suggest that further research on factors other than egg deposition which control smolt production will not necessarily improve our ability to predict returns of adults. In other words, we can predict numbers of smolts from eggs fairly accurately. By contrast, more research on the factors which influence sea survival might prove to be fruitful.

Attempts to improve salmon production through enhancement may increase the number of fish, but not in a predictable manner. The variation in fry to smolt survival rates from stocking on the Pollett River was greater than the variation observed for sea survival (Figure 16.5). Hence, if our goal is to improve and stabilise yield in fisheries, then enhancement activities in healthy rivers will only increase variation in survival, or introduce more noise into the system. Therefore enhancement might have more predictable results if it focused on the marine environment. Cage rearing of Atlantic salmon is one example of marine enhancement which is successful because it reduces the variation in marine survival.

Although the variation in sea survival cannot be separated into natural mortality and fishing mortality, it is possible to estimate crudely which source of mortality is most important. Mortality estimates based on data collected at counting fences, tagging of adult salmon, and the theoretical inverse relationship between size and mortality of salmon indicate that natural mortality is eight to ten times greater than fishing mortality (Table 16.5). I have not included information from smolt tagging studies because the survival of tagged smolts is only one-tenth that of untagged smolts (Table 16.6). It is probable then that most of the observed variation in sea survival is because of natural causes.

It is also probable that most of the natural mortality occurs during the first year at sea. This conclusion can be inferred from the small size of smolts. It can also be inferred from the better relationships that we find between 1SW and 2SW salmon compared to those between smolts and 1SW salmon. This contrast is easily seen by comparing the correlation coefficients in Figures 16.2 and 16.3 versus Figure 16.4.

Table 16.5: Estimates of natural and fishing mortality rates for 1SW salmon (grilse) and 2SW salmon (salmon) in Canadian rivers. Estimates were made using data in Murray (1968b), Pratt et al. (1974) and Chadwick, Reddin and Burfitt (1985)

	Numbers of smolts	Losses Natural	Fishing	Returns to river
Salmon	100	90	9	1
Grilse	100	84	12	4
% Grilse	50		55	80

Table 16.6: Comparison of sea survival rates for tagged and untagged smolts which returned as 1SW salmon. Data are from Big Salmon River, New Brunswick

Year	Tagged Smolts	% returns	Untagged Smolts	% returns
1967	3,900	0.9	7,250	16.7
1968	5,120	0.06	16,640	1.0
1969	3,792	0.1	9,638	4.4
1970	9,864	0.05	19,766	2.4
1971	7,786	0.5	18,384	5.4
Mean		0.3		6.0

Source: Jessop (1986)

Straying should also be considered as a source of mortality. Straying is probably most important in small rivers where salmon are unable to enter because of low water levels. This problem was discussed by A.R. Murray (Anon., 1962) on Little Codroy River where a significant correlation was found between numbers of adults which entered the river and water discharge on a large tributary (Figure 16.6). This phenomenon is well known in small Newfoundland rivers, where adults will remain in pools near the river mouth waiting for sufficient water discharge to attract them up river. Salmon held up in this manner are particularly vulnerable to the sport fishery, whether or not they are caught legitimately, and accounts for the significant inverse relationships between sport catch and water discharge (Figure 16.7). On Western Arm Brook, sea survival was lowest during the two years of lowest water discharge during August and September (1984 and

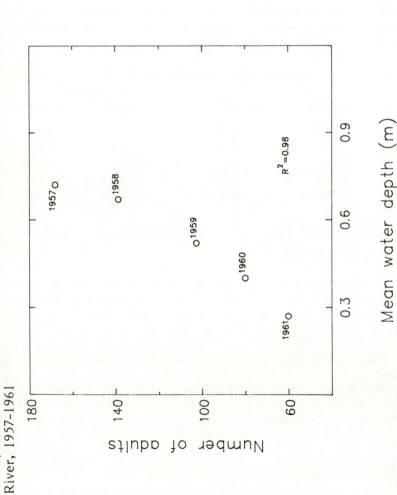

Figure 16.6: Relationship between the June-September mean water depth (feet) as measured at Northern Brook, and the number of Atlantic salmon mature adults recorded at the counting fence, Little Codroy River, 1957-1961

Source: Anon. (1962)

312

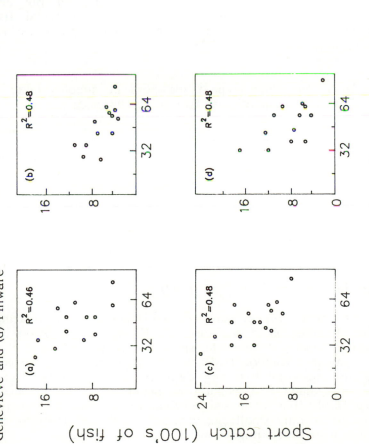

Figure 16.7: Correlations between mean daily water discharge for July and August recorded on St Genevieve River and sport catch of salmon in four nearby rivers, (a) River of Ponds, (b) Castors, (c) St Genevieve and (d) Pinware

313

1985) which are the months salmon enter this river.

Strays are not necessarily lost to regional production because they probably enter other rivers. On Little Codroy River there was a stray rate of about 0.5 per cent probably into the nearby and much larger Grand Codroy River (A.R. Murray personal communication and Murray, 1968b). The strays, however, are lost to local production and become part of the variation in estimates of sea survival.

BENEFITS OF REPEAT SPAWNING

One obvious means of adapting to the large variation in sea survival is the life history tactic of multiple spawning. That multiple spawners can be an important part of a population is attested by two facts: a large sample of Newfoundland commercial fishery in 1931 indicated that over 30 per cent of 1SW and 3SW salmon and 15 per cent of 2SW salmon in the catch (Table 16.7) were repeat spawners; secondly, on Big Salmon River, whose stock was exposed to little commercial fishing, there was an average 50 per cent repeat spawners in the spawning migration (Table 16.8).

Table 16.7: Percentage repeat spawners in samples taken from the Newfoundland commercial fishery in 1931

Sea age	Sample size	Repeat spawners (%)
1SW	275	31
2SW	3,597	15
3SW	132	31
4SW	2	100

Source: Lindsay and Thompson (1932)

As a result repeat spawners are important in mitigating the effects of variation in sea survival. First, the variation in egg deposition based on multiple spawners is almost half that of virgin spawners (Figure 16.8). Secondly, with multiple spawners there are no very low egg depositions and more high egg depositions than if a population depends on virgin spawners (Figure 16.8). Thus when sea survival of one smolt migration is low, its potentially small egg deposition can be offset by repeat spawners from smolt migrations in previous years.

Table 16.8: Percentage repeat spawners in Big Salmon River, New Brunswick, for adults belonging to the same smolt migration

Year of smolt migration	No. of adults counted	Repeat spawners (%)
1966	1,984	61
1967	2,730	54
1968	429	56
1969	534	19
1970	705	47
Average		47

Source: Jessop (1986)

Figure 16.8: Relative variation in egg depositions from virgin spawners and virgin plus repeat spawners on Big Salmon River, New Brunswick

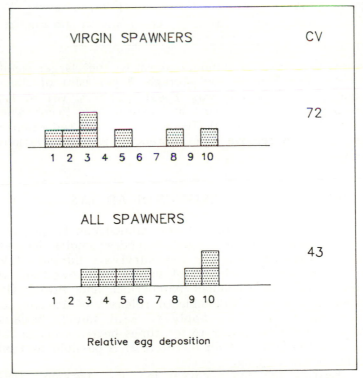

Source: Jessop (1986)

Finally, repeat spawners are less common than they used to be. The proportion of repeat spawners harvested in the commercial fisheries of Statistical Area A, Newfoundland, has declined from 15-20 per cent in the 1930s to < 5 per cent in the early 1970s (Table 16.9). This decline was probably because of

Table 16.9: Percentage repeat spawners for Atlantic salmon sampled in the commercial fisheries of Statistical Area A, Newfoundland

Year	Sample size	% repeat spawners	Reference
1931	3,161	16	Lindsay and Thompson (1932)
1938	380	12	Belding and Préfontaine (1961)
1939	728	20	Blair (1943)
1969	446	4	Lear and May (1972)
1970	578	3	Lear and May (1972)
1971	554	4	Lear (1973)
1973	649	1	Lear, Batton and Burfitt (1976)

greater selection of gillnets towards the larger-sized repeat spawners. For example, on average, 8 per cent of the catch of salmon in St Barbe Bay was repeat spawners, yet < 1 per cent repeat spawners were found in Western Arm Brook which flows into the bay (Table 16.10). The effect of this reduction in repeat spawners is to remove egg insurance against variations in sea survival.

PREDICTING CHARACTERISTICS OF ADULTS

The size and maturity of Atlantic salmon can be predicted from smolts. Evidence from tagged hatchery smolts indicates that larger smolts have better sea survival: Larsson (1977) has summarised data on releases of smolts from Swedish hatcheries; Ritter (1977) for Canadian hatcheries; and Bilton (1984) for hatcheries in British Columbia of releases of Pacific salmonids. These results may not apply to wild smolts because wild, untagged smolts have four to ten times better survival rates than tagged smolts (Table 16.6); therefore it is possible that only large smolts survive tagging.

Bearing the above caveat in mind, Larsson's (1977) data provide convincing evidence that large smolts fare better at sea.

Table 16.10: Repeat spawning 1SW salmon sampled in the commercial fishery of St Barbe Bay and Western Arm Brook as a percentage of all 1SW salmon (virgins and repeat spawners)

| Year | Commercial fishery of St Barbe Bay | | Western Arm Brook | |
	No. of repeat spawners	% of total 1SW salmon	No. of repeat spawners	% of total 1SW salmon
1977	36	9.8	0	0
1978	4	4.2	0	0
1979	4	2.0	0	0
1980	3	1.3	2	3.2
1981	56	18.0	3	4.9
Total	103	8.6	5	1.3

Source: Chadwick (1982b)

Between 23,000 and 109,000 smolts were released each year from 1955-1966. The average sea survival of 14 cm smolts was 8 per cent compared to 15 per cent for 17 cm smolts (Table 16.11). Thus, 17 cm smolts had twice the sea survival rate as 14 cm smolts and variation in sea survival was also less for the larger smolts. Thus, smolt size may explain why sea survival of Western Arm Brook smolts, with an average fork length of 17 cm (Chadwick, 1982a) was more than twice that of smolts from Big Salmon River (Table 16.4) which had an average fork length of 14 cm (Jessop, 1975).

Despite the apparently dramatic variation in sea survival for different sizes of smolts, there are some contradictions. For example, there is evidence that sea survival is not very size dependent. On Western Arm Brook the size of smolts could be used to predict the size of returning adults; in other words, survivors of smolt migrations where the average size of the smolts was small returned as smaller adults (Figure 16.9). For this type of relationship to exist, the survival of average-sized smolts must be greater than that of very small and very large smolts. In fact, there is good evidence of strong selection at sea against extreme sizes. The coefficient of variation in mean fork length of adults is about half that for smolts (Figure 16.10), that is adults are more similar in size than smolts.

Relationships between smolts and adults are complicated by the inability to identify the sea age that smolts are destined to become. There are potentially five types or sea ages of virgin

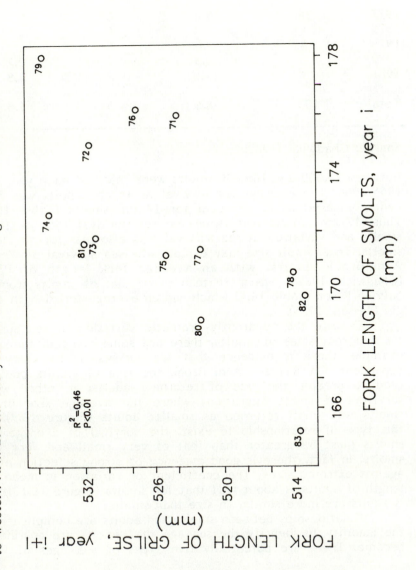

Figure 16.9: Relationship between fork length of smolts and of 1SW salmon returning the following year to Western Arm Brook. The year of the smolt migration is indicated

Table 16.11: Sea survival of smolts released in Swedish rivers at two different sizes: 14 cm and 17 cm. Survival rates were calculated from data presented in Larsson (1977). The numbers of smolts released ranged from 23,000 to 109,000

| | % survival | |
Year	14 cm smolts	17 cm smolts
1955	10	15
1956	3	10
1957	9	16
1958	7	13
1959	15	27
1960	12	23
1961	6	14
1962	4	10
1963	7	12
1964	5	11
1965	6	11
1966	8	15
Mean	8	15

spawners that a given smolt can develop into and it is unlikely that any salmon stock is a pure strain of only one type; rather it will contain a mixture of sea ages. Jack salmon or 0 sea-winter (0SW) salmon spend one summer at sea before first spawning. This type is rare, but it has been documented in Little Codroy River (Murray, 1968a), Western Arm Brook (Chadwick, 1982a) and in rivers of Ungava Bay (Robitaille, Côté, Shooner and Hayeur, 1986). Grilse or 1SW salmon spend one winter at sea and are the predominant type of salmon in rivers of insular Newfoundland and the inner Bay of Fundy (Figure 16.1). Most of the rivers in Nova Scotia, New Brunswick, Prince Edward Island and Labrador support populations of 2SW salmon. The two remaining types of salmon are those which first spawn after three and four years at sea. These types of salmon stocks are found throughout Quebec and Atlantic Canada.

The final relationship that I will discuss is between ovarian development of smolts and sea age of returning adults. Our research group has just completed two years of research on this topic. Recently we compared ovarian development of Atlantic salmon in 14 groups of smolts and parental sea age of smolts explained most of the variation in ovarian development (Figure 16.11). Smolts from 1SW parents had > 10 per cent stage 5 and 6

Figure 16.10: Annual changes in the coefficient of variation of fork length for smolts and 1SW salmon (grilse), in the following year sampled in Western Arm Brook. The year of the smolt migration is indicated

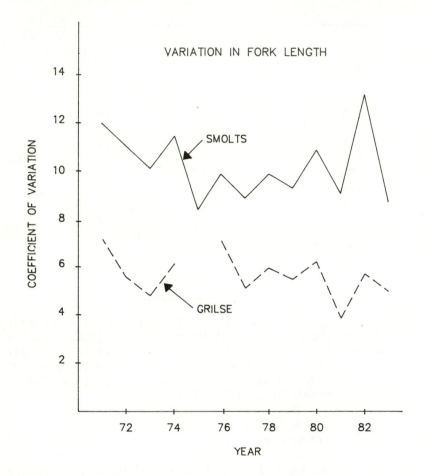

Figure 16.11: Mean ovarian development for smolts sampled in various 1SW, 2SW and 3SW salmon stocks in Atlantic Canada in 1984-85. The ordinate indicates the mean oocyte stage of smolts; the abscissa indicates the sea age smolt parents. See Chadwick et al. (1986) for details

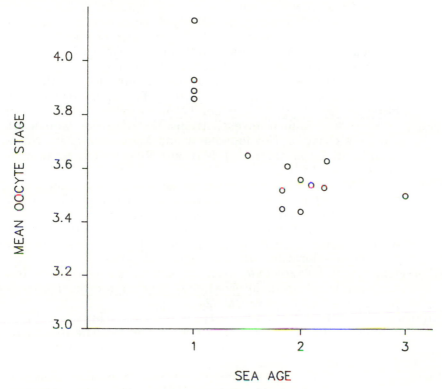

oocytes in their ovaries, while smolts from 3SW parents had none. Annual within-stock variation and covariance with freshwater age were not significant. Hatchery-reared smolts had similar ovarian development to their wild counterparts. There was also a significant, positive correlation between fork length of smolts and ovarian development within groups. These results indicate that sea age of Atlantic salmon can be at least partly determined from close examination of the smolts.

In summary, I have shown briefly that numbers of smolts can be used to predict returns of adults. In order to understand variation and sea age in Atlantic salmon it is necessary to understand the variation in smolts which can best be accomplished by obtaining complete counts of smolts and adults at representative rivers.

REFERENCES

Anon. (1962) Report of the Biological Station, St John's New-foundland, for 1961-62. W. Templeman, Director. Fisheries Research Board of Canada Atlantic Salmon, Appendix 18, p. 98-103 by A.R. Murray

Belding, D.L. and Préfontaine, G. (1961) A report on the salmon of the north shore of the Gulf of St Lawrence and of the northeastern coast of Newfoundland. Contribution from the Department of Fisheries, Québec, 82, 104 pp

Bilton, H.T. (1984) Returns of Chinook salmon in relation to juvenile size at release. Canadian Technical Report of Fisheries and Aquatic Science, 1245, 33 pp

Blair, A.A. (1943) Salmon investigations. 2: Atlantic salmon of the east coast of Newfoundland and Labrador, 1939. Newfoundland Department of Natural Resources, Research Bulletin No. 13 (Fisheries), 21 pp

Chadwick, E.M.P. (1982a) Dynamics of an Atlantic salmon stock (Salmo salar) in a small Newfoundland river. PhD Thesis, Memorial University of Newfoundland, 267 pp

Chadwick, E.M.P. (1982b) Recreational catch as an index of Atlantic salmon spawning escapement. International Council for the Exploration of the Sea CM 1982/M: 43, 5 pp

Chadwick, E.M.P. (1982c) 1SW Atlantic salmon harvests predicted one year in advance. International Council for the Exploration of the Sea CM 1982/M: 20, 10 pp

Chadwick, E.M.P. (1984) Prediction of 1SW Atlantic salmon returns Statistical Area N, 1984. Canadian Atlantic Fisheries Scientific Advisory Committee Research Document 83/84, 8 pp

Chadwick, E.M.P. (1985) Fundamental research problems in the management of Atlantic salmon, Salmo salar L., in Atlantic Canada. Journal of Fish Biology, 27 (supplement A), 9-25

Chadwick, E.M.P., Randall, R.G. and Léger, C. (1986) Ovarian development of Atlantic salmon (Salmo salar) and age at first maturity. In D.J. Meerburg (ed.), Salmonid age at maturity. Canadian Special Publication of Fisheries and Aquatic Science, 89, pp 15-23

Chadwick, E.M.P., Reddin, D.G. and Burfitt, R.F. (1985) Fishing and natural mortality rates for 1SW Atlantic salmon. International Council for the Exploration of the Sea CM 1985/M, 18, 11 pp

Jessop, B.M. (1975) Investigation of the salmon (Salmo salar) smolt migration of the Big Salmon River, New Brunswick, 1966-72. Technical Report Series. No. MAR T-75-1. Resource Development Branch. Maritimes Region, 57 pp

Jessop, B.M. (1986) Atlantic salmon (Salmo salar) of the Big Salmon River, New Brunswick. Canadian Technical Report of Fisheries and Aquatic Science, 1415, 50 pp

Larsson, P.O. (1977) Size dependent mortality in Salmon smolt plantings. International Council for the Exploration of the Sea CM 1977/M, 43, 8 pp

Lear, W.H. (1973) Size and age composition of the 1971 Newfoundland - Labrador commercial salmon catch. Fisheries Research Board of Canada Technical Report, 392, 43 pp

Lear, W.H. and May, A.W. (1972) Size and age composition of the Newfoundland and Labrador commercial salmon catch. Fisheries Research Board of Canada Technical Report, 353

Lear, W.H., Batton, W.N. and Burfitt, T.R. (1976) Seasonal trends in the biological characteristics of salmon from three selected areas along the southeast and northeast Newfoundland coasts. Fisheries Marine Service Technical Report, 601, 75 pp

Lindsay, S.T. and Thompson, H. (1932) Biology of the salmon (Salmo salar L.) taken in Newfoundland waters in 1931. Report Newfoundland Fisheries Research Commission, 1(2), 80 pp

Marshall, T.L. (1984) Status of Saint John River, New Brunswick, Atlantic salmon in 1984 and forecast of returns in 1985. Canadian Atlantic Fisheries Scientific Advisory Committee Research Document 84/84, 24 pp

Murray, A.R. (1968a) Numbers of Atlantic salmon and brook trout captured and marked at the Little Codroy River, Newfoundland, counting fence and auxiliary traps, 1954-63. Fisheries Research Board of Canada Technical Report, 84, 135 pp

Murray, A.R. (1968b) Smolt survival and adult utilization of Little Codroy River, Newfoundland, Atlantic salmon. Journal of Fisheries Research Board of Canada, 25(10), 2165-218

Pippy, J. (1982) Report of the Working Group on the interception of mainland salmon in Newfoundland. Canadian MS Report on Fisheries Aquatic Sciences, 1654, 196 pp

Pratt, J.D., Hare, G.M. and Murphy, H.P. (1974) Investigation of production and harvest of an Atlantic salmon population, Sandhill River Labrador. Resource Development Branch, Newfoundland Region Technical Report Series No. NEW/T-74-1, 27 pp

Randall, R.G., Chadwick, E.M.P. and Schofield, E.J. (1985a) Status of Atlantic salmon in the Miramichi River, 1984. Canadian Atlantic Fisheries Scientific Advisory Committee Research Document, 85/2, 21 pp

Randall, R.G., Chadwick, E.M.P. and Pickard, P.R. (1985b) Status of Atlantic salmon in the Restigouche River, 1984. Canadian Atlantic Fisheries Scientific Advisory Committee Research Document, 85/1, 18 pp

Ritter, J.A. (1977) Relationship between smolt size and tag return rate for hatchery-reared Atlantic salmon (Salmo salar). International Council for the Exploration of the Sea CM 1977/M: 27, 6 pp

Robitaille, J.A., Côté, Y., Shooner, G. and Hayeur, G. (1986) Growth and maturation patterns of Atlantic salmon Salmo salar in the Kiksoak River, Ungava, Quebec. In D.J. Meerburg (ed.), Salmonid age at maturity. Canadian Special Publication of Fisheries and Aquatic Science, 89, pp. 62-9

Chapter Seventeen

MEASUREMENT OF ATLANTIC SALMON SPAWNING ESCAPEMENT[1]

P. Prouzet[2] and J. Dumas[3]
2 IFREMER, France
3 INRA, France

Laboratoire d'Ecologie des Poissons, Station d'Hydrobiologie, BP 3, Saint-Pée-Sur-Nivelle, 64310 ASCAIN

17.1   INTRODUCTION

In common with all migratory salmonids, the adult Atlantic salmon always returns to the place where it was born (the 'homing' phenomenon). This characteristic makes it possible to establish, for any given river, a relationship between spawning stock size and the progeny (recruits) produced. This link which is generally described as the 'stock recruitment relationship', is peculiar to each stock and depends amongst other things on the characteristic dynamics of the population and territorial productivity of the environment (particularly in freshwater).

Although all salmonid populations present different patterns in their stock recruitment relationship, they all have the same need for an optimal number of spawners on their initial spawning grounds in order to achieve an increase in production of the stock in question in the future and to maintain it at the highest possible level.

Thus, the acknowledged aim of the manager will be to allow the necessary number of spawners into the spawning grounds to ensure a maximum production of smolts. To achieve this aim, he will first have to calculate the optimal number. After this, he will need to estimate the density of stock during the terminal phase of their migration.

It can be appreciated that the estimation of these two parameters is not at all easy. It calls for the use of a certain number of techniques which can be divided into two categories: direct and non-direct evaluation.

The purpose of this paper is to describe them and to compare their respective advantages and disadvantages, as well as their degree of accuracy.

1.   The French version of this paper may be obtained from Dr. P. Prouzet at the above address.

The debate will deal with a possible choice of technique and replace in a more general framework the role of the manager whose decisions will depend not only on the accuracy of these estimations but also on their forecasts.

17.2   ESTIMATION OF REQUIRED NUMBER OF SPAWNERS

Direct evaluation

The production of Atlantic salmon is closely linked to the amount of space and food in the salmonid production area (Symons, 1968, 1971; Chapman, 1966; Slaney and Northcote, 1974; Kalleberg, 1958; Heland, 1977). Thus, for these areas, the optimal density of eggs or smolts necessary to saturate the territory can be determined (Le Cren, 1970; Elson, 1975; Symons, 1979; Buck and Hay, 1984).

Because of this, the calculation of the number of spawners required will take into account these two restrictions:

$$Ns = \text{Area} - \text{OED} / \text{FEC} \qquad (17.1)$$

where Ns, = number of spawners required; Area, spawning area in m^2; OED, optimal egg density per m^2; FEC, fecundity per spawner.

The use of formula (17.1) involves the calculation of three variables.

The production area, the measurement of which requires a long inspection of cartographic or photographic documents and field observations. The surface area of the nursery and production zones is calculated from criteria defined by various authors such as Elson (1975), Symons and Heland (1978), Bagliniere and Champigneulle (1982).

The density of eggs laid, the importance of which usually follows certain norms. For example, the norm used for Canadian rivers is that established by Elson (1975): 2.4 eggs/m^2 (Chadwick, 1985), although Chadwick (1982a) points out that this is inadequate for the rivers of Newfoundland. Elson and Tuomi (1975) estimate it at 1.68 eggs/m^2 for the River Foyle in Ireland, whereas Symons (1979) suggests modifying these estimations according to the age of the smolts produced (reduction of the optimal rate of eggs laid in ratio to the increased longevity of Atlantic salmon in freshwater). It is therefore advisable to be very conservative as to the extrapolation of a set norm in a given environment, and, even for neighbouring rivers it will be necessary to check that any local disturbance of the environment does not lead to significant divergence from this norm. In which case, it would be advisable to base the estimation on observations carried out on fish population and completed analysis of territorial needs and behaviour (Kalleberg, 1958; Allen, 1969; Symons and Heland, 1978).

The fecundity of the spawning population, which requires some knowledge of the sex ratio for each type of salmon (this can be obtained in a reliable way from newly ascended individuals using serodiagnosis (Prouzet, Le Bail and Heydorff, 1984)), the break-down of the different sea-age groups within the ascending population, and lastly the average weight and relative fecundity (number of eggs/kg/female) of each of the stock constituents (salmon and early or late grilse). Obtaining these variables requires accurate observations either using a sample catch (on condition that the fishing period covers the whole of the ascent period) or using sample individuals from traps.

Using this approach it is possible to estimate the required number of spawners. However, this evaluation is subject to an uncertainty which can be significant, bearing in mind the errors which might be made in the assessment of the principal variables involved in the calculation of this number (cf. appendix).

Non-direct evaluation

The required number of spawners can also be calculated from a stock recruitment relationship established with the aid of statistics bearing on catches, or catches by unit effort (CPUE) carried out by commercial or sport fisheries, as well as on their production rate. Amongst the models used to establish this connection, those developed by Beverton and Holt (1957) and Ricker (1958) are the most widely known.

For the migratory salmonids, the basic model which is most used is Ricker's exponential model (1958, 1980). It is laid out in the following general manner

$$R = P.e \ a \ (1 - P/Pr)$$

where
- R number of recruits;
- P abundance of parental stock;
- a parameter of the form of the curve, equal to Pr/Pm;
- Pr abundance of spawners required to replace stock;
- Pm abundance of spawning stock to ensure maximum level of recruitment

Thus by using this rule of thumb model it is possible to estimate the size of the parental stock (Pm) which maximises the catches at long term (ultimate aim of the manager). Hilborn (1985) also shows that the parameter Pm can be obtained quite simply:

$$Pm = Pr (0.5 - 0.07a) \text{ when } 0 < a < 3$$

The interest of working out such a curve lies in the overall diagnosis which can be deduced as to the condition of the stock concerned and in particular to its level of productivity. This takes into account the Maximum Sustainable Yield (MSY) which brings to light the excess production per spawner and the maximum recruitment per spawner (Mr).

This type of model was currently used by several authors for the study of Atlantic salmon stocks (Ricker, 1958; Van Hyning, 1973; Dahlberg, 1973) as well as by Dempson (1980) for the Labrador Atlantic Fishery.

It should be pointed out that to obtain a certain degree of accuracy in the stock recruitment curve it is necessary to have a long series of fishing statistics for a stock, which if possible, has been subject to important fluctuations in abundance. Finally, the use of this deterministic model leads to the acquisition of average curves. These estimations can be improved by the use of a stochastic version obtained by a simple modification of the basic model (Hilborn, 1985).

17.3   ESTIMATION OF NUMBER OF SPAWNERS PRESENT

Direct evaluation

The direct methods of estimating the abundance of the number of potential spawners present are extremely varied, ranging from a simple visual count to enumeration using structures of different degrees of sophistication, automated or not. These techniques have been developed both in North America and Europe (Hellawell, 1973; Lawson, 1974; MacGrath, 1975; Ruggles, 1975, 1980; Cousens, Thomas, Swann and Healey, 1982).

These methods can be divided into three major categories: (1) visual counts without use of traps; (2) counts with traps which permit manipulation of the fish; (3) automatic counts.

Visual counts

Several techniques have been perfected: observation of holding areas or spawning grounds on foot, by divers, from boats or from planes, or by visualisation of fish passes. They all require collection of additional data to estimate the required number of spawners, and in particular data on the characteristics of the stock (size of the different age groups, as well as their respective average weights and sex ratio), the relation between the number of females and the number of redds, the difference between salmon spawning grounds and those of other salmonids (particularly sea trout).

Observation of spawning grounds from the bank

This is generally done on foot. Redd counts (characteristic light patches on the river bed) are made from the banks once or twice a week during the whole of the spawning season.

This method, widely used in North America for the Pacific anadromous salmonid species (Cousens, et al., 1982) is also used in Europe for Atlantic salmon (Bagliniere, Champigneulle and Nihouarn, 1979; Hay, 1984). It applies to rivers which are not very turbid, shallow and not too wide (20m maximum).

This type of evaluation is easy, inexpensive, but requires skilled personnel. Carried out under adequate conditions (clear water, single species migratory stock) the estimation can be quite accurate. Hay (1984) draws attention to a maximum fluctuation of - 10 per cent to + 15 per cent in the Girnock Burn in Scotland between the counts made using this method and those obtained by downstream traps. However, it should be noted that according to Cousens et al. (1982) this method tends to underestimate the number of redds.

Counts made by divers or from a boat

These are usually reserved for narrow rivers with difficult access (Cousens et al., 1982). They turn out to be as simple and inexpensive as the counts made on foot. They are subject to the same constraints and lead to the same underestimation of the number of redds. The method is used in France for the spawning grounds of the Haut Allier and the Gave d'Oloron (Cuinat and Bousquet, personal communication) as well as in Gaspesie for the spawners grouped in the pools in September, where according to Côté and Bastien (1982) the count obtained made it possible to estimate their number as closely as 10 or 20 per cent.

Aerial counts

Reserved for wide rivers without steep embankments and where the preceding techniques are ineffectual. This method is widely used for census of the spawners of Pacific anadromous salmonids and their spawning grounds (Nielson and Geen, 1981; Cousens et al., 1982). It is used in Norway for enumeration of the spawning grounds of migratory salmonids (Heggberget, Haukebo and Voie-Rosvoll, 1986). These authors obtained abundance ratings of salmon population by associating the aerial observations with observations made on the spawning grounds (identification of the eggs in the redds). Obviously, this is an expensive method, but one which enables the coverage of large expanses of clear-water rivers.

Enumeration in fish passes

This process is widely used on the coast of the United States (Cousens et al., 1982) and Canada (Prouzet, 1983). The same technique is used for Atlantic salmon on the east coast of Canada (Bagliniere, 1983). It is also used, for example, to calibrate the automatic fish counters used in the British Isles (MacGrath, 1975; Bussell, 1978; Dunkley and Shearer, 1983) and at fish passes in Scotland (Mills and Graesser, 1981). The type of information obtained as to the abundance of spawners is more accurate than the preceding methods (particularly as regards identification of species) but in order to assess the number of spawners, it also needs an estimate of the angling and natural mortality rate after the count in the pass.

Trap counts

Trapping can be used either on the whole of the ascending population, or a part of it. In this case, it becomes a method of indirect approach which calls for the use of additional techniques to estimate the number of available spawners (e.g. marking and recapture).

Interception trapping of the whole ascending population

The principle is to integrate a capture device in a construction which deflects the ascending population.

A variety of devices are used (Blair, 1957; Clay, 1961; Pyefinch and Mills, 1963; Murray, 1968; Kerswill, 1971; MacGrath, 1974, 1975; Ruggles, 1975, 1980; Cousens et al., 1982). Whichever device is used its design should avoid any delay in the access of the fish to their spawning grounds as this might have harmful consequences on the distribution of the spawners (Ruggles, 1975, 1980; Craig, 1980). Furthermore, maintenance (raising the trap, cleaning) should be rigorous so that the trapped fish do not suffer too much stress. This technique is simple and gives a very reliable enumeration of individuals as well as an understanding of their different characteristics. It is used as a standard for the testing of other counting methods. An estimation of the number of spawners can only be made if the angling and natural mortality rate after trapping is known. The cost of the construction used depends on the degree of sophistication of the structure, its size and whether or not a deflection barrier is built.

330

Partial trapping

The structures used are generally inspired by the machines used in commercial fisheries. Ruggles and Turner (1973) and Ruggles (1975, 1980) described those of the Miramichi River in New Brunswick. They are generally used for sampling and for gathering data on catches, but also for evaluation of the production rate in order to estimate the abundance of population returns (marking-recapture). It is a technique which is relatively inexpensive and which can be integrated in the non-direct evaluation techniques developed.

Automatic counts

Several types of automatic counter have been perfected:

(1)   electro-mechanical (Jackson and Howie, 1967).
(2)   photo-electric (Bell and Armstrong, 1970).
(3)   resistivity counters (Hellawell, 1973; Lawson, 1974; Hellawell, Leatham and Williams, 1974; MacGrath, 1975; Bussel, 1978; Cousens et al., 1982; Dunkley and Shearer 1983; Mann, Hellawell, Beamont and Williams, 1983; Struthers and Stewart, 1984, 1985).
(4)   sonar (Braithwaite, 1974; Hellawell, 1973; Cousens et al., 1982)

Only the last two have known any real development for fish counts. Note that principally these counters only furnish a gross abundance which should be completed by information obtained by additional studies of the migratory population (sex-ratio, age and size structures).

Resistivity counters

These operate on a simple principle: a fish passing through a weak electric field created by submerged electrodes unbalances a Wheatstone bridge. The signal emitted is amplified and then processed. From this signal, it is possible to enumerate either fish which are over a certain size, or distinguish two or three different sizes. The majority of counters show the direction of the fish and eliminate the impulses due to the passage of drifting objects.

The interest of this type of counter is the possibility of enumerating not only the ascending fish, but also those descending. On the other hand, frequent adjustment is necessary to maintain them in good working order because the sensitivity of the apparatus is affected by changes in the resistivity of the water, especially if this is only slightly mineralised. It is therefore important to readjust the machine with either a video camera or by simultaneous trapping.

Counts made with tubular electrodes are the most accurate: between 90 and 95 per cent of the individuals are accounted for by units installed in Ireland (MacGrath, 1975), between 90 per cent at Cluny (Struthers and Stewart, 1985) and 100 per cent at Pitlochry on the River Tummel in Scotland (Bussell, 1978). Cousens et al. (1982) point out, however, that it is possible to underestimate the number of individuals by as much as 25 per cent when too many spawners pass through the tube at the same time.

Electrode counters made up of bands fastened down on the top of dams can, under optimal conditions, provide accurate results: 95 per cent of salmon and trout in the River Frome in England are accounted for in this way (Hellawell et al., 1974; Mann et al., 1983).

Sonar

This technique has been used very little for Atlantic salmon (Hellawell, 1973; Braithwaite, 1974). It has been mainly developed for counting Pacific salmon (Cousens et al., 1982). Its principle resides in the reception and analysis of an echo produced by a fish as it passes through a vertical ultrasonic beam coming from a transmitter placed across the river bed. In optimum conditions, according to Cousens et al. (1982) the accuracy of the counts is of the order 90 to 95 per cent.

The only difficulty in the use of this type of machine seems to reside in the choice of a suitable place for its installation, but also its adjustment and the interpretation of the data recorded.

Non-direct evaluation

The fishing statistics gathered from the commercial and sport fisheries can be used as such (CPUE, comparative abundance rating completed by an estimation of the productor rate) or in conjunction with marking campaigns which give an idea of the size of the population produced (marking-recapture techniques).

Use of CPUE

As a rule, catch per unit effort (CPUE) is used to measure the comparative abundance of a stock. This comparative abundance rating makes it possible to follow the fluctuations in the size of a fish population, on condition that certain basic theories are complied with.

As a matter of fact, the expression of the catch (C) as defined by Gulland (1969):

$$C = qf \, N/A$$

where q =   the constant expressing the stock capture potential or vulnerability,

      f =   fishing effort per operating unit,

     N =   average abundance of stock produced,

     A =   area inhabited by stock produced,

shows that the size of the catch is dependent on three parameters: density of stock N/A, the fishing effort deployed and the capture potential of the stock. Taking CPUE as the comparative abundance rating one is independent of the variation in effort from one year to the next but not yearly fluctuations in vulnerability (q).

If the theories of constancy of vulnerability and constancy in the zone investigated (in time and space) are respected, then the CPUE establishes a rating enabling the study of the fluctuations in abundance of the fish populations.

In spite of these reservations, the CPUE is widely used for an abundance rating of anadromous salmonids (Cousens et al., 1982). For Atlantic salmon, rod-fishing catches are usually used for the parental abundance rating (Dempson, 1980) working from the additional theory that fishing effort does not vary from one year to another. In several cases, this seems to be confirmed because a reasonable balance between rod-fishing catches and stock abundance has been observed (Elson, 1974; Pomerleau, Côté and Migneault, 1980; Chadwick, 1982b), although Piggins (1976) remarks on a slight correlation between catches and stock in Lough Feeagh in Ireland (low average production rate 8.8 per cent).

The relationship between CPUE or catches, and the number of spawners is carried out in different ways: either from models, taking into account the reference data linking the CPUE with the abundance of spawners on the spawning grounds (Cousens et al., 1982) or from the assessment of the production rate of the fishery as defined by different mark-recapture techniques (Ruggles and Turner, 1973; Elson, 1974) or by counting the spawners on the spawning grounds (Pomerleau et al., 1980; Jensen, 1981b; Cousens et al., 1982).

It will be easily understood that an estimation of the abundance of spawners by this method is based on the accuracy of the fishing statistics used, as well as the conditions under which the mark-recapture techniques are used to determine the production rate (see below).

Mark-Recapture Estimates

The estimation of the size of a population by marking is based on the fact that the number of individuals in a population can be calculated by marking sample individuals which are then released amongst the population. Given the number of marked individuals (M), the number of individuals examined to find the marks (C) and the number of recaptured individuals (R) it is possible to estimate the size (N) of the population:

$$N = MC/R$$

This basic method due to Petersen was improved on by different authors (cf. Ricker, 1980).

For ascending migratory salmonids, marking takes place during one season only. For this type of marking, different methods have been proposed (Petersen, Petersen modified, Chapman, Bailey, Schnabel, Ricker or Schaefer). Amongst these, the Schaefer method seems particularly appropriate because it can be carried out on a cross-section of the population in cases where there is not a homogeneous mixture of marked and unmarked fish (Ricker, 1980; Cousens et al., 1982). In addition, certain expedients produced by different factors (size of sample, selection, recruitment effect) can be estimated by different methods (Parker, 1955; Robson and Regier, 1964; Jensen, 1981a).

It should be pointed out, however, that the accuracy of the estimates obtained by these techniques is dependent on the basic theories (homogeneous distribution of the marked animals, identical vulnerability of the marked animals, constancy in the balance between marked and unmarked animals during the recapture period), which are very rarely respected. For this reason, the method is rarely used for abundance estimations, except for big streams and rivers where no other method can be used (Cousens et al., 1982). It is used for Atlantic salmon to determine the production rate of a fishery (Ruggles and Turner, 1973; Elson, 1974). The estimation of the production rate using this technique is obviously subject to the same constraints as those controlling the reliability of abundance rating. Any departure from the basic theories leads to a considerable lack of precision, as pointed out by Elson (1974) on the Miramichi.

17.4 DISCUSSION AND CONCLUSION

Choice of a technique for enumerating spawners

Table 17.1 attempts a comparison between different methods of estimating the abundance of individuals, with, it is true, all the

Table 17.1: Comparison between different estimation methods

■ :yes; ` :no   +:low; ++:medium; +++:high

METHODS	APPLICATION FIELD	PRECISION	FISHING MORTALITY	NATURAL MORTALITY	EXPLOITATION RATE	STOCK CHARACTERISTICS	RELATIONSHIP Nber REDDS /Nber SPAWNERS	DISTINCTION BETWEEN SPECIES	COST
1/DIRECT EVALUATION									
1.1 VISUAL COUNTS WITHOUT TRAPPING									
# ON FOOT	ACCESSIBLE RIVERS NOT VERY TURBID SHALLOW AND NOT TOO WIDE EXPERIMENTED TEAM	++				■	■	■	+
# BY DIVING OR FROM A BOAT	HARDLY ACCESSIBLE RIVERS NOT VERY TURBID SHALLOW AND NOT TOO WIDE EXPERIMENTED TEAM	++				■	■	■	+ A ++
# IN A FISHPASS	GLASS PANEL OR CLEAR BOTTOM CHANNEL CLEAR WATER	++ A +++	■	■		■			++ A +++
1.2 TRAP COUNTS									
# PARTIAL TRAPPING	MODIFIED OR NOT FISHING GEARS TRAP IN A FISHWAY USED FOR ESTIMATION OF EXPLOITATION RATE	+ A ++	■	■					+ A ++
# TOTAL TRAPPING	FIXED STRUCTURES: LADDERS, SLUICE GATES, TRAPS,..	+++	■	■					+++
1.3 AUTOMATIC COUNTS									
# RESISTIVITY COUNTER	FIXED STRUCTURES: LADDERS, SLUICE GATES, TRAPS,.. STABLE WATER RESISTIVITY	++ A +++	■	■		■	■	■	++
# SONAR	BOTTOM OF THE RIVERS UPPER PART OF DAM	++ A +++	■	■		■		■	+++
2/ NON DIRECT EVALUATION									
2.1 CPUE	RELIABLE FISHING STATISTICS ESTIMATION OF EXPLOITATION RATE STATISTICS CONSTRAINTS	++		■	■				+
2.2 MARKING/ RECAPTURE	RELIABLE FISHING STATISTICS GOOD RETURN RATE FOR THE TAGS STATISTICS CONSTRAINTS	++	■	■	■				++

imperfections and approximations involved in this type of exercise. However, from the analysis of this table, certain points become evident, and in particular the following.

There is no 'miracle technique'; all the methods evoked need, to different degrees, additional sources of information which necessitate bringing into operation a variety of techniques.

If total trapping seems to be the most adequate solution, it is not always possible to use it, for several reasons (technical, financial or even political). In addition, it requires important maintenance work which can only be carried out to the detriment of other tasks such as watch-keeping and collecting the fishing statistics indispensable to the evaluation of the potential number of spawners.

The use of fishing statistics in conjunction with the setting up of marking schemes makes it possible to obtain estimates of the number of spawners, certainly not as accurate as those obtained by the use of certain direct evaluation methods, but which contribute to an improvement in the network of statistics on the gathering of fishing data. In addition, these methods of non-direct evaluation make it possible to gather an important amount of additional data which is not obtained by the first category of methods, or only at the price of an increase in stress (as in the case of trapping).

Need for an abundance forecast

Whatever degree of accuracy is obtained concerning the estimation of the abundance of spawners, it should not be forgotten that good management is only possible when a more or less long-range forecast can be made of the abundance of future generations, in order to adapt production to the possible increases in stock, by the use of appropriate regulations.

In other words, it is better to lose a degree of accuracy, than not to make the effort of a middle-range forecast. Thus, as noted by Paulik and Greenough (1969), accuracy as to the composition of a stock increases as data gathering improves, but inversely, the manager's freedom of action is diminished as his decisions will affect only a limited part of the stock because they will become effective so much later. This type of short-range estimated rating (CPUE of commercial fisheries on the coast for the current year), is used for the Pacific salmon populations (which ascend in a more closely grouped fashion than the Atlantic salmon) and allows partial management of the terminal fisheries.

A middle-range forecast is therefore desirable. For Atlantic salmon, the abundance of grilse can be used to forecast a season in advance the yields of two sea-winter salmon (Scarnecchia, 1984; Chadwick, 1985).

Over a longer range, the estimated survey ties up with the conceptions set forth by Wroblewski (1983) who underlines that researchers have recourse to the model more often for exploratory purposes (improving on a theory) than for forecast purposes. Several authors have attempted an indirect approach using stochastic models (Larkin and Hourston, 1964; Ward and Larkin, 1964; Allen, 1973; Gros and Prouzet, 1984, 1986). Certain workers such as Evans and Dempson (1986), have, using a simplified model of the biological cycle of Atlantic salmon, made a study of the sensitivity of this model to the change in different factors (development, environment). These latter authors have thus shown that measures taken to protect the habitat have more lasting effects on the stability of the population than modification of the regulations.

In addition, the basis for the regulations does not only take into account biological requirements (very often complicated by mixing of species in the fisheries) but also the socio-economical or political imperatives.

This leads to the advice of fishery biologists being taken into consideration in a variety of ways.

All things considered, the manager can obtain a certain number of estimates, the problem is not to receive them too late to make a minimum of decisions. Moreover, these will be partly impeded by events outside his sphere of intervention (fisheries in coastal or international waters). Without wishing to be too pessimistic like Larkin (1981) in thinking that 'Salmon management is a cul de sac of frustration' there is no alternative but to acknowledge that the setting up of medium- or long-term management which will permit a maximum number of spawners to colonise the spawning grounds year after year, is still a long way off, chiefly when the resources are actively exploited.

REFERENCES

Allen, K.R. (1969) Limitations on production in salmonid populations in streams. Symposium on salmon and trout in streams, H.R. MacMillan Lectures in Fisheries, University of British Columbia, 3-18

Allen, K.R. (1973) The influence of random fluctuations in the stock-recruitment relationship on the economic return from salmon fisheries. Fish Stocks and Recruitment. Rapports et Procès Verbaux du Conseil International pour l'Exploration de la Mer, 164, 350-9

Bagliniere, J.L. (1983) Connaissance et gestion des populations de Saumon Atlantique (Salmo salar L.) au Quebec. Bulletin scientifique et tecnique INRA, Hydrobiologie, 14, 23 pp

Bagliniere, J.L. and Champigneulle, A. (1982) Densité des populations de Truite Commune (Salmo trutta L.) et de juvéniles de Saumon Atlantique (Salmo salar L.) sur le cours principal du Scorff (Bretagne): Préférendums physiques et variations annuelles (1976-1980). Acta Oecologica Applicata, 3 (3), 241-56

Bagliniere, J.L., Champigneulle, A. and Nihouarn, A. (1979) La fraie du Saumon Atlantique (Salmo salar L.) et de la Truite Commune (Salmo trutta L.) sur le bassin du Scorff. Cybium, série (3), 7, 75-96

Bell, W.H. and Armstrong, M.C. (1970) A photoelectric fish counter. Fisheries Research Board Canadian Technical Report, 215, 23 pp

Beverton, R.J.H. and Holt, S.J. (1957) On the dynamics of exploited fish populations, UK Ministry for Agriculture and Fisheries, Fisheries Investigations, Series 2, 19, 533 pp

Blair, A.A. (1957) Counting fence of netting. Transactions of American Fisheries Society, 86, 199-207

Braithwaite, H. (1974) Sonar fish counting. EIFAC 74/I/Symposium, 17, 9 pp

Buck, R.J.G. and Hay, D.W. (1984) The relation between stock size and progeny of Atlantic salmon, Salmo salar L., in a Scottish stream. Journal of Fish Biology, 23, 1-11

Bussell, R.B. (1978) Fish counting stations. Notes for guidance, their design and use. Department of Environment, London, 97 pp

Chadwick, E.M.P. (1982a) Stock-recruitment relationship for Atlantic salmon (Salmo salar) in a Newfoundland river. Canadian Journal of Fisheries and Aquatic Sciences, 39, 1496-501

Chadwick, E.M.P. (1982b) Recreational catch as an index of Atlantic salmon spawning escapement. International Council for the Exploration of the Sea, C 1982 M: 43, 4 pp

Chadwick, E.M.P. (1985) Fundamental research problems in the management of Atlantic salmon, Salmo salar L., in Atlantic Canada. Journal of Fish Biology, 27, 9-25

Chapman, D.W. (1966) Food and space as regulators of salmonid populations in streams. American Naturalist, 100, 345-57

Clay, C.H. (1961) Design of fishways and their facilities. Queen's Printer, Ottawa, 301 pp

Coté, Y. and Bastien, Y. (1982) Fishing success and angling mortality rates of Gaspe salmon stocks as estimated by creel census and spawner counts. American Fisheries Society Atlantic Chapter Annual Meeting, 1981, 22 pp

Cousens, N.B.F., Thomas, G.A., Swann, G.G., Healey, M.C. (1982) A review of salmon escapement estimation techniques. Canadian Technical Report of Fisheries and Aquatic Sciences, 1108, 122 pp

Craig, J.F. (1980) 5. Sampling with traps In Backiel, T. and Welcomme, R.L. (eds) Guidelines for sampling fish in inland waters, EIFAC, Technical papers, 33, 143-58

Dahlberg, M.L. (1973) Stock and recruitment relationships and optimum escapements of Sockeye salmon stocks of the Chignik lakes, Alaska. Fish stocks and Recruitment, Rapport et Procès Verbaux des Réunions du CIEM, 164, 98-105

Dempson, J.B. (1980) Application of a stock recruitment model to assess the Labrador Atlantic salmon fishery. International Council for the Exploration of the Sea, C 1980 M: 28, 15 pp

Dunkley, D.A. and Shearer, W.M. (1983) The reliability of population data obtained from the use of a resistivity counter in a major salmon river. International Council for the Exploration of the Sea, C 1983 M: 25, 11 pp

Elson, P.F. (1974) Impact of recent economic growth and industrial development on the ecology of Northwest Miramichi Atlantic Salmon (Salmo salar). Journal of the Fisheries Research Board of Canada, 31 (5), 521-44

Elson, P.F. (1975) Atlantic salmon rivers, smolt production and optimal spawning: an overview of natural productions. International Atlantic Salmon Foundation Special Publications Series, 6, 96-119

Elson, P.F. and Tuomi, A.L.W. (1975) The Foyle fisheries: new bases for rational management. Special Report to the Foyle Fisheries Commission, Londonderry, Northern Ireland, pp. 1-194

Evans, G.T. and Dempson, J.B. (1986) Calculating the sensitivity of a salmonid population model. Canadian Journal of Fisheries and Aquatic Sciences, 43, 863-8

Gros, Ph. and Prouzet, P. (1984) Modèle stochastique de l'évolution des captures de Saumon de Printemps sur l'Aulne de

1973 à 1980. Conseil International pour l'Exploration de la Mer, Comité Anadrome-Catadrome, M, 3, 20 pp

Gros, Ph. and Prouzet, P. (1986) Modèle stochastique prévisionnel des captures de Saumons de Printemps (Salmo salar) dans l'Aulne (Bretagne): Eléments d'aménagement de la pêcherie. Conseil International pour l'Exploration de la Mer C 1986 M: 3, 20 pp

Gulland, J.A. (1969) Manuel des méthodes d'évaluation des stocks d'animaux aquations. Première partie - Analyse des populations. Manuel FAO de Science halieutique, 4, 160 pp

Hay, D.W. (1984) The relationship between redd counts and the numbers of spawning salmon in the Girnock Burn (Scotland). International Council for the Exploration of the Sea, C 1984 M: 22, 4 pp

Heggberget, T.C., Haukebo, T. and Veie-Rosvoll, B. (1986) An aerial method of assessing spawning activity of Atlantic salmon, Salmo salar L., and brown trout, Salmo trutta L., in Norwegian streams. Journal of Fish Biology, 8, 335-42

Heland, M. (1977) Recherches sur l'ontogénèse du comportement territorial chez l'alevin de Truite Commune, Salmo trutta L. Thèse de troisième cycle en Biologie Animale, Université de Rennes, 239 pp

Hellawell, J.M. (1973) Automatic methods of monitoring salmon populations. Special publications Atlantic Salmon Foundation, 4, 317-37

Hellawell, J.M., Leatham, and Williams, G.I. (1974) The upstream migratory behaviour of salmonids in the River Frome, Dorset. Journal of Fish Biology, 6, 729-44

Hilborn, R. (1985) Simplified calculation of optimum spawning stock size from Ricker's stock recruitment curve. Canadian Journal of Fisheries and Aquatic Sciences, 42, 1833-4

Jackson, P.A. and Howie, D.I.D. (1967) The movement of salmon (Salmo salar) through an estuary and fish pass. Journal of Fisheries Investigations, (A)2, 1-28

Jensen, A.L. (1981a) Sample sizes for single mark and single recapture experiments. Transactions of American Fisheries Society, 110, 455-8

Jensen, K.W. (1981b) On the rate of exploitation of salmon from two Norwegian rivers. International Council for the Exploration of the Sea, C 1981 M: 11, 7 pp

Kalleberg, H. (1958) Observations in a stream tank of territoriality and competition in juvenile salmon and trout (Salmo salar and Salmo trutta). Report of Institute for Freshwater Research Drottingholm, 39, 55-98

Kerswill, C.J. (1971) Relative rates of utilisation by commercial and sport fisheries of Atlantic salmon (Salmo salar) from the Miramichi river, New Brunswick. Journal of the Fisheries Research Board of Canada, 28(3), 351-63

Larkin, P.A. (1981) A perspective on population genetics and salmon management. Canadian Journal of Fisheries and Aquatic Sciences, 38, 1469-75

Larkin, P.A. and Hourston, A.S. (1964) A model for simulation of the population biology of Pacific salmon. Journal of the Fisheries Research Board of Canada, 21(5), 1245-65

Lawson, K.M. (1974) The electronic monitoring of salmon in Lancashire, England. EIFAC 74/1 Symposium, 15, 11 pp

Le Cren, E.D. (1970) The population dynamics of young trout (Salmo trutta) in relation to density and territorial behaviour. International Council for the Exploration of the Sea, Symposium on Stock and Recruitment, 11, 12 pp

MacGrath, C.J. (1974) The fish trapping installation at Salmon Leap, Furnace, Co. Mayo. EIFAC 74/1, 9 pp

MacGrath, C.J. (1975) A report on fish countng installations in Ireland. EIFAC, Technical report 23, 447-65

Mann, R.H.K., Hellawell, J.M., Beamont, W.R.C. and Williams, G.I. (1983) Records from the automatic fish counter on the river Frome, Dorset. 1970-1981. Freshwater Biological Association, Occasional Publication, 19, 100 pp

Mills, D.H. and Graesser, H.W. (1981) The Salmon Rivers of Scotland, Cassell, London, 339 pp

Murray, A.R. (1968) Estuarine net counting fence for trapping Atlantic salmon. Transactions of American Fisheries Society, 97(3), 282-6

Nielson, J.D. and Geen, G.H. (1981) Enumeration of spawning salmon from spawner residence time and aerial counts. Transactions of American Fisheries Society, 110,(4), 554-6

Parker, R.A. (1955) A method for removing the effect of recruitment on Petersen-type population estimates. Journal of Fisheries Research Board of Canada, 12, 447-50

Paulik, G.J., Greenough Jr, J.W. (1969) Management analysis for a salmon resource system. In K.E. Watt (ed.) Systems Analysis in Ecology, New York, 215-50

Piggins, D.J. (1976) Exploitation of grilse stocks by rod fishing in a lake system and subsequent spawning escapements, smolt productions and adult returns, 1970-75. International Council for the Exploration of the Sea, C 1976 M: 10, 6 pp

Pomerleau, C., Coté, Y. and Migneault, J.G. (1980) Répertoire de données relatives aux populations de Saumon Atlantique (Salmo salar) des rivières de la région du Bas Saint-Laurent et de la Gaspésie. 1- Guide méthodologique. Direction de la Recherche Faunique. Ministère du Loisir, de la Chasse et de la Pêche, 331 pp

Prouzet, P. (1983) Aménagement et gestion des populations de Salmonidés Anadromes en Colombie Britannique. Départe-ment Biologie Aquaculture Pêche, COB, rapport interne, 23 pp

Prouzet, P., LeBail, P.Y. and Heydorff, M. (1984) Sex ratio and potential fecundity of Atlantic salmon (Salmo salar L.) caught by anglers on the Elorn river (Northern Brittany, France) during 1979 and 1980. Fishery Management, 15,(3), 123-30

Pyefinch, K.A. and Mills, D.H. (1963) Observations on the move-ments of Atlantic salmon (Salmo salar L.) in the River Conon and the River Meig, Ross-shire. Salmon and Fresh-water Fisheries Research, 31, 24 pp

Ricker, W.E. (1958) Handbook of Computations for biological statistics of fish populations. Bulletin of the Fisheries Research Board of Canada, 119, 300 pp

Ricker, W.E. (1980) Calcul et interpretation des statistiques biologiques des populations de poissons. Bulletin de l'Office des Recherches sur les Pêcheries du Canada, 191F, 409 pp

Robson, D.S. and Regier, H.A. (1964) Sample size in Petersen mark-recapture experiments. Transactions of American Fisheries Society, 93, 215-26

Ruggles, C.P. (1975) The use of fish passes, traps and weirs in Eastern Canada for assessing populations of anadromous fishes. EIFAC, Technical Report, 23, 466-89

Ruggles, C.P. (1980) 10. Sampling migrating salmon. In T. Backiel and R.L. Welcomme (Eds) Guidelines for sampling fish in inland waters, EIFAC, Technical Paper, 33, 143-58

Ruggles, C.P. and Turner, G.E. (1973) Recent changes in stock composition of Atlantic salmon (Salmo salar L.) in the Miramichi River, New Brunswick. Journal of Fisheries Re-search Board of Canada, 30(6), 779-86

Scarnecchia, D.L. (1984) Forecasting yields of two sea winter Atlantic salmon (Salmo salar) from Icelandic rivers. Canadian Journal of Fisheries and Aquatic Sciences, 41, 1234-40

Slaney, P.A. and Northcote, T.G. (1974) Effects of prey abund-ance on density and territorial behaviour of young rainbow trout (Salmo gairdneri) in laboratory stream channels. Journal of Fisheries Research Board of Canada, 31, 1201-9

Struthers, G. and Stewart, D. (1984) A report on the composition of the adult salmon stock of the upper river Tummel, Scot-land and the evaluation of the accuracy of a closed channel resistivity counter. International Council for the Explora-tion of the Sea, C 1984 M: 20, 19 pp

Struthers, G. and Stewart, D. (1985) The composition and migra-tion of the adult salmon stock in the upper river Tummel, Scotland, in 1984, with observations of the accuracy of resistivity counters at two fish passes. International Council for the Exploration of the Sea, C 1985 M: 14, 18 pp

Symons, P.E.K. (1968) Increase in aggression and in strength of the social hierarchy among juvenile Atlantic salmon deprived of food. Journal of the Fisheries Research Board of Canada, 25(11), 2387-401

Symons, P.E.K. (1971) Behavioural adjustment of population density to available food by juvenile Atlantic salmon. Journal of Animal Ecology, 40(3), 569-87

Symons, P.E.K. (1979) Estimated escapement of Atlantic salmon (Salmo salar) for maximum smolt production in rivers of different productivity. Journal of Fisheries Research Board of Canada, 36, 132-40

Symons, P.E.K. and Heland, M. (1978) Stream habitats and behavioral interactions of underyearling and yearling Atlantic salmon (Salmo salar). Journal of the Fisheries Research Board of Canada, 35(2), 175-83

Van Hyning, J.M. (1973) Stock-recruitment relationships for Columbia river Chinook Salmon. Fish Stocks and Recruitment. Rapports et Procès Verbaux du Conseil International pour l'Exploration de la Mer, 164, 89-97

Ward, F.J. and Larkin, P.A. (1964) Cyclic dominance in Adams River sockeye salmon. International Pacific Salmon Fisheries Commission, Progress Report, 11, 116 pp

Wroblewski, J.S. (1983) The role of modeling in biological oceanography. Ocean Science and Engineering, 8(3), 245-85

APPENDIX

Assessment of the accuracy in the estimation of the required number of spawners

Given a function $\phi(x,y,z)$, such as $\phi(x,y,z) = x.y/z$. If $v(x)$, μx; $v(y),\mu y$; $v(z),\mu z$ are respectively the variance and the average of the independent variables x,y and z, the variance of the function $\phi,[v(\phi)]$, can be determined by the Delta method as follows:

$$v(\phi) = v(x).(\mu y/\mu z)^2 + v(y).(\mu x/\mu z)^2 + v(z).(-\mu x.\mu y/\mu z^2)^2 \qquad (17.2)$$

The variables being independent, the covariance terms are non-existent.

Example of a calculation:

> spawning area (area) = 100 hectares
> variation coefficient 10%
>
> optimal egg density per m^2 (Oed) = 2
> variation coefficient 25%
>
> fecundity per spawner (Fec) = 4,000 eggs
> variation coefficient 25%

If the formula Ns = Area.Oed/Fec is applied, the average estimation of the required number of spawners (Ns) is 500 individuals. The estimation of the variation coefficient using formula (17.2) is 37.7% (standard deviation: s(Ng) = 183).

Chapter Eighteen

STOCK ENHANCEMENT OF ATLANTIC SALMON (SALMO SALAR L.)

G.J.A. Kennedy

Fisheries Research Laboratory, Department of Agriculture, 38 Castleroe Road, Coleraine, Co Londonderry, N. Ireland

18.1 INTRODUCTION

The practice of artificial rearing of salmonids for stocking has been carried out for over 100 years, with the first commercial hatchery having been developed on the Rhine in 1852 (Harris, 1978). Hatcheries proliferated in the present century, fuelled by the belief that increases in the numbers of juveniles produced must result in increases in the abundance of adult stocks. Hatchery propagation was considered to be an improvement over the natural condition, where high juvenile mortalities were seen as wasteful of resources. Certainly hatchery techniques improved rapidly, and large numbers of juvenile salmonids became available for stocking. However, doubts about the efficiency of artificial propagation for increasing adult numbers began to grow, and various pieces of scientific evidence from the 1930s onwards confirmed that little benefit had accrued from the stocking practices then in vogue (e.g. Foerster, 1936, 1938; Hobbs, 1948).

In relation to stock enhancement in the British Isles, Harris (1978) noted that up until the 1970s little scientific assessment of artificial rearing and stocking of Atlantic salmon (Salmo salar L.) had been carried out. He rightly considered this disturbing since about 50 organisations professed to pursue policies directed at enhancing their salmon stocks, and at least 38 operated their own salmon hatcheries at this time. He calculated the total potential adult return from all stocking policies combined in Britain, and estimated that this contributed < 2.5 per cent of the total commercial catch. Harris's conclusion was that 'the very heavy commitment to artificial propagation in the British Isles was difficult to justify in terms of resultant benefits ... largely because it was carried out on too small a scale'.

On the American continent recent enhancement programmes for Pacific salmonids have been subject to considerable cost-benefit analyses (as reviewed by Solomon, 1983; Shepherd,

1984). The need for a similarly justifiable economic rationale was stressed in proposals put forward for Atlantic salmon enhancement in Canada (e.g. Anon, 1982; Morse and DeWolf, 1973, 1979). However, economic costings are not widely applied to Atlantic salmon enhancement programmes and from the published literature only limited progress appears to have been made on the quantitative evaluation of stock enhancement programmes in terms of additional adult returns.

18.2   STOCKING TECHNIQUES

On the more positive side, the techniques utilised in habitat evaluation (Kennedy, 1984a) and for salmon stocking at different stages of the life cycle have now been widely investigated and recommendations made for optimising stocking survival (e.g. Harris, 1978; Egglishaw, Gardiner, Shackley and Struthers, 1984). The survival rates have varied considerably, depending on the habitat, the stocking densities and the techniques of the investigators. A range of the published values for survival to the end of the first growing season and estimated smolt production rates from various stocking methods has been summarised in Table 18.1.

Planting ova
There are two commonly recognised stages at which eggs can be planted out in streams.

(1)   As 'green' ova - between 24 and 48 hours from stripping. The main advantage of planting at this stage is that no hatchery is required and the eggs can be planted out utilising only limited overnight holding facilities. However, there are also several disadvantages to this method of stocking. The time period suitable for planting is very short before they become sensitive to damage by movement as the embryo begins a period of rapid development. This may result in eggs being planted out in less than ideal water flow and habitats - producing low survival rates. Furthermore the eggs must remain in the bed of the stream for long periods of time before their emergence from the gravel, thus rendering the artificial redds vulnerable to damage by siltation and floods.
(2)   As 'eyed' ova - this covers a period of about three weeks following the development of pigmented eyes, but before the alevins hatch. Stocking at this stage has the advantage that the eggs have a shorter period to remain in the gravel prior to emergence, thus improving their chances of survival. Also, the stocking programme will be more flexible because of the longer

Table 18.1: Survivals and estimated smolt production from various salmon stocking methods and densities

Stage of stocking	Reference	Density stocked (m^{-2})	% survival to end of first growing season	Estimated smolt production per 1,000 stocked[a]
green ova	Kennedy and Strange (1981)	6.2	4.0	10.0
	Shearer (1961)	52.3 - 59.0	1.7 - 2.7	4.3 - 6.8
eyed ova	Egglishaw and Shackley (1973)	3.7 - 11.1	13.0 - 13.9	32.5 - 34.8
	Elson (1957a)	0.4 - 0.7	6.0 - 8.0	15.0 - 20.0
	Kennedy and Strange (1981, 1986) and Kennedy (1982)	6.2	16.9 - 19.4	42.3 - 48.5
	McCarthy (1980)	5.0 - 6.0	3.5 - 7.5	8.8 - 26.3
unfed fry	Ayton (1973)	3.6	-	32.0
	Egglishaw (1984)	2.0 - 6.2	14.9 - 26.2	37.3 - 65.5
	Egglishaw and Shackley (1980)	3.6 - 29.3	9.4 - 31.0	24.0 - 77.5
	Kennedy (1982)	6.2	16.7	41.8
	Kennedy and Strange (1986)	6.2[b]	38.6	96.5
		6.2[c]	23.1	57.8
	Mills (1964, 1969)	1.7 - 15.5	1.3 - 30.3	3.3 - 75.8
	McCarthy (1980)	5.0 - 10.0	2.2 - 11.3	5.5 - 28.3
	O'Connell, Davis and Scott (1983)	0.3 - 0.9	-	> 36.2[d]

Table 18.1 (Cont'd)

Stage of stocking	Reference	Density stocked (m^{-2})	% survival to end of first growing season	Estimated smolt production per 1,000 stocked[a]
fed fry	Ayton (1973)	1.2	-	26.6 - 49.4
	Elson (1957b)	0.1 - 1.8	6.7 - 22.7	16.8 - 56.8
	MacCrimmon (1954)	0.8	10.7 - 14.6	26.8 - 36.5
	Rogers (1968)	-	-	2.5 - 17.0

Notes:

a. Where smolt production was not quantified by the authors it has been estimated for the purposes of comparison in this table by assuming that (i) all smolts are 2+, and (ii) annual mortality to the smolt stage is 50%.

b. Stocking here was carried out after removal of all the other fish from the stocked section.

c. Stocking here was carried out in the presence of salmon parr, but in the virtual absence of all other fish.

d. Partial smolt counts only were carried out, but adult returns were quantified (see Section 18.5).

time period over which eyed ova are available - thus optimising planting time and site distribution. Furthermore, eyed ova are fairly robust and easy to transport. The main disadvantage in stocking these as opposed to green ova is the requirement for hatchery facilities.

Two methods are commonly employed for planting eggs; either directly into an artificially created gravel redd using a plastic tube (Sedgewick, 1960) or in plastic slatted Vibert boxes anchored to the stream bed and covered with stones and gravel. The former method was used by Kennedy and Strange (1981) for comparing the survival rates from green eggs and eyed ova in a small brown trout (Salmo trutta L.) stream at Altnahinch in Northern Ireland. The results indicated that survival from eyed egg stocking was almost five times better than that from green eggs. This was attributed to a high level of washout of the artificial redds by winter floods. Work reported by Harris (1978) indicated that up to 30 per cent of natural redds can suffer destruction by floods, and it seems likely that artificially constructed ones may be even less robust than those produced by spawning fish. Evidence for the destruction of redds by floods during the shorter time period following stocking with eyed eggs was also found at Altnahinch, where the survival of these planted in a high gradient area was only about 20 per cent of that in the equivalently stocked area of a low gradient stream (Kennedy, 1982).

The method adopted by Shearer (1961) for stocking with green eggs in a Scottish stream was the anchoring of Vibert boxes containing the eggs to the stream bed before burying with gravel. This presumably reduced the chances of them being washed away by floods. However, the survival rates to the end of the first summer reported by Shearer were even lower than those at Altnahinch (see Table 18.1). Closer examination of the data reveals that the stocking densities employed by Shearer were about nine times higher than those utilised at Altnahinch, and were well above the holding capacity of the stream. In fact many fry were captured emigrating from the stream, and the remaining summer densities were equivalent to those found using eyed ova at Altnahinch. This suggests that stocking with green eggs at lower densities may in fact be a satisfactory stocking method, and emphasises the dangers of interpreting results on the basis of percentage survivals alone.

Unfed fry stocking

This is probably the most common method of salmon stocking, having several advantages over eggs. Fry can be more rapidly distributed over a system using buckets or watering cans (or even by helicopter (O'Connell, Davis and Scott, 1983)), without the need for redd construction. This can be of particular benefit in areas where no suitable gravel is available for building redds. Egglishaw et al. (1984) consider that twice as many fry as eggs can be stocked by a ground team within the same time period. The disadvantages of unfed fry stocking are that the fry are less robust and require more oxygen for transport than ova. There is also the disadvantage that there is a restricted time period prior to complete absorption of the yolk sac when unfed fry can be planted out (Marr, 1966; McCarthy, 1980; Prouzet, 1983).

Comparative stocking experiments in the Altnahinch stream indicated that there was no significant difference in the survival of unfed fry compared with eyed ova stocked at the same densities (Kennedy, 1982). This is at variance with an unpublished study (Sturge, 1968) cited by O'Connell et al. (1983), where there was only a 20 per cent egg to fry survival rate. This was used as a basis for multiplying swim up fry stocking rates on the Exploits River by a factor of five to equate with ova deposition rates. Similarly, recent work by Berg, Abrahamsen and Berg (1986) suggests that only about 28 per cent of the eggs laid by spawning salmon are satisfactorily retained in the redd and survive. This figure can apparently drop to as low as 8 per cent when the spawning salmon have been previously injured by fishing gear. Certainly hatchery survival rates between green ova and swim up fry are usually much higher than this (Kennedy and Johnston, 1986), and Shearer (1961) reported hatching rates of about 85 - 90 per cent from green eggs in Vibert boxes anchored to the stream bed. However, this aspect requires further research under a variety of natural conditions as this area is fundamentally important both for the comparison of stocked and natural survival rates and for the provision of meaningful stocking recommendations.

As Table 18.1 illustrates, there is considerable variation in the reported survival rates from stocking with green and eyed eggs and unfed fry by different authors. This may be attributed partly to differences in the habitats being stocked and partly to variation in the stocking densities used. Mills (1969) not only found that the lowest survival rates tended to be recorded at the highest stocking densities, but that the latter resulted in high emigration rates of fry from the stocked areas. It is not clear whether these high emigration rates result in greater colonisation of unstocked areas downstream, or whether these fry are the exhausted losers in territorial disputes and are subject to high mortalities. Work by Ottoway and Clarke (1981) in flume tanks suggested that higher flows may aid downstream dispersal of salmon fry planted at high densities. However, this work was carried out in short artificial channels and it cannot be used as a guideline for the distances over which salmon fry can disperse in the wild, where they must establish a territory before their yolk sac is used up. Work on the downstream dispersal of fry emerging from redds (Egglishaw and Shackley, 1973) and fry stocked as unfed 'swim-ups' (Kennedy, 1982) suggest that movement is normally very limited, with the majority of fry not moving further than about 100 m from the stocking site and none being found more than 400 m downstream. This emphasises the need for careful distribution of both eggs and fry over a river system during a stocking programme to eliminate clumping.

The variation in stocking densities employed by different authors appears to have been partly determined by the availability of stock and partly by estimating the productivity of the areas being stocked. Symons (1979) pointed out that the carrying capacity of a river for juvenile salmon depends on the survival rates from egg to smolt, the growth rates to the smolt stage (and therefore the smolt ages produced by the system) and the space requirements of the young salmon at all stages between fry and smolt. From the published data available Symons estimated that the egg depositions required for various systems in the range 0.75 m^{-2} for 4+ smolt production, $1.5 - 2.0$ m^{-2} for 3+ smolt production, 2.2 m^{-2} for 2+ smolt production and > 5.0 m^{-2} for 1+ smolt production.

Whether these are minimum or optimum recommendations is the subject of much debate as it is still unclear whether the freshwater stock-recruitment relationship of juvenile salmonids follows a flat-topped or dome-shaped type curve. Discussions of the available evidence are given by Symons (1979), Le Cren (1984), Chadwick (1985), Elliott (1985) and Solomon (1985).

Symons (1979) also points out that the recommended densities relate to whole rivers, which have a wide range of habitat types, not all of which are suitable for juvenile salmonids. Therefore in rivers such as those in N. Ireland, which produce mostly 1+ and 2+ smolts, overall stocking densities of between 2.2 and 5.0 ova per m^2 are recommended - with higher values than this possible in localised areas. The experiments carried out to date in the Altnahinch streams suggest that 6.2 ova or unfed fry per m^2 in good habitat is close to the optimum in the presence of trout, but below the optimum for salmon fry on their own or in an area containing no older age classes of salmon (Kennedy and Strange, 1980, 1986).

Egglishaw et al. (1984) recommend stocking densities for Scottish streams varying between 2 and 10 ova or unfed fry per m^2, depending on the productivity, altitude and size of the streams. In earlier work in Scotland, Egglishaw and Shackley (1980) found that a higher stocking density of 13.9 m^{-2} gave improved survival and an increased size range of salmon if this was carried out as a 'double stocking' separated by 19 days. However, investigations of this technique do not appear to have been followed up in Scotland or elsewhere. In the Republic of

Ireland, McCarthy (1980) recommended an overall stocking density of 5 m^{-2} for ova and unfed fry. Previous recommendations for the R. Foyle in Ireland by Elson (1975) were based on densities of 1.7 ova m^{-2}. These were derived from studies in Canadian streams, and as recognised by Symons (1979), now appear to be too low for the higher productivity of rivers in the British Isles. Further work on this aspect is required in areas of differing carrying capacity before local recommendations for stocking densities can be defined.

Stocking with fed fry and parr
Fed fry and parr are released at various times after first feeding through to the pre-smolt stage. This long period of time over which stocking may take place permits a choice of stocking conditions and is considered to be one of the main advantages of this form of stocking (Harris, 1978). However, Shustov, Shchurov and Smirnov (1980) emphasised that stocking in the autumn or winter does not give good survival. They attributed this to the need for a long adaption period for hatchery-reared parr to feeding on wild food. However, some work on smolt releases in the R. Bush (Kennedy, Strange, Anderson and Johnston, 1984) indicated that hatchery fish which had been reared in an unfiltered river water supply (and were therefore familiar with invertebrate drift), did not show any reluctance to feed on wild food following release. Similarly Solomon (1983) noted that Pacific salmonids raised in the semi-natural conditions of rearing channels benefited from learning to feed on some of the natural food available and recommended that this rearing technique might be extended to Atlantic salmon.

Smirnov, Shustov and Shchurov (1983) also noted high mortalities from a number of stockings of hatchery-reared parr released at other times of the year in the USSR. They considered that some of these losses may have been the result of an inability of hatchery fish to take up territories. It was hypothesised that the high densities under which these fish had been reared in fish farms had conditioned them to schooling behaviour which they maintained when stocked, and consequently they drifted many miles with the current due to their low stamina, before succumbing to predation or exhaustion, or simply dropping into the sea. However, it seems likely that stocking at very high densities into inadequate habitat exacerbated this behaviour in these studies, as work by Symons (1969) indicated that hatchery-reared parr tended not to move very far from their release point when introduced to tributaries of the Miramachi in small numbers. Similarly Elson (1957b) found that while hatchery-reared underyearlings released at high densities in the Pollet river distributed themselves further than the parr in

Symons' work, most did not move further than about one mile from the release point. In a tributary of the R. Usk, Bulleid (1973) also found limited movement of small numbers of stocked parr, but noted that this tended to be greater in a year with higher flows.

Elson (1957b) recommended stocking densities for parr in Canadian rivers and Symons (1979) made extrapolations for areas of differing productivity. However, the published information on stocking of fed fry and parr does not present a consistent picture with respect to survival, extent of movement, feeding and optimum stocking densities.

Stocking with smolts

Juvenile salmon may be reared throughout their whole freshwater life and released as they are entering their migratory phase as smolts. This has the advantage of improving the survival so that the number of eggs required is much reduced from that needed for wild smolt production. Since large numbers of smolts can be produced in a limited volume of water this technique permits the stocking of rivers well beyond their rearing capacity. This form of 'sea ranching' is particularly important in rivers where large areas of spawning and nursery habitats have been lost through impoundment. Reared smolts are easy to release since this can be done in bulk without the need for dispersal in either their home river or a stocked river - although specialised release ponds have been found to improve survival (Isaksson, Rasch and Poe, 1978). Smolt release programmes also lend themselves to fuller evaluation of return and exploitation rates through counts and tagging programmes than other stocking techniques.

The main disadvantages are the extensive hatchery facilities that are required, with their attendant high running costs. Also, poorer and variable returns have been achieved from hatchery-reared smolts than from wild fish. This problem and other aspects of smolt stocking and sea ranching are subject to widespread research programmes, which are outside the remit of this paper (e.g. see Anon. (1979); Eriksson, Ferranti and Larsson (1983) Thorpe (1980) for collected papers).

Similarly, the genetic implications of stocking and the constraints on the selection of broodstock for ranching and enhancement schemes will be discussed elsewhere in this symposium.

18.3   ALTERNATIVE ENHANCEMENT TECHNIQUES

Stream remedial measures

From the evidence available on the stock-recruitment relationship of Atlantic salmon Solomon (1985) concluded that natural reproduction at low stock densities is a very efficient process, and he suggested that 'management of spawning escapement should be a most effective means of enhancement of depleted stocks'. Enhancement by stream remedial work is aimed at improving fish passage at natural or man-made obstructions to permit part of the natural spawning escapement access to spawning and nursery habitat which is suitable but normally inaccessible. This includes removal of natural or man-made obstacles, creation of channels and resting pools and construction of fish passes. Evaluation of the rate of natural recolonisation of such opened areas by counts at fishways in Canada suggests that about 10 per cent of the available downstream stock will stray upstream through such facilities (Anon., 1978). However, maximum benefits can be obtained more quickly if a stocking programme is combined with this type of remedial work, as for the Great Rattling Brook in Newfoundland (Taylor and Bauld, 1973). Stream remedial work to improve the nursery habitat, spawning gravel and adult holding areas can also be undertaken (e.g. following river drainage schemes, see Kennedy (1984b) and Solomon (1983)).

Adult transfers

The transfer of adult fish from below impassable barriers or from overstocked areas to artificial spawning channels (Pratt, Farwell and Rietveld, 1974) or other under-utilised parts of the same river system has been used with some success in Canada, as in the East River, Nova Scotia (Ducharme, 1972) and Great Rattling Brook, Newfoundland (Farwell and Porter, 1976). This technique has not been widely adopted in the British Isles although schemes have been described on the River Towy (Howells and Jones, 1972) and the River Foyle (Kennedy, Hadoke and Sheldrake, 1975). Trials involving transfer of adult salmon to an artificial spawning channel have also been carried out on the Nivelle River in France (Beall and Marty, 1983).

The method can also be used with success for transfer to a different river system if several criteria are met (Anon, 1978).

(1)   The donor stock should have similar biological characteristics to stocks in the recipient system.

(2)   The mouth of the donor stream should be in the same general area as the mouth of the recipient stream with similar directional orientation of outflows.

354

(3)   Both streams should be in the same geological formation
        and the water chemistry of both streams should be similar.

These criteria were followed successfully in transfer from
Western Arm Brook 112 km south to the Torrent River in
Newfoundland (Anon., 1978). The evidence suggests that if
sufficient donor stock is available, the areas stocked can be self
supporting in five or six years.

One of the advantages of this technique is that the progeny
of transferred adults will be of 'wild quality' compared with
juveniles produced in hatcheries and released in streams. Also,
adult transfers can be undertaken more cheaply than hatchery
rearing ventures, although the usual limiting factor is the
availability of brood stock. However, from the success
experienced with this technique in Canada there may well be
scope for further investigations into its applicability elsewhere.

Lake rearing

This is also referred to as semi-natural rearing. Extensive work
has been carried out on this and a review of the results of
studies in the British Isles was presented at a previous Atlantic
Salmon Symposium (Harris, 1973). His conclusion was that
although the technique successfully produced salmon smolts there
was a need for further study of problems associated with
non-migration, smolt viability, predation by birds and older age
classes of salmon.

More recent investigations in Newfoundland have indicated
that some juvenile salmon utilise lake habitat naturally
(Chadwick, 1985; Chadwick and Green, 1985; Pepper, 1976), but
that large deep lakes are not utilised to the same extent as
shallow ponds. In a study of three ponds ranging in area from 27
to 66 hectares, Pepper, Oliver and Blundon (1985) investigated
the results of swim-up fry stocking at a density of about 1,000
per hectare. Smolt trapping over a three year period indicated a
survival range of 0.9 - 20.0 per cent from fry to smolt, with the
best pond averaging 11.5 per cent. However, survival of fry
showed a decreasing trend with time - concomitant with an
increase in mean smolt size and age. The authors interpreted this
as indicative of increasing intraspecific competition between
stocked year classes. Problems with predation by brook trout
(Salvelinus fontinalis Mitchell) and non-migration due both to low
flows and residualism were also encountered. The latter was
associated with high growth rates and precocious maturation.
However, micro-tagging indicated that marine survivals of
lake-produced smolts were equivalent to those from riverine
nursery areas. Conversely, preliminary results from tagging
experiments in Norway indicated that lake-reared smolts showed

a return rate of only 0.3 per cent compared to 12.5 per cent from riverine smolts (Hansen, 1983). This author suggested that this may be related to the irregular timing of migration of the lake-reared compared with naturally produced smolts.

In general terms therefore the conclusions of Harris (1973) are still valid, and this area of research into salmon enhancement has not yet produced definitive recommendations for a lake stocking strategy.

18.4 A REVIEW OF SALMON ENHANCEMENT WORK IN PROGRESS

In order to collate some of the unpublished information on salmon stocking operations presently in progress I circulated a questionnaire to most of the Atlantic salmon producing countries and to each Water Authority in Britain. The aim was to determine not only the extent of enhancement schemes, but the rationale behind the work and the level of scientific assessment of the outcome. Not everyone replied and not all the replies were comprehensive. The list is therefore incomplete but a summary of the information provided is shown in Table 18.2.

It is apparent that stocking of salmon for a variety of purposes is widespread, with somewhere in the region of 38 million juveniles of varying age classes stocked worldwide annually. The rationale behind stocking with different age classes varies with the rearing facilities available and the particular requirements of each situation, including smolts for sea ranching, micro-tagging research and in mitigation for impoundment or pollution kills. However, some of the variation also appears to be related to the particular experience and perceptions of the workers involved as to the efficiency of each stocking method. For example, fry stocking was originally carried out in Iceland from 1932 onwards with several hundred thousand unfed fry annually. However, this has now ceased as no correlation between fry releases and catch records could be detected (Gudjonsson, 1978). This author comments that the failure of fry stocking was probably due to poor handling and release techniques. However, despite improvements in the general understanding of these aspects, all Icelandic stocking is presently carried out with pre-smolts and smolts. In other areas unfed fry stocking is favoured, and reportedly can give better survival rates to smolt and adult than natural spawning (O'Connell et al., 1983, and see Section 18.5). Other constraints, particularly financial, influence stocking decisions, and the views of various fishery managers who are not in favour of allocating stock enhancement a high priority were summed up by one respondent to the questionnaire as follows:

Table 18.2: Summary of stocking activities in various Atlantic salmon producing countries

Country/Region	Number of rivers stocked	Type and maximum numbers of salmon stocked annually	Source of broodstock	Reasons for stocking
England:				
Northumbrian Water Authority	7	525,000 ova 500,000 parr and smolts	Local and Scottish	Mitigation for impoundment Enhancement Pollution kills Research
North-West WA	18	267,000 ova 722,000 unfed fry 789,000 fed fry 25,000 smolts	Local and Scottish	Enhancement Pollution kills Mitigation for effects of acid rain
Thames WA	1	70,565 parr 38,688 smolts	Scottish and Welsh	Replace extinct run
Yorkshire WA	1	40,000 unfed fry 30,000 smolts	Scottish	Enhancement 'Opportunistic stocking'
Severn-Trent WA	4	8,000 unfed and fed fry	–	Research purposes
Southern WA	1	25,000 smolts	Local and 'foreign'	Research purposes (micro-tagging)

Table 18.2 (Cont'd)

Country/Region	Number of rivers stocked	Type and maximum number of salmon stocked annually	Source of broodstock	Reasons for stocking
England:				
South-West WA	-	'some juveniles'	Local	Pollution kills
Wessex WA	None	-	-	-
Wales	14	250,000 eyed ova 540,000 unfed and fed fry 75,000 parr and smolts	Local and Scottish	Mitigation for impoundment Enhancement Pollution kills
Scotland	> 50	> 13 million eyed ova and unfed fry 200,000 parr 60,000 smolts	'Mostly local'	Enhancement Research
N. Ireland	7	1 million eyed ova and unfed fry 40,000 parr 37,000 smolts	Local and Republic of Ireland	Enhancement Research Pollution kills Mitigation for drainage

Table 18.2 (Cont'd)

Country/Region	Number of rivers stocked	Type and maximum numbers of salmon stocked annually	Source of broodstock	Reasons for stocking
Republic of Ireland	13	2.7 million eyed ova and unfed fry 320,000 parr 330,000 smolts	Local and N. Ireland	Mitigation for impoundment Enhancement Mitigation for drainage Research
Canada	Potentially 175	1.9 million unfed fry 1.3 million fed fry 200,000 parr 800,000 smolts	Local	Mitigation for impoundment Enhancement Research
Faroes	5	200,000 'fry' 75,000 smolts	Icelandic and now 'local' Danish, Norwegian and now 'local'	Enhancement

Table 18.2 (Cont'd)

Country/ Region	Number of rivers stocked	Type and maximum numbers of salmon stocked annually	Source of broodstock	Reasons for stocking
France	6+	unfed fry, parr and smolts. Details not known - but see Brunet (1980)	Irish Scottish Norwegian Canadian	Enhancement
Iceland	50	300,000 smolts	Local	Enhancement
Norway	'Wide-spread'	12 million unfed and fed fry 500,000 smolts	Local	Enhancement Mitigation for impoundment Research
USA	3+	Details not known - but see Stolte (1980) and Jones (1983)	-	Enhancement Replace extinct run

> [We] no longer pursue a policy of attempting to enhance salmon stocks by means of either artificial propagation or restocking. Following a recent review of management practices we concluded that our resources could be more efficiently deployed on controlling the human exploitation of stocks as well as monitoring and mitigating the effect of man's industrial activities on the freshwater habitat of salmon.

Obviously views differ widely, not just on the relative merits of different stocking methods, but on the value of salmon stocking per se. This may be related to the paucity of published information in terms of adult returns for enhancement schemes analysed on a cost-benefit basis. Assessment of the results of stocking is not carried out at all in some cases, and frequently does not extend beyond electrofishing surveys. In many cases smolt and adult counting facilities are sited so that they include runs not attributable to enhancement. Similarly, reliance on rod and commercial catches in areas with a natural salmon run is difficult to justify as these not only include fish not attributable to enhancement, but are also influenced by environmentally induced fluctuations in catch per unit effort.

The next question must therefore be; how successful have stock enhancement schemes been in terms of additional adult returns?

18.5 ADULT RETURNS FROM STOCKING PROGRAMMES

Enhancement as discussed in this paper covers a much wider range of activities than that defined by the EIFAC working party on stock enhancement (Anon., 1983): 'Enhancement: To keep the production of stocks above the level that would be naturally sustained in the presence of heavy exploitation.'

This limited view of the term has not yet been adopted by most fishery managers and scientists engaged in salmon enhancement. A broader definition might be that all salmon enhancement schemes are ultimately aimed at improving adult returns beyond the levels that would be expected if no action was taken. This would apply equally to schemes designed to halt stock declines, replace eliminated stocks, increase depleted stocks, return stocks to historical levels or improve stocks beyond their natural limits.

Precise quantification of the numbers of returning adult salmon directly attributable to enhancement schemes is still very rare due either to the lack of facilities or availability of only partial facilities for trapping or counting adults. For this reason assessment of the relative contribution of restocking is often difficult to separate from other factors influencing salmon numbers. For example, in the R. Rheidol in Wales no adult counts were available following a restocking scheme with ova and fry, although both the redd counts and angler catches showed dramatic increases during the 1950s and 1960s (Jones and Howells, 1969). These were taken as indicative of some success of restocking although not conclusive proof.

In the R. Thames, restocking with salmon could not have been contemplated without improvement in water quality. Although some natural recolonisaton of this cleaner environment by straying cannot be dismissed, it is assumed that most of the recent adult returns can be attributed to the stocking programme. From an average annual stocking of 58,000 parr and 12,300 smolts, the recapture rate to date has been in the order of 100 adults per year from partial trapping, electrofishing and angling (Anon., 1985). It has been estimated that this represents an actual return of 200 - 400 adults per year. The costs of the scheme are presently running at about £65,000 per year, suggesting a production cost of £163 to £325 for each returning adult.

A much larger restoration scheme has been undertaken in the Connecticut river in the USA over the last two decades. As in the Thames, the run here had also been completely eliminated. An update on this enhancement scheme is given elsewhere in these proceedings, but up to 1982, over half a million smolts had produced returns of about 900 adult salmon (Jones, 1983). In addition to smolt costs, the construction of fishways and hatcheries in this scheme pushed the costs per returning adult some orders of magnitude higher than on the Thames.

The high costs of enhancement schemes in relation to the relatively low adult returns has led to considerable controversy in some areas. For example in France the depletion of the salmon resource was reported to have been widespread with the total catches falling to about 5,000 fish annually in the 1970s (Dumas, Prouzet, Porcher and Davaine, 1979). The main causes were identified as river obstructions, water abstraction, pollution and overfishing, and recommendations were made by these authors and by the Interdepartmental Committee for Nature and the Environment (Brunet, 1980) for both environmental improvement schemes and for restocking. Encouraging survivals from fry stocking and habitat improvement were recorded in four Scorff tributaries in south Brittany (Baglinière, 1979) and adult returns have been quantified from stocking of unfed fry and smolts in the Elorn river (Prouzet and Gaignon, 1982). The success of this work is leading to the development of larger hatchery facilities and an extension of the restocking and sea ranching programme in this area (Prouzet, 1983). An experimental restocking programme with fry, parr and smolts in conjunction with the installation of fish passes and trapping facilities has also been initiated in the R Nivelle in the southwest of France (Dumas, 1979, 1986; Dumas and Casaubon, 1985). Other rivers such as the Dordogne, Aulne, Allier-Loire rivers and streams in Brittany, Normandy, the Auvergne and Basque-Bearne region have also been earmarked for improvement schemes and enhancement - with rearing stations and research centres being established (Brunet, 1980). However, the traditional view of the past abundance of the salmon resource, the value of the high investment costs in enhancement schemes and the objectives of such schemes have been questioned by Thibault (1983) and Rainelli and Thibault (1985). These authors consider that the scientific assessment of salmon stock enhancement has been inadequate and the results unconvincing to date, and they call for a halt to stocking until a full scientific and financial evaluation of the potential contribution of hatchery reared fish to the returns has been undertaken.

In Canada, consideration of the economic management of the salmon resource and assessment of the relative cost-benefits of management and enhancement of Atlantic salmon stocks to various users have been undertaken (e.g. Morse and DeWolf, 1973, 1979). Full assessments of the potential additional salmon returns from enhancement schemes were calculated from habitat surveys, production potentials and survival and exploitation rates (Anon., 1978). Within Atlantic Canada the potential for enhancement was identified in 175 of the 350 major watersheds, and the additional adult production was estimated at 562,000 adults. It was further estimated that an additional 360,000 adults could be produced from smolt rearing and sea ranching. The relative allocation of these additional fish to regional fisheries within Canadian waters was also assessed as were the cost benefits of this potential harvest. (The total projected costs for a large scale enhancement programme in the Maritimes Provinces and Newfoundland were about $250 million in 1978.)

A full assessment of the additional adult production from enhancement has been undertaken on the Exploits River in Newfoundland (O'Connell, et al., 1983). Here, stocking with from 0.14 million to 1.83 million unfed fry was carried out annually from 1968 onwards. Adult returns to a downstream fishway have built up from zero to a mean of 3,100 from 1980 to 1982. This represents an average egg-adult survival of 0.17 per cent for this period, which is 3.4 times better than that quantified from natural spawning in an adjacent river (Great Rattling Brook). The authors suggest that this justifies the technique of fry stocking as an effective and efficient use of brood stock. Furthermore, this is additional evidence that natural ova deposition rates cannot be equated to artificial stocking densities, see Section 18.2. In addition there is considerable evidence from this study that there was a marked increase in the commercial catch and catch per unit effort in the local coastal salmon fisheries following the enhancement programme.

This concept of the value of enhancement in terms of return to the national catch is shared by workers elsewhere, e.g. the Republic of Ireland (Browne, 1981; Twomey, 1982). Here the returns from micro-tagging experiments have indicated that commercial exploitation rates by coastal drift nets are in the range of 52 - 88 per cent from different smolt releases (Browne and Piggins, 1986; Piggins, 1979). Thus for every adult salmon returning to the stocked river or hatchery of origin up to eight additional fish may be taken by local commercial nets. This may justify the principle of enhancement on a national basis, but the exploitation of a resource by users other than those developing the resource leads to a variety of economic management discussions beyond the scope of this paper (see McKernan, 1980; Mills, 1983; Morse and DeWolf, 1973, 1979; and various contributions to Grover (ed.), 1980).

18.6   SUMMARY

The efficacy of different enhancement techniques is evaluated under two main headings:

(1)   Stocking techniques. The relative merits and drawbacks of stocking with green and eyed ova, unfed and fed fry, parr and smolts is discussed, along with an assessment of appropriate stocking densities. Recommendations for further research are made in a number of areas:

   (a)   Stocking with green ova in anchored plastic slatted boxes has been inadequately evaluated. There is some evidence that this technique may produce survival rates which are comparable with those from eyed ova and may exceed the survival rates from natural spawning.
   (b)   The survival rates from natural spawning to emerging fry require precise quantification if meaningful stock-recruitment relationships and comparisons with stocked survival rates are to be made, or recommendations formulated for stocking densities.
   (c)   Further studies on optimum stocking densities of different stages in different habitats should be undertaken - including 'double stocking' techniques.
   (d)   The extent of dispersal of salmon stocked at different stages, with varying hatchery experience and in different habitats requires further quantification.

(2)   Alternative enhancement techniques are discussed under
the headings of stream remedial measures, adult transfers
and lake rearing. It is concluded that more research is
required into problems associated with the latter before
definitive recommendations can be produced. Successes
with adult transfers in Canada suggest that there is scope
for investigation of the applicability of this technique
elsewhere.

The results of a survey of the extent of enhancement
programmes worldwide indicate that over 38 million juvenile
Atlantic salmon of various stages may be stocked annually.
However, some of the responses to a questionnaire suggest that
there is great disparity between various workers' views, not just
on the efficacy of different stocking methods, but in the value of
enhancement per se. This is discussed in terms of the paucity of
data available to quantify the returns from enhancement projects
both as increased adult numbers and from cost-benefit analyses.
Some of the available literature on adult returns from ongoing
enhancement programmes is evaluated.

REFERENCES

Anon. (1978) Atlantic salmon review, Government of Canada, Fisheries and Oceans Resource Development Sub-Committee Report, November 1978, 55 pp

Anon. (1979) (Ocean Ranching and Restocking) Aquaculture Extensive et Repeuplement. Centre National Pour l'Exploitation des Oceans, (France). Publ. by : CNEXO, COB, Brest (France) (Actes Colloques) No 12

Anon. (1982) Management of the Atlantic salmon in the 1980s. A discussion paper. Department of Fisheries and Oceans, Ottawa, Canada, May 1982, 21 pp

Anon. (1983) Report of the EIFAC working party on stock enhancement. Hamburg, Federal Republic of Germany, 16-19 May 1983. EIFAC Tech. Pap./Doc. Tech. CECPI, (44) 22 pp

Anon. (1985) Salmon rehabilitation scheme. Phase 1 review. Thames Water, Rivers Division, Special Publication, 46 pp

Ayton, W.J. (1973) Smolt trapping in the Severn River Authority area, Journal of the Institute of Fisheries Management, 4, 52-5

Baglinière, J.L. (1979) Production des juveniles de saumon Atlantique (Salmo salar L.) dans quatre affluents du Scorff, Rivière de Bretagne-Sud. Annales de Limnologie, 15(3), 347-66

Beall, E. and Marty, C. (1983) Reproduction du saumon Atlantique Salmo salar L. en milieu semi-natural controlé. Bulletin Francais de Pisciculture, 289, 77-93

Berg, M., Abrahamsen B. and Berg, O.K. (1986) Spawning of injured compared to uninjured female Atlantic salmon, Salmo salar L. Aquaculture and Fisheries Management, 17, 195-9

Browne, J. (1981) First results from a new method of tagging salmon - the coded wire tag. Department of Fisheries and Forestry, Dublin, Fishery Leaflet 114, 6 pp

Browne, J. and Piggins, D.J. (1986) The Burrishoole as an index river. Working Paper for ICES Working Group on North Atlantic Salmon, Copenhagen, 17-26 March 1986, 4 pp

Brunet, A.R. (1980) Present status of the Atlantic salmon stocks in France and environmental constraints on their extension. In Atlantic salmon : its future. Proceedings of the second International Atlantic Salmon Symposium, Edinburgh. Fishing News Books, Farnham, Surrey, pp 128-34

Bulleid, M.J. (1973) The dispersion of hatchery-reared Atlantic salmon (Salmo salar) stocked into a fishless stream. International Atlantic Salmon Foundation Special Publication Series, 4(1), 169-80

Chadwick, E.M.P. (1985) Fundamental research problems in the management of Atlantic salmon, Salmo salar L., in Atlantic

Canada. Journal of Fish Biology, 27 (Supplement A), 9-25

Chadwick, E.M.P. and Green, J. (1985) Atlantic salmon (Salmo salar L.) production in a largely lacustrine Newfoundland watershed. Verhandlugen der Internationalen Vereinigung fuer Theoretische und Angewandte Limnologie, 22(4), 2509-15

Ducharme, L.J.A. (1972) Atlantic salmon (Salmo salar) rehabilitation in the East River, Sheet Habour, Nova Scotia. Project description and initial results. Department of the Environment, Canada, Fisheries Service Halifax, Progress Report, No. 4, 29 pp

Dumas, J. (1979) Les saumons (Salmo salar L.) adultes de la Nivelle (Pyrénées-Atlantiques) en 1977. Debut de restauration avec des smolts d'élevage d'origine Ecossaise. Annales de Limnologie, 15(2), 223-38

Dumas, J. (1986) La population de saumons de la Nivelle Bilan 1977-1985. Cent. Rech. Hydrobiol, INRA, St Pée-sur-Nivelle, 7 pp

Dumas, J. and Casaubon, J. (1985) Connaissance et restauration de la population de Saumon Atlantique (Salmo salar L.) de la Nivelle (Pyrénées Atlantiques). Communication au Colloque Franco-Québécois, Restauration des Rivières à Saumons, Bergerac, Mai 1985, INRA Publ., 15 pp

Dumas, J., Prouzet, P., Porcher, J.P. and Davaine, P. (1979) État des connaissances sur le saumon en France. In Centre National pour l'Exploitation des Oceans, (France). Publ. du CNEXO, série : Actes de Colloques, No 12, 153-70

Egglishaw, H.J. (1984) Guidelines for restocking. Atlantic Salmon Trust Workshop on Stock Enhancement, University of Surrey, 9-11 April 1984, 8 pp

Egglishaw, H.J. and Shackley, P.E. (1973) An experiment on faster growth of salmon Salmo salar (L) in a Scottish stream. Journal of Fish Biology, 5, 197-204

Egglishaw, H.J. and Shackley, P.E. (1980) Survival and growth of salmon, Salmo salar (L), planted in a Scottish stream. Journal of Fish Biology, 16, 565-84

Egglishaw, H.J., Gardiner, W.R., Shackley, P.E. and Struthers, G. (1984) Principles and practices of stocking streams with salmon eggs and fry. Department of Agriculture and Fisheries for Scotland. Scottish Fisheries Information Pamphlet, No. 10, 22 pp

Elliott, J.M. (1985) Population dynamics of migratory trout, Salmo trutta, in a Lake District stream, 1966-83, and their implications for fisheries management. Journal of Fish Biology, 27, (Supplement A), 35-43

Elson, P.F. (1957a) Number of salmon needed to maintain stocks. Canadian Fish Culturist, 21, 19-23

Elson, P.F. (1957b) Using hatchery-reared Atlantic salmon to best advantage. Canadian Fish Culturist, 21, 1-17

Elson, P.F. (1975) The Foyle Fisheries : New basis for rational management. Foyle Fisheries Commission, Co Londonderry, N. Ireland, Special Publication, 224 pp

Eriksson, C., Ferranti, M.P. and Larsson, P.O. (1982) Sea Ranching of Atlantic salmon. COST 46/4 Workshop, EEC, Lisbon, 26-29 October, 1982

Farwell, H.K. and Porter, T.R. (1976) Atlantic salmon enhancement techniques in Newfoundland. FAO Technical Conference on Aquaculture, Japan 1976, 7 pp

Foerster, R.E. (1936) Sockeye salmon propagation in British Columbia. Bulletin of the Biological Board of Canada, 53, 1-16

Foerster, R.E. (1938) An investigation of the relative efficiencies of natural and artificial propagation of sockeye salmon (Oncorhynchus nerka) at Cultus Lake, British Columbia. Journal of the Fisheries Research Board of Canada, 4 (3), 151-61

Grover, J.H. (ed.) (1980) Allocation of fishery resources. Proceedings of the Technical Consultation on Allocation of Fishery Resources, Vichy, France, 20-23 April 1980, FAO, 623 pp

Gudjónsson, T. (1978) The Atlantic salmon in Iceland. Journal of Agricultural Research in Iceland, 10(2), 11-39

Hansen, L.P. (1983) Stocking streams and lakes with eggs and juveniles of Atlantic salmon, Salmo salar L. In N. Johansson (ed.), Salmon Symposium, Lulea, Sweden 4-6 October, 1983, 19 pp

Harris, G.S. (1973) Rearing smolts in mountain lakes to supplement salmon stocks. International Atlantic Salmon Foundation, Special Publication Series, 4(1), 237-52

Harris, G.S. (1978) Salmon propagation in England and Wales. A Report by the Association of River Authorities/National Water Council Working Party. National Water Council, London, 62 pp

Hobbs, D.F. (1948) Trout fisheries in New Zealand : their development and management. New Zealand Marine Department, Fisheries Bulletin No 9, 75 pp

Howells, W.R. and Jones, A.N. (1972) The R. Towy regulating reservoir and fishery protection scheme. Journal of the Institute of Fisheries Management, 3 (3), 5-19

Isaksson, A., Rasch, T.J. and Poe, P.H. (1978) An evaluation of smolt releases into salmon and non-salmon producing streams using two release methods. Journal of Agricultural Research in Iceland, 10(2), 100-13

Jones, R.A. (1983) The restoration of Atlantic salmon in the Connecticut River. Proceedings of the Institute of Fisheries Management 14th Annual Study Course, 19th-22nd September 1983, City University, London, pp. 9-13

Jones, A.N. and Howells, W.R. (1969) Recovery of the River Rheidol. Effluent and Water Treatment Journal, November, pp 70-6

Kennedy, G.J.A. (1982) Factors affecting the survival and distribution of salmon (Salmo salar L.) stocked in upland trout (Salmo trutta L.) streams in N. Ireland. Symposium on Stock Enhancement in the Management of Freshwater Fish, Budapest, 31 May - 2 June 1982. EIFAC Technical Papers (42), Suppl. Vol. 1, 227-42

Kennedy, G.J.A. (1984a) Evaluation of techniques for classifying habitats for juvenile Atlantic salmon (Salmo salar L.). Atlantic Salmon Trust Workshop on Stock Enhancement, University of Surrey, 9th-11th April 1984, 23 pp

Kennedy, G.J.A. (1984b) The ecology of salmonid habitat re-instatement following river drainage schemes. Institute of Fisheries Management (N. Ireland Branch) Institute of Continuing Education, New University of Ulster, Fisheries Conference Proceedings, pp 1-13

Kennedy, G.J.A., Hadoke, G.D.F. and Sheldrake, D.R. (1975) Transplanting of adult Atlantic salmon (Salmo salar L.) in the River Foyle as a viable method of supplementing the spawning stock. Fisheries Management, 8(4), 120-7

Kennedy, G.J.A. and Johnston, P.M. (1986) A review of salmon (Salmo salar L.) research on the River Bush. Proceedings of the Institute of Fisheries Management 17th Annual Study Course, University of Ulster, 9th-11th September 1986, pp

Kennedy, G.J.A. and Strange, C.D. (1980) Population changes after two years of salmon (Salmo salar L.) stocking in upland trout (Salmo trutta L.) streams. Journal of Fish Biology, 17, 577-86

Kennedy, G.J.A. and Strange, C.D. (1981) Comparative survival from salmon (Salmo salar L.) stocking with eyed and green ova in an upland stream. Fisheries Management, 12(2), 43-8

Kennedy, G.J.A. and Strange, C.D. (1986) The effects of intra- and inter-specific competition on the survival and growth of stocked juvenile Atlantic salmon, Salmo salar L., and resident trout, Salmo trutta L., in an upland stream. Journal of Fish Biology, 28, 479-89

Kennedy, G.J.A., Strange, C.D., Anderson, R.J.D. and Johnston, P.M. (1984) Experiments on the descent and feeding of hatchery reared salmon smolts (Salmo salar L.) in the River Bush. Fisheries Management, 15 (1), 15-25

Le Cren, E.D. (1984) Report of a workshop on salmon stock enhancement, University of Surrey, 9th-11th April 1984. Atlantic Salmon Trust Special Publication, 20 pp

MacCrimmon, H.R. (1954) Stream studies on planted Atlantic salmon. Journal of the Fisheries Research Board of Canada, 11, 362-403

McCarthy, D.T. (1980) Stocking of salmon fry and ova in rivers. Department of Fisheries and Forestry, Dublin, Fisheries Handbook No 2, 12 pp

McKernan, D.L. (1980) The future of the Atlantic salmon - an international issue. In Atlantic Salmon : its future. Proceedings of the second International Atlantic Salmon Symposium, Edinburgh. Fishing News Books, Farnham, Surrey, pp. 18-29

Marr, D.H.A. (1966) Factors affecting the growth of salmon alevins and their survival and growth during the fry stage. Association of River Authorities Year Book, 1965, 133-41

Mills, D.H. (1964) The ecology of the young stages of the Atlantic salmon (Salmo salar L.) in the River Bran, Ross-shire, Department of Agriculture and Fisheries for Scotland, Freshwater and Salmon Fisheries Research, 32, 58 pp

Mills, D.H. (1969) The survival of hatchery reared salmon fry in some Scottish streams. Department of Agriculture and Fisheries for Scotland, Freshwater and Salmon Fisheries Research, 39, 12 pp

Mills, D. H. (1983) Problems and solutions in the management of open seas fisheries for Atlantic salmon. Atlantic Salmon Trust, Special Publication, 22 pp

Morse, N.H. and DeWolf, A.G. (1973) Economic principles for the management of Atlantic salmon - St John River system. International Atlantic Salmon Foundation, Special Publication Series, 4(1), 355-64

Morse, N.H. and DeWolf, A.G. (1979) Options for the management of the Atlantic salmon fisheries of the Saint John River, New Brunswick. Fisheries and Marine Service Technical Report No. 819, 57 pp

O'Connell, M.F., Davis, J.P. and Scott, D.C. (1983) An assessment of the stocking of Atlantic salmon (Salmo salar L.) fry in the tributaries of the middle Exploits River, Newfoundland. Canadian Technical Report of Fisheries and Aquatic Sciences No. 1225, 142 pp

Ottoway, E.M. and Clarke, A. (1981) A preliminary investigation into the vulnerability of young trout (Salmo trutta L.) and Atlantic salmon (Salmo salar L.) to downstream displacement by high water velocities. Journal of Fish Biology, 19, 135-45

Pepper, V.A. (1976) Lacustrine nursery areas for Atlantic salmon in insular Newfoundland. Fisheries and Marine Service Technical Report No 671, 61 pp

Pepper, V.A., Oliver, N.P. and Blundon, R. (1985) Evaluation of an experiment in lacustrine rearing of juvenile anadromous Atlantic salmon. North American Journal of Fisheries Management, 5(4), 507-25

Piggins, D.J. (1979) Atlantic salmon sea ranching in Ireland. In Centre National pour l'Exploitation des Oceans, (France). Publ. du CNEXO, série : Actes de Collegues, No 12, 127-33

Pratt, J.D., Farwell, M.K. and Rietveld, H.J. (1974) Atlantic salmon production using a spawning channel. Resource Development Branch, Newfoundland Region. Internal Report Series No. NEW/1-74-7, 29 pp

Prouzet, P. (1983) Salmon rehabilitation and management on the River Elorn, Northern Brittany, France. Proceedings of the Institute of Fisheries Management 14th Annual Study Course, 19th-22nd September, City University, London, pp 28-43

Prouzet, P. and Gaignon, J.L. (1982) Production de saumon Atlantique (Salmo salar L.) juveniles et adultes sur un ruisseau pepinière de Bretagne Nord (France) a partir d'une souche Irlandaise. Revue des Travaux-Institut des Pêches Maritimes, 45(2), 155-74

Rainelli, P. and Thibault, M. (1985) La surabondance de consommation de saumon autrefois : une surabondance véritablement ... fabuleuse. Cahiers de Nutrition et de Dietetique, 20(4), 292-7

Rogers, A. (1968) Salmon fry survival in a small west of Ireland stream. Salmon and Trout Magazine, 183, 107-12

Sedgewick, D.S. (1960) Planting salmon. Salmon and Trout Magazine, 160, 204-10

Shearer, W.M. (1961) Survival rate of young salmonids in streams stocked with 'green' ova. International Council for the Exploration of the Sea, Salmon and Trout Committee 1961, No. 98, 3 pp

Shepherd, B.G. (1984) The biological design process used in the development of federal government facilities during phase 1 of the salmonid enhancement programme. Canadian Technical Report of Fisheries and Aquatic Sciences, No. 1275, 188 pp

Shustov, Yu. A., Shchurov, I.L. and Smirnov, Yu. A. (1980) Adaptation times of hatchery salmon, Salmo salar, to riverine conditions. Journal of Ichthyology, 20(4), 156-9

Smirnov, Yu. A., Shustov, Yu. A. and Shchurov, I.L. (1983) Behaviour of hatchery salmon parr, Salmo salar, in relation to stocking. Journal of Ichthyology, 23, 85-95

Solomon, D.J. (1983) Salmonid enhancement in North America. Description of some current developments and their application to the UK. The Atlantic Salmon Trust/Atlantic Salmon Federation, Bensinger-Liddell Memorial Atlantic Salmon Fellowship 1983, 40 pp

Solomon, D.J. (1985) Salmon stock and recruitment, and stock enhancement. Journal of Fish Biology, 27 (Supplement A), 45-58

Stolte, W. (1980) Planning as related to the restoration of Atlantic salmon in New England. In Atlantic salmon : its future. Proceedings of the Second International Atlantic Salmon Symposium, Edinburgh. Fishing News Books, Farnham, Surrey, 135-45

Sturge, C.C. (1968) Production studies on the young stages of Atlantic salmon (Salmo salar L.) in an experimental area of Indian River, Notre Dame Bay, Newfoundland. M. Sc. Thesis, Dept of Biology, Memorial University of New-Foundland, 134 pp

Symons, P.E.K. (1969) Greater dispersal of wild compared with hatchery-reared juvenile Atlantic salmon released in streams. Journal of the Fisheries Research Board of Canada, 26(7), 1867-76

Symons, P.E.K. (1979) Estimated escapement of Atlantic salmon (Salmo salar) for maximum smolt production in rivers of different productivity. Journal of the Fisheries Research Board of Canada, 36, 132-40

Taylor, V.R. and Bauld, B.R. (1973) A programme for increased Atlantic salmon (Salmo salar) production in a major Newfoundland river. International Atlantic Salmon Foundation, Special Publication Series, 4(1), 339-47

Thibault, M. (1983) Les transplantations de salmonides d'eau courant en France: saumon Atlantique (Salmo salar L.) et truite commune (Salmo trutta L.) Compte Rendu Sommaire des Seances Societé du Biogeographie, 59 (3c), 405-20

Thorpe, J.E. (ed.) (1980) Salmon ranching. Academic Press, London, 441 pp

Twomey, E. (1982) The contribution of hatchery-reared smolts to the Irish drift net fishery. Department of Fisheries and Forestry, Dublin, Fishery Leaflet No. 118, 9 pp

Chapter Nineteen

SALMON ENHANCEMENT: STOCK DISCRETENESS AND
CHOICE OF MATERIAL FOR STOCKING

J.E. Thorpe

Freshwater Fisheries Laboratory, Pitlochry, PH16 5LB, Scotland

19.1   INTRODUCTION

Reduction of wild stocks of Atlantic salmon, Salmo salar L.,
through heavy exploitation or through loss of habitat, has
encouraged managers to try to compensate for these decreases by
stock enhancement. When catches of salmon in a river are less
than expected, or of a quality which differs from the fisherman's
anticipations, it is commonly assumed that the river is
underpopulated (or will be), and that the simple remedy must be
to increase future numbers by stocking that river with more
juveniles. Is this assumption valid?

To obtain a representative index of changes in numbers,
sizes and ages in a salmon population, it needs to be fished at
the same intensity on all its component parts over a period of
years. Close times and close seasons restrict fishing to an
arbitrary, and thus unrepresentative, fragment of the population.
Even this fragment varies, since whether fish are there to be
caught at a particular time depends on both ocean and river
conditions, which vary from year to year. So, by themselves,
numbers in the catch are not normally a reliable indicator of the
numbers in the stock (Browne, 1986; Dunkley, 1986; Shearer,
1986). Likewise the age and size composition of the catch are
characteristics which do not necessarily reflect accurately those
same features of the stock (Dunkley, 1986).

Hence assumptions about stock based on the partial
samples normally provided by catches are not sound. How, then,
can one assess the status of a particular stock?

The manager's objective is to ensure that there is a
continuing adequate supply of harvestable adult fish returning
from the sea. He will achieve this only if sufficient juvenile fish
enter the sea from the nursery grounds. Future catches, as with
all fisheries, depend on successful production of juveniles. Hence
predictions about the quantity and quality of future catches

373

depend on knowledge of the quantity and quality of the juveniles which will yield the adults that make up those catches.

Egglishaw, Gardiner, Shackley and Struthers (1984) have set out simple rules for survey of river systems to assess the status of juvenile fish stocks in terms of the densities of young fish present in relation to the productivity of those streams, and thus to their carrying capacity, and Mills and Tomison (1985) give an example of such a survey. If the results of such a survey suggested that there was spare productive capacity in the river, there would be grounds to consider stock enhancement. What fish should then be stocked?

19.2   STOCK DISCRETENESS

Natural selection ensures the success of a species by favouring reproduction among animals which are adapted harmoniously with their environment, and by eliminating those not so well adapted. Every river system is different. It imposes physical and biological constraints on the animals developing within it, which differ in subtle ways from those imposed in other rivers. Hence the pressures of natural selection differ between rivers, and it is therefore not surprising to find that the genetic composition of the salmon population in one river differs from that in the next (Moller, 1970; Cross, Healy and O'Rourke 1978; Ryman, 1981; Stahl, 1981, 1983; Slynko, Semenova and Kazakov, 1981; Cross and King, 1983; Altukhov and Salmenkova, 1987). This close match between genotype and environment is preserved efficiently through high homing precision, so that these stocks remain discrete (Behnke, 1972; Power, 1981; Saunders, 1981; Thorpe and Mitchell, 1981). So the stock of salmon of a particular river is likely to possess a genetically determined range of patterns of development, tailored by natural selection, which ensure success in that specific river. If salmon are to be stocked into a system to increase the numbers of an existing population, then ideally they should be of the same genotype as the recipient population. If they differ, subsequent interbreeding could modify the genetic spectrum of that recipient population. Taggart and Ferguson (1986) have documented just such changes in the brown trout S. trutta population of L. Erne, following the stocking of over 8 million Movanagher trout over the period 1968-83. It has been shown that species such as salmonids whose populations are subdivided into discrete stocks are particularly vulnerable to directional changes in genetic composition (Thorpe and Koonce, 1981). Such changes could impair the ability of the population to respond to environmental change.

That this is more than a theoretical possibility has been illustrated by the records of a stocking exercise with chum salmon, <u>Oncorhynchus keta</u>, in Sakhalin. Altukhov (1981) reported that 350,000,000 fertilised eggs were transferred from the Kalininka River to the Naiba River between 1964 and 1971. Before the transfer, the Naiba river carried a spawning population of about 650,000 chum salmon. By 1969-70 the genetic characteristics of the stock returning to the Naiba had shifted towards those of the Kalininka fish, and by 1980 the returning population had decreased to 30,000-40,000 spawners. By 1985 the population was virtually extinct (Yu. Altukhov, personal communication, 1986). He concluded that this local disaster was the result of the massive genetic migration of non-adapted genotypes.

Similarly but less dramatically, Smoker (1987) recorded that the return of native chum salmon to a site in Alaska was twice as great as that of transplanted stock, but from hybrids between the two the return was very small indeed.

Apart from differences in biochemical genetic markers (of unknown functional significance), the differences in biology between all these various chum stocks are unknown. It is not known, therefore, at what stage of life history the hybrids failed through lack of appropriate adaptations.

These examples are all the more important for Atlantic salmon managers, since the genetic variation between stocks of <u>Salmo</u> species is known to be very much greater than that between stocks of <u>Oncorhynchus</u> species (Ryman, 1983; Altukhov and Salmenkova, 1987). If transfers of the relatively similar chum salmon stocks can interfere with one another so adversely, how much more damaging could be transfers of Atlantic salmon stocks? When ranched Atlantic salmon in New Brunswick were hybridised between stocks, Bailey (1987) recorded a <u>decrease</u> in return rate in the second generation. He suggested that these inter-stock pairings may have interfered with the genetic components of navigation, and so impaired the fishes homing ability.

Atlantic salmon have been stocked repeatedly in many river systems containing supposed depleted populations, throughout their natural range for a very long time, and the fish are still there. This is encouraging, but current scarcity of evidence for adverse effects (e.g. Moller, 1970; but also Bailey, 1987) similar to those seen in the Naiba River chums is not evidence of total absence of such effects in Atlantic salmon enhancement. The real long-term success of augmentation in this species has not been measured, and would require a study similar to that made in Kamchatka over a period of 20 years. However, some predictive pointers do exist:

(1)   stockings have often been total failures;

(2)   natural strays make negligible impact on recipient stocks; and

(3)   multiple distinct stocks do coexist in some systems.

     To take these points of evidence in turn, firstly, failures: Saunders (1981) pointed out that many salmon stocks have been eliminated due to human competition for the use of their native rivers. Restoration of these vanished populations has been exceedingly difficult to attain, for example in New England, and he suggested that important genetically based behavioural traits had been lost, especially in relation to appropriate run timing of smolting and spawning fish. Bailey's (1987) data would add navigational control to this list. Similarly, Altukhov (1981) reported many instances of complete failure of chum salmon transplants. Taggart, Ferguson and Mason (1981) suggested that since similar genotypes often existed in contiguous geographic areas, failure of stockings of 'generalised or mixed' strains of hatchery trout could have been due to lack of local genetic adaptation.

     Secondly, straying: Rasmuson (1968) predicted that a straying rate of only 2 per cent per annum would be enough to prevent genetic differentiation between Atlantic salmon populations. The few estimates of straying rate in Atlantic salmon suggest that it is greater than this, at about 2-7 per cent per annum (Thorpe and Mitchell, 1981; Browne et al., 1983), but Hansen and Lea (1982) have reported up to 17.6 per cent in some Norwegian rivers. However, Stahl (1981) demonstrated clear genetic differentiation between stocks from biochemical evidence in a group of six Swedish rivers. In these rivers, 2 per cent straying represented 50-200 fishes: but Stahl (1981) calculated from his samples that in fact less than one stray salmon per year had contributed to the progeny of subsequent generations. Hence differences of developmental timing between stocks, or behavioural inhibition, may have diminished the chances of natural hybridisation in these river stocks.

     Thirdly, multiple stocks: Riddell, Leggett and Saunders (1981) noted the separate nature of the Atlantic salmon populations of different tributaries of the Miramichi. Stahl (1981) found distinct biochemical differences between fish of the River Lainio, and of the River Torne, to which it is a tributary. Altukhov (1981) reported the existence of 30 different spawning stocks of sockeye salmon, Oncorhynchus nerka, in Lake Azabachye, Kamchatka, defined by biochemical genetic differences. Ryman and Stahl (1981) reported many examples of stock differentiation in Arctic charr, Salvelinus alpinus, within Scandinavian lakes, and in brown trout in lakes and even within small streams. Ferguson (1980), and Ferguson and Mason (1981) found evidence for coexistence of three reproductively isolated and genetically distinct stocks of brown trout in Lough Melvin. Crozier and Ferguson (1986) have found two distinct trout stocks in Lough Neagh, each of which is made up of components associated with several spawning streams, such that there is stock differentiation even within small streams. Many other examples were given in the STOCS Symposium (1981).

To summarise these findings, much 'foreign' material would be useless for stocking into a river since it would fail to return; some might be of value only once, if after return it failed to breed at all; and some could be useful, as long as its breeding behaviour and occupancy of the river environment complemented, but did not interfere with, those of existing stocks, so that there was no reduction or dilution of essential genetic information within the population.

19.3   NATURE OF STOCK DIFFERENCES

The developmental programme for an organism is genetically determined, but it runs under environmental instruction. Major changes in development, such as smolting, and emigration from freshwater, are influenced environmentally by changes in daylength and temperature. It is likely that these indicators of season (especially daylength) act as synchronisers of physiological change (Saunders and Henderson, 1970; Lundqvist, 1983), and so of the changed behaviour patterns which ensue (Wagner, 1974). Natural selection will have determined the optimal time for such changes for any given stock: those fish that make their changes at the optimum time will survive to grow and reproduce; those which change at other times will be eliminated. Timing of events such as emigration from rivers must relate to optimal protection from predators, optimal physical conditions for downstream transport, and optimal physical and biological conditions for entering the sea (cf. Lundqvist and Eriksson, 1985). In large southern systems like the Connecticut River this would include arrival in the ocean before sea temperatures reach lethally high levels. Such a timetable will be specific not only to individual river systems, but to separate regions of those systems.

A probable instance of this has been documented by Struthers and Stewart (1986) for Atlantic salmon in Scotland. Five streams devoid of wild salmon, draining into Loch Rannoch, were stocked with unfed fry derived from eggs taken from adults trapped at least 35-50 km downstream, and some from other branches of the watershed (Figure 19.1). In subsequent years, descending emigrants were trapped at Dunalastair Dam 15-30 km below the stocking sites. Progeny of wild fish from 4-10 km upstream emigrated through the trap between weeks 9-21 (March-May): stocked fish arrived there during weeks 22-24 (June) (Figure 19.2). On evidence from within the experiment, and from

Figure 19.1: Struthers and Stewart's Atlantic salmon stocking experiment. Streams with native populations: N; Stocked streams: S; Trap for smolting emigrants: T

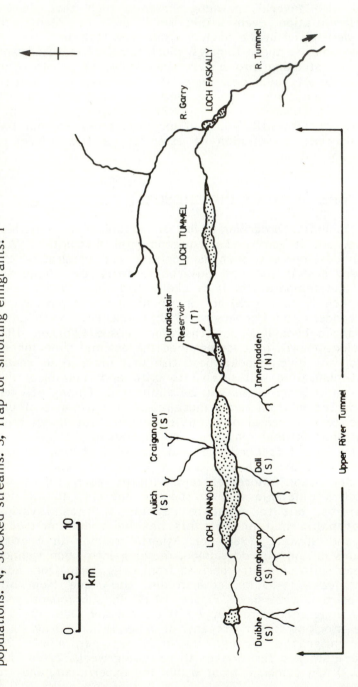

Source: After Struthers and Stewart (1986)

Figure 19.2: Weekly cumulative percentage catches of emigrant juvenile salmon at the Dunalastair trap. Native stocks: ; introduced stocks: o

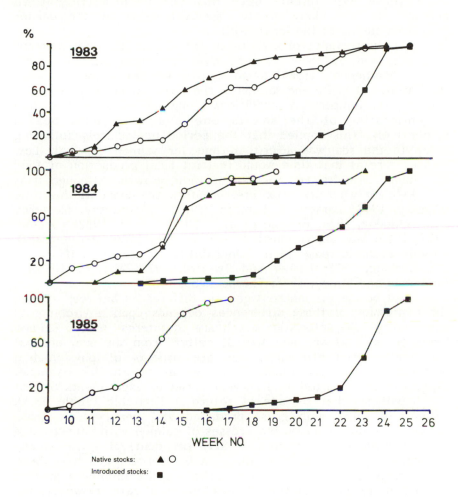

Source: After Struthers and Stewart (1986)

other studies of movement of smolting fish through rivers and lakes (Thorpe, Ross, Struthers and Watts 1981), the difference in migration distance was not enough to account for the almost total separation of migration periods between the wild and stocked fish. The authors attribute the delay among stocked fish to a difference in developmental timing implying differences in adaptation between the local and planted stocks. (Adult returns from these experiments have not yet been sufficient to demonstrate the likely greater survival value of the earlier migration timing of the local fish.)

The importance of arriving in the sea at the right moment is emphasised by some indirect evidence from wild Icelandic stocks. Scarnecchia (1984) found significant correlations between the catches of grilse and salmon in the rivers of northern Iceland, and the sea temperature conditions along the coast at the time of emigration of the smolts one and two years before respectively. This implied that the period immediately following entry to the sea was a critical one for salmon survival. Low temperature at that time would restrict local production of food for the post-smolts, and so restrict their growth. Although it is not yet clear what predation pressure they are under at this time (Piggins, 1958; Larsson & Larsson, 1975; Bakshtansky, Nesterov and Nekludov, 1976; Browne et al., 1983; Hansen, 1982; Larsson, 1982, 1985; Morgan, Greenstreet and Thorpe, 1986), a priori slow growth would increase their vulnerability to predation (Horwood and Cushing, 1978; Lasker and Zweifel, 1978; Walters, Hilborn, Peterman and Staley, 1978; Thorpe 1980, 1987).

But beside the inherent genetic differences between stocks, the expression of those differences depends upon environmental opportunity. Characteristics which are of interest to the salmon manager, such as age and size at return from the sea, are not genetically fixed attributes, but are aspects of physiological performance (Thorpe, 1986). Take age at return, for example: Piggins (1983) reported on a 16-year series of experiments to test the hypothesis that sea-age at return is heritable (Table 19.1). Crosses of grilse with grilse produced 98.2 per cent of grilse among returning progeny: but crosses of salmon with salmon have also produced high percentages (87.1 per cent) of grilse among the returning progeny. That these latter percentages have been consistently less than those from grilse x grilse implied a genetic component to the regulation of developmental rate. However, the fact that even among the progeny of the slower developers (salmon) there is a preponderance of grilse, implies that the environmental opportunities have been so good that even fish of relatively low developmental rate can mature quickly.

Further examples of this have been demonstrated recently. From the same St John River stock, 2+ smolting salmon reared on in sea cages in the Bay of Fundy showed grilse:salmon ratios of approximately 1:100, whereas those released to sea produced ratios of approximately 1:1 (Saunders, Henderson, Glebe and Loudenslager, 1983; Saunders, 1986). 1+ smolting fish of this same stock produced no grilse at all in the cages, and a ratio of 0.21:1

Table 19.1: Atlantic salmon: sea-age at return of progeny of grilse and 2-sea-winter salmon

Parental	Sea-age at return		
Sea-age	0[a]	1	2
1	8 (0.2%)	4283 (98.2%)	70 (1.6%)
1 x 2	0	147 (96.1%)	6 (3.9%)
2	0	183 (87.1%)	27 (12.9%)

Note: a, Pre-grilse: fish returning to freshwater in the year of smolt emigration

Source: Piggins (1983).

among released fish. Salmon of the Neva stock (USSR) emigrate from the river into the Gulf of Finland mostly (76-77 per cent) at age 2+, and most (58.8 per cent) return to spawn after 3 or more years at sea (Melnikova, 1980). When material from this stock was reared at Ims, southwest Norway, the fish grew fast, smolted after 1 year, and returned predominantly as grilse (Hansen, personal communication). In both these cases, as in Piggins' experiments, environmental opportunity has dictated the age at maturity, within the developmental limits set by the respective genotypes.

While rearing experiments have shown that developmental rates are heritable characteristics in salmon (Thorpe 1975; Thorpe and Morgan, 1978; Bailey, Saunders and Buzeta, 1980; Thorpe, Morgan, Talbot and Miles, 1983), the expression of those different genetic capacities is heavily dependent on local environmental conditions. Hence, the life-history strategy adopted by successful stocked fish is likely to resemble that of the native stock.

19.4 CHOICE OF MATERIAL FOR STOCKING

If assessment of the population status indicates that stocking would be a valuable management undertaking, then the evidence set out above would suggest that the first measure should be to augment with the existing native stock. By rearing the progeny of pairings of wild fish in a hatchery, and then on to maturity in cages, a large broodstock could be built up quickly, from which the progeny would be available for planting out. To maintain the genetic diversity of such a stock, as much information as possible should first be gained about the genetic structure of that stock.

If distinct components exist in the population, then each should be treated separately, and for each the age groups of spawners should be selected in proportion to their occurrence in the stock, from over the whole seasonal range of return to the river. To avoid reducing the fitness of the population through inbreeding, at least 30 males and 30 females should be used for each stock component in the initial spawning (Falconer, 1964; Ryman and Stahl, 1980).

In the case of environments which have lost their salmon altogether, the choice must depend on whatever can be determined about the genetic structure of neighbouring populations. As Taggart, Ferguson and Mason, (1981) noted, there are often very similar genotypes in geographically close but isolated populations, so that there is greater probability of successful establishment of a 'foreign' stock in a vacant system, if that stock comes from a nearby river. However, since there are now no native fish at risk of genetic damage from such an introduction, there may be advantage in using fish from several neighbouring stocks, to broaden the genetic range initially, and so provide more scope for rapid natural selection of suitable combinations in subsequent generations.

As evidenced above, the life-history characteristics of such planted fish cannot be predicted with certainty from their performance in their native rivers, since such characteristics will be strongly influenced by the new environment.

19.5 SUMMARY

Catch data for Atlantic salmon are unreliable indices of strength of stocks. Stream surveys, to determine the quantity and quality of the population present, are necessary before decisions are taken on augmentation of those populations.

Salmonid fish species are composed of genetically distinct, reproductively isolated stocks, finely adapted to the specific features of the environments in which they develop. Interference with this close match of genotype and environment through introduction of 'foreign' stocks is potentially disastrous, as in the destruction of the Naiba River stock of chum salmon, and the reduced success of inter-stock crosses of Atlantic salmon. Many stockings have failed, probably through lack of appropriate adaptation of the introduced stocks. At low levels of straying, barriers exist to the effective reproduction of these strays. From this, and the discovery of sympatric non-interbreeding stocks, it is possible that some stockings can succeed, provided that the breeding behaviour and occupancy of the river environment by the introduced fish complement but do not interfere with those of the native populations.

Behavioural differences exist between stocks, as seen in the different migration timing of native and introduced Atlantic salmon in the Loch Rannoch area of Scotland. As the developmental programme for organisms is genetically fixed but environmentally controlled, the performance of salmon transferred into a new habitat will depend on the nature of both the fish and the habitat. The same genetic stock of Atlantic salmon reared in two different environments in Canada matured at different ages in those two environments. Conversely, in one environment in Ireland, the progeny of both grilse and 2 sea-winter salmon matured predominantly as grilse. Thus the life-history pattern adopted by successfully stocked fish would be heavily dependent on local conditions.

Consequently the ideal material for stocking a system is that which is native to it. Since each salmon stock consists of diverse components, augmentation should aim to increase all those components in proportion to their occurrence in the initial population. To reduce the possibility of impairment of genetic diversity by inbreeding, at least 30 males and 30 females should be used as the founding broodstock of each stock component. In stocking systems empty of salmon, neighbouring stocks are the most likely to provide successful material. In such cases there may be advantage in using several stocks initially, to broaden the genetic range on which selection may act.

REFERENCES

Altukhov, Y.P. (1981) The stock concept from the viewpoint of population genetics. Canadian Journal of Fisheries and Aquatic Sciences, 38, 1523-38

Altukhov, Y.P. and Salmenkova, E.A. (1987) Population genetics of coldwater fish. In: K. Tiews (Ed) EIFAC/FAO Symposium on Selection, Hybridisation and Genetic Engineering in Aquaculture. Heenemann, Berlin

Bailey, J.K. (1987) Canadian sea ranching program (East Coast). In: K. Tiews (Ed) EIFAC/FAO Symposium on Selection, Hybridisation and Genetic Engineering in Aquaculture. Heenemann, Berlin

Bailey, J.K., Saunders, R.L. and Buzeta, M.I. (1980) Influence of parental smolt age and sea age on growth and smolting of hatchery reared Atlantic salmon (Salmo salar). Canadian Journal of Fisheries and Aquatic Sciences, 37, 1379-86

Bakshtansky, A.L., Nesterov, V.D. and Nekludov, M.N. (1976) Predators' effect on the behaviour of Atlantic salmon smolts in the period of downstream migration. International Council for the Exploration of the Sea CM 1976/M:3

Behnke, R.J. (1972) The systematics of salmonid fishes of recently glaciated lakes. Journal of the Fisheries Research Board of Canada, 29, 639-71

Browne, J. (1986) The data available for analysis on the Irish salmon stock. In: D. Jenkins and W.M. Shearer (eds), The status of the Atlantic salmon in Scotland. ITE Symposium No. 15, Institute of Terrestrial Ecology, Abbots Ripton, pp. 84-90

Browne, J., Eriksson, C., Hansen, L.P., Larsson, P.O., Lecomte, J., Piggins, D.J., Prouzet, P., Ramos, A., Sumari, O., Thorpe, J.E. and Toivonen, J. (1983) COST Project 46/4 on ocean ranching of Atlantic salmon: final report. Action COST 46 Mariculture: Rapport Final. Comm. des Comm. Europ., Brussels, pp. 16-94

Cross, T.F., Healy, J.A. and O'Rourke, F.J. (1978) Population discrimination in Atlantic salmon from Irish rivers, using biochemical genetic methods. International Council for the Exploration of the Sea, CM 1978/M.2

Cross, T.F. and King, J. (1983) Genetic effects of hatchery rearing in Atlantic salmon. Aquaculture, 33, 33-40

Crozier, W.W. and Ferguson, A. (1986) Electrophoretic examination of the population structure of brown trout, Salmo trutta L., from the Lough Neagh catchment, Northern Ireland. Journal of Fish Biology, 28, 459-77

Dunkley, D.A. (1986) Changes in the timing and biology of salmon runs. In: D. Jenkins and W.M. Shearer (eds) The status of the Atlantic salmon in Scotland. ITE Symposium No.15, Institute of Terrestrial Ecology, Abbots Ripton, pp. 20-7

Egglishaw, H.J., Gardiner, W.R., Shackley, P.E. and Struthers, G. (1984) Principles and practice of stocking streams with salmon eggs and fry. Scottish Fisheries Information Pamphlet, 10, 1-22

Falconer, D.S. (1964) Introduction to quantitative genetics. Oliver & Boyd, Edinburgh

Ferguson, A. (1980) Biochemical systematics and evolution. Wiley, New York

Ferguson, A. and Mason, F.M. (1981) Allozyme evidence for reproductively isolated sympatric populations of brown trout Salmo trutta L. in Lough Melvin, Ireland. Journal of Fish Biology, 18, 629-42

Hansen, L.P. (1982) Salmon ranching in Norway. In: C. Eriksson, M.P. Ferranti and P.O. Larsson (eds), Sea ranching of Atlantic salmon. Comm. des Comm. Europ., Brussels, pp. 95-108

Hansen, L.P. and Lea, T. (1982) Tagging and release of Atlantic salmon smolts (Salmo salar L.), in the River Rana, Northern Norway. Report of the Institute for Freshwater Research Drottningholm, 60, 31-8

Horwood, J.W. and Cushing, D.H. (1978) Spatial distribution and ecology of pelagic fish. In: J.H. Steele (ed.), Spatial pattern in plankton communities. Plenum Press, New York, pp. 355-83.

Larsson, P.O. (1982) Salmon ranching in Sweden. In: C. Eriksson, M.P. Ferranti and P.O. Larsson (eds), Sea ranching of Atlantic salmon. Comm. des Comm. Europ., Brussels, pp. 127-37

Larsson, P.O. (1985) Predation on migrating smolt as a regulating factor in Baltic salmon, Salmo salar L., populations. Journal of Fish Biology 26, 391-7

Larsson, H.O. and Larsson, P.O. (1975) Predation pa nyutsatt odlad smolt, Lulealven 1974. (Predation on hatchery-reared smolts after release into the River Lule 1974). Laxforskningsinstituted Meddelande, 9

Lasker, R. and Zweifel, J.R. (1978) Growth and survival of first feeding northern anchovy larvae (Engraulis mordax) in patches containing different proportions of large and small prey. In: J.H. Steele (ed.), Spatial pattern in plankton communities. Plenum Press, New York, pp. 329-54

Lundqvist, H. (1983) Precocious sexual maturation and smolting in Atlantic salmon (Salmo salar L.): photoperiodic synchronisation and adaptive significance of annual biological cycles. PhD Thesis, University of Umea, Sweden

Lundqvist, H. and Eriksson, L.O. (1985) Annual rhythms of swimming behaviour and seawater adaptation in young Baltic salmon, Salmo salar, associated with smolting. Environmental Biology of Fisheries, 14, 259-67

Melnikova, M.N. (1980) Sovremennoye sostoyaniye stada Nevskogo lososya Salmo salar (L.) (Present state of the Neva stock of Salmo salar (L.)). In: O.A. Skarlato (ed.) Lososevidnye ryby. Akademiya Nauk USSR, Leningrad, pp. 101-5

Mills, D.H. and Tomison, A. (1985) A survey of the salmon and trout stocks of the Tweed basin. Tweed Foundation, Edinburgh, 39 pp

Møller, D. (1970) Transferrin polymorphism in Atlantic salmon (Salmo salar). Journal of the Fisheries Research Board of Canada, 27, 1617-25

Morgan, R.I.G., Greenstreet, S.P.R. and Thorpe, J.E. (1986) First observations on distribution, food and fish predators of post-smolt Atlantic salmon, Salmo salar, in the outer Firth of Clyde. International Council for the Exploration of the Sea CM 1986/M:27

Piggins, D.J. (1958) Investigations on predators of salmon smolts and parr. Annual Report of the Salmon Research Trust of Ireland, 5

Piggins, D.J. (1983) Census work on fish movements: reared salmon: summary of selective breeding programme. Annual Report of the Salmon Research Trust of Ireland, 27, 38

Power, G. (1981) Stock characteristics and catches of Atlantic salmon (Salmo salar) in Quebec, and Newfoundland and Labrador in relation to environmental variables. Canadian Journal of Fisheries and Aquatic Sciences, 38, 1601-11

Rasmuson, M. (1968) (Quoted in Stahl, 1981: see below).

Riddell, B.E., Leggett, W.C. and Saunders, R.L. (1981) Evidence of adaptive polygenic variation between two Atlantic salmon (Salmo salar) populations within the S.W. Miramichi River, New Brunswick. Canadian Journal of Fisheries and Aquatic Sciences, 38, 321-33

Ryman, N. (ed.) (1981) Fish Gene Pools. Ecological Bulletin, 34, 5-111

Ryman, N. (1983) Patterns of distribution of biochemical genetic variation in salmonids: differences between species. Aquaculture, 33, 1-21

Ryman, N. and Stahl, G. (1980) Genetic changes in hatchery stocks of brown trout (Salmo trutta). Canadian Journal of Fisheries and Aquatic Sciences, 37, 82-7

Ryman, N. and Stahl, G. (1981) Genetic perspectives of the identification and conservation of Scandinavian stocks of fish. Canadian Journal of Fisheries and Aquatic Sciences, 38, 1562-75

Saunders, R.L. (1981) Atlantic salmon (Salmo salar) stocks and management implications in the Canadian Atlantic Provinces and New England, USA. Canadian Journal of Fisheries and Aquatic Sciences, 38, 1612-25

Saunders, R.L. (1986) The scientific and management implications of age and size at sexual maturity in Atlantic salmon (Salmo salar). In: D.J. Meerburg (ed.) Salmonid age at maturity. Canadian Special Publication on Fisheries and Aquatic Sciences, 89

Saunders, R.L. and Henderson, E.B. (1970) Influence of photoperiod on smolt development and growth of Atlantic salmon (Salmo salar). Journal of the Fisheries Research Board of Canada, 27, 1295-311

Saunders, R.L., Henderson, E.B., Glebe, B.D. and Loudenslager, E.J. (1983) Evidence of a major environmental component in determination of the grilse:larger salmon ratio in Atlantic salmon. Aquaculture, 33, 107-18

Scarnecchia, D. (1984) Climatic and oceanic variations affecting yield of Icelandic stocks of Atlantic salmon. Canadian Journal of Fisheries and Aquatic Sciences, 41, 917-35

Shearer, W.M. (1986) An evaluation of the data available to assess Scottish salmon stocks. In: D. Jenkins and W.M. Shearer (eds), The status of the Atlantic salmon in Scotland. ITE Symposium No.15, Institute of Terrestrial Ecology, Abbots Ripton, pp. 91-111

Slynko, V.I., Semenova, S.K. and Kazakov, R.V. (1981) Izucheniye populyatsionno-geneticheskoyi struktury atlanticheskogo lososya v svyazi s zadachami ego razvedeniya. Soobshch. II. Analiz chastot fenotipov malikenzima v populyatsiyakh lososya rek Nevy i Narovy. (Study of the population-genetic structure of Atlantic salmon in relation to problems of its culture. II. Analysis of frequencies of malic enzyme phenotypes in the salmon populations of the Neva and Narova Rivers.) Sbornik nauchn. trudov GosNIORKh, 163, 124-8

Smoker, W.W. (1987) Survival of transplant, native and hybrid chum salmon (Oncorhynchus keta) at an ocean ranch. In: K. Tiews (Ed) EIFAC/FAO Symposium on Selection, Hybridisation and Genetic Engineering in Aquaculture. Heenemann, Berlin

Stahl, G. (1981) Genetic differentiation among natural populations of Atlantic salmon (Salmo salar) in northern Sweden. Ecological Bulletin, 34, 95-105

Stahl, G. (1983) Differences in the amount and distribution of genetic variation between natural populations and hatchery stocks of Atlantic salmon. Aquaculture, 33, 23-32

STOCS Symposium. (1981). Stock Concept International Symposium. Canadian Journal of Fisheries and Aquatic Sciences, 38, 1457-921

Struthers, G. and Stewart, D. (1986) Observations on the timing of migration of smolts from natural and introduced juvenile salmon in the upper River Tummel, Scotland. International Council for the Exploration of the Sea, CM 1986/M:4

Taggart, J.B. and Ferguson, A. (1986) Electrophoretic evaluation of a supplemental stocking programme for brown trout, Salmo trutta L. Aquaculture and Fisheries Management, 17, 155-62

Taggart, J.B., Ferguson, A. and Mason, F.M. (1981) Genetic variation in Irish populations of brown trout (Salmo trutta L.): electrophoretic analysis of allozymes. Comparative Biochemistry and Physiology, 69B, 393-412

Thorpe, J.E. (1975) Early maturity in male Atlantic salmon. Scottish Fisheries Bulletin, 42, 15-17

Thorpe, J.E. (1980) Salmon ranching: current situation and prospects. In: J.E. Thorpe (ed.) Salmon ranching. Academic Press, London, pp. 395-405

Thorpe, J.E. (1986) Age at first maturity in Atlantic salmon, Salmo salar: freshwater period influences and conflicts with smolting. In: D.J. Meerburg (ed.), Salmonid age at maturity. Canadian Special Publication on Fisheries and Aquatic Sciences, 89, 7-14

Thorpe, J.E. (1987) Some biological problems in ranching salmonids. Report of the Institute of Freshwater Research, Drottningholm, 63, 83-95

Thorpe, J.E. and Koonce, J.F. (with Borgeson, D., Henderson, B., Lamsa, A., Maitland, P.S., Ross, M.A., Simon, R.C. and Walters, C.) (1981) Assessing and managing man's impact on fish genetic resources. Canadian Journal of Fisheries and Aquatic Sciences, 38, 1899-907

Thorpe, J.E. and Mitchell, K.A. (1981) Stocks of Atlantic salmon (Salmo salar) in Britain and Ireland: discreteness and current management. Canadian Journal of Fisheries and Aquatic Sciences, 38, 1576-90

Thorpe, J.E. and Morgan, R.I.G. (1978) Parental influence on growth rate, smolting rate and survival in hatchery reared juvenile Atlantic salmon, Salmo salar. Journal of Fish Biology, 13, 549-56

Thorpe, J.E., Morgan, R.I.G., Talbot, C. and Miles, M.S. (1983) Inheritance of developmental rates in Atlantic salmon, Salmo salar L. Aquaculture, 33, 119-28

Thorpe, J.E., Ross, L.G., Struthers, G. and Watts, W. (1981) Tracking Atlantic salmon smolts, Salmo salar L., through Loch Voil, Scotland. Journal of Fish Biology, 19, 519-37

Wagner, H.H. (1974) Photoperiod and temperature regulation of smolting in steelhead trout (Salmo gairdneri). Canadian Journal of Zoology, 52, 219-34

Walters, C., Hilborn, R., Peterman, R.M. and Staley, M.J. (1978) Model for examining early ocean limitation of Pacific salmon production. Journal of the Fisheries Research Board of Canada, 35, 1303-15

Chapter Twenty

ATLANTIC SALMON IN AN EXTENSIVE FRENCH RIVER SYSTEM: THE LOIRE-ALLIER

Robin Cuinat*

Regional Delegate of the Conseil Superieur de la Peche, Auvergne-Limousin

20.1 THE LOIRE-ALLIER SYSTEM - DESCRIPTION

The Loire is the largest river in France, 1,000 km in length, with a catchment of 117,480 km^2. Its flow-rate at Montjean (an area of 109,930 km^2 or 94 per cent of the catchment) is 841 cumecs (mean) but varies between 257 cumecs (August average) to 1,520 cumecs (February average), with extremes of from 50 to more than 6,000 cumecs, derived in the main, from precipitation.

The salmon holding area of the Loire

From the mouth of the Loire to its confluence (called 'Bec d'Allier') with its most important tributary (the Allier) is a distance of 536 km, of which about 30 km can be termed estuarial. The gradient is very flat, at an average of 0.34/1,000 and the river bed is largely sandy. Water temperatures fall to 0-5°C in winter but can increase to over 30°C in summer.

Water quality is generally 'good' (Quality 1B, where quality 1A is very good; 1B is good; 2 is moderate; 3 is bad; 4 is very polluted) from the Béc d'Allier as far as Tours but then becomes progressively worse, especially downstream of the larger towns, eventually being categorised as 'bad' (Quality 3) in the final hundred kilometres to the sea.

The quality of this part of the river does not allow of reproduction or growth of salmonids.

There are five obstacles to fish migration: Chinon, Blois, St-Laurent-des-Eaux, Dampierre-en-Burly and Belleville. All except Blois (a mobile, leisure-sports dam) are for conventional

---

* Copies of the French version of this paper can be obtained from the author at Conseil Superieur de la Peche, 6e Delegation Regionale, Auvergne-Limousin.

or nuclear power stations. These dams are from 1 to 2 metres high and all except that at Belleville, have fish passes.

Dredging and extraction of sand and gravel has been practised intensively since about 1950 and has brought about a significant lowering of the river bed, which in turn has caused various problems during the past twenty years (Clavel, Cuinat, Hamon and Romaneix, 1977; Cuinat, 1981):

(1)   the penetration further up-river of brackish water and a 'silt plug',
(2)   a relative increase in the height of all dams along the course of the Loire, causing delays to the upstream passage of migratory fish, especially in conditions of low flow and cold water.

The principal affluent streams in this portion of the Loire are the Maine, the Vienne and the Cher. The latter two were salmon rivers until the nineteenth century when the construction of dams without efficient fish passes caused their disappearance. A restoration scheme was begun in 1980 on the Vienne-Creuse-Gartempe system.

In the portion of the Loire upstream of the Bec d'Allier, salmon became progressively less numerous, from the same causes, from 1836 onwards, finally disappearing around 1940 (Bachelier, 1963-1964). Since then, further dams have been built, some of them very high (Grangent, Villerest, Bapallisse); part of the flow of the upper Loire has been diverted by Electricité de France into the Rhone catchment and water quality is moderate or bad over the greater part of this length of river. The reintroduction of salmon was thus made extremely difficult.

Apart therefore, from salmon derived from recent restockings of the Gartempe, all the salmon of the Loire system reproduce in the Allier.

The Allier

The flow in this tributary is almost as great (average: 156 cumecs) as that of the Loire immediately above the junction (average at Nevers: 190 cumecs). The length of the Allier is also of the same order as that of the upper Loire (404 km and 464 km respectively) The Allier can be divided into seven sections, according to the physical characteristics of the river and the degrees to which salmon can grow and reproduce.

The lower part (174 km) has a low gradient (0.8%), is relatively polluted and plays no part in the reproduction of salmon. It contains six dams, of which five have fish passes (Pont-du-Guetin (no fish pass), dam at Laurins, bridge apron at Moulins, leisure-sports dam at Vichy, rubble dam at Madeleines, dam at Pont-du-Chateau). The dam at Vichy causes the most frequent delays to the upstream migration of salmon. As in the Loire, the lowering of the river bed by sand and gravel extraction (forbidden since 1979) has aggravated this problem.

The zone where salmon reproduction is possible begins upstream of Pont-du-Chateau (near Clermont-Ferrand, 44 km), but siltation and the presence of predatory fish militate against successful salmon reproduction.

From Issoire to Langeac (75 km) conditions are better and there are significant numbers of salmon juveniles in the upper area.

Above Langeac, the gradient is steeper, (4, 5 and 6.5%), the water is of very good quality and has moderate summer temperatures. This area constitutes ideal salmon spawning and nursery areas. However, the lower area is affected by considerable variations in water flow, resulting from the operation of sluice gates at the hydroelectric station at Poutès and salmon cannot reach the upper part (35 km) since the construction of the dam at Poutès in 1941. In an attempt to open up this excellent nursery area, a fish-lift has just (1986) been built. Surveys and management studies have been undertaken to achieve satisfactory upstream and downstream migrations (Bomassi and Travade, personal communication).

In order to reclaim the uppermost area (38 km) it would be necessary to find a means of passing fish over the Barrage de St-Etienne, which would involve very considerable technical problems. There are nine weirs or dams between Pont-du-Chateau and Poutès, of which four have working fish passes. (Weirs at Vic-le-Comte and Vézezoux (no fish pass), dams at Brioude, Vielle-Brioude and Chilhac (inadequate fish passes), Cerzat, Langeac; Poutès (fish pass under construction) St Etienne-du-Vigan (totally impassable at the present time).

The system, in general, from the sea to the spawning grounds
Following their entry into the estuary, salmon must swim 750-860 km (and soon, 900 km) before they find themselves in the uplands of the Massif Central. They are found at altitudes ranging from 370 to 860 m, in areas where the stream gradient is adequate (1.4 - 7%) and where water quality and summer conditions are good to very good for growth and reproduction.

On their journey, salmon must surmount 12 - 19 obstacles, of which about a third are difficult in low water conditions or at low water temperatures (less than 5°C according to Baril and Gueneau, 1986).

Thus, not only is their upstream migration endangered, but they are also subject to mortalities from fishing in localised areas where fish are concentrated at the foot of these obstacles.

Water pollution, in certain areas of the Loire and Allier can cause 'biological blocks' at certain times, thereby increasing mortalities. This is especially true in the lower area of the Allier during prolonged droughts; since the Naussac reservoir was built (1983) on an affluent stream near Langogne, the maintenance of dry weather flows (this reservoir maintains a flow rate of about 6 cumecs at Brioude and 14.3 cumecs in the lower region when the natural flow has fallen to 2 and 5 cumecs respectively, during drought periods) has eased this problem; at the same time, it has probably improved living conditions for young salmon.

20.2   BIOLOGY, BEHAVIOUR AND LIFE-CYCLES OF SALMON OF THE LOIRE-ALLIER

For studies relating to this section see Cuinat, Bomassi, Bousquet, Joberton and Marty (1980), Cuinat (1980), Cuinat and Bomassi (1985), Cuinat (1986) and Prevost (1985).

Juveniles in freshwater
Spawning normally takes place between 5 November and 15 December. Embryonic stages of the life-history have not been studied but the alevins probably emerge from the gravel during March/April. Complete ice-cover of the Allier is extremely rare.
The parr are found mostly in fast-flowing stretches (0.2-1 m/s); their population density in September varies from several hundred to 4,000 per hectare and exceptionally, up to 10,000/ha Population density of 0+ fish, 1+ parr are normally ten times less numerous.
Growth is rapid - mean (total) length in September is 9-10 cm; in winter, the population structure becomes bimodal.
Smolt migration takes place mainly in April and May in the Allier; 80-95 per cent of the smolts are 1+ years old and virtually all the remainder are 2+. Their mean length (both age groups) varies between 15.5 and 18.0 cm in different years (SD - 2.5). This growth is probably more closely linked with the productivity of the rivers than to any genetic characteristics of the Loire/Allier stock. This is confirmed by the slower growth of artificially reared smolts in the CSP salmon hatchery at Augerolles, where most of the fish smoltify at 2+ years of age.
The rate of smolt migration seems to vary between 10 and 30 km per day (exceptionally 70 km/ day) depending on water flow rates in the spring and obstacles encountered. Thus the complete journey to the estuary probably involves a period of 1 - 3 months, during which the migrating fish continue feeding. We have no information on losses due to predators or fishing during this migratory phase. Delays and injuries are caused at certain

nuclear power stations during low water conditions in the spring (Cuinat et al., 1980). The number of smolts moving downstream in the Allier would seem to vary between 50,000 and 200,000 in different years but these estimates should be treated with caution.

Marine life and return to the estuary

The recovery of tags (external, dorsal) is made mostly in NW Greenland, with some from the Faeroes. It is possible, however, that some salmon (mostly those rare specimens that return to the Loire after one year in the sea) make most of their growth in less distant waters. These fish probably run up the Allier after the fishing season has closed and are sub-samples only, the scales and tags being provided in the main, by fishermen.

The majority of the Loire-Allier salmon spend two summers (small salmon) or three summer (large salmon) in the sea and a few stay away for four summers (Table 20.1). By reason of their lengthy stay in the estuary, the Allier salmon do not normally show a final winter band on their scales, in contrast to other French river systems. In order not to confuse the situation, therefore, we count summers in the sea rather than winters. In the Loire, the average weight of small salmon is about 4.5 kg and that of large salmon 7.5 kg, but may range up to 12 kg and occasionally, even more. In the Allier, weights are generally 10-20 per cent less, these salmon being thinner.

They enter the estuary at intervals from October (sometimes September) up to the end of spring, a period of 6 - 8 months. Usually, the earliest salmon are the large fish and the later running are small salmon. The returns are delayed during periods of particularly low water in the Loire, such as occurred in the winter of 1985-86.

Adult upstream migration in freshwater

The speed of upstream migration of salmon in the Loire and then the Allier, is very variable (probably by factors of 1-3), depending on conditions of water flow and temperature. These factors, it will be remembered, affect the passage over some of the weirs and dams.

Migration is interrupted during a period in the summer (duration varying between years); fish movement recommences in the first floods of September but is then less vigorous, with reduced swimming speeds.

Fishing effort come from:

(1)   estuary nets, which are poorly supervised and from 'accidental' captures (at present, one cannot rely on the

Table 20.1: Percentages of the different sea-age groups in salmon catches on the Loire-Allier system (catches on the Allier mainly by rod and line)

Years	Time spent at sea (summers in the sea)				Sample size
	1	2	3	4	
1973	0	14.3	85.7	0	35
1974	0	37.5	50.0	12.5	8
1975	-	-	-	-	0
1976	0	23.5	70.6	5.9	34
1977	0	21.4	75.0	3.6	28
1978	0	23.8	76.2	0	21
1979	0	42.0	58.0	0	69
1980	0	20.6	78.9	0.3	286
1981	0	57.0	42.0	1.0	100
1982	0	14.6	85.4	0	82
1983	0	10.9	89.1	0	46
1984	0	26.0	74.0	0	27
1985		6.6	92.4	1.0	105
Average for the period 1978-1985	0	24.6	75.0	0.4	841

      data of salmon captures in the estuary or in the sea, especially those which derive from the period outside the legal fishing season); also from 'palisade nets', flatfish nets and trammels, in the Loire.

(2)   rod-fishing in the Allier (and occasionally in the Loire), the largest proportion of which comes from the lower half of the Allier. The salmon are less susceptible to rod-fishing effort at the end of the season or when they have reached their spawning areas.

      Sanctuaries have been set up below most of the obstacles but they are not always large enough (especially on the Loire) to ensure adequate protection for stocks of fish waiting for more favourable conditions for upstream passage.

      Four extremes of conditions are found in the upstream run:

(1)   Early salmon encounter the following conditions over the whole system:

(a) Favourable: few losses to the fishery and successful early return to the good spawning areas, high up in the system, assuming an efficient 'homing mechanism'.
(b) Bad: when they suffer heavy losses to the fishery, especially to the estuary nets and when held up below obstacles. However, any survivors reach the spawning areas in time to reproduce.

(2) Late salmon encounter:

(a) Favourable conditions: when they suffer only moderate losses from fishing and the survivors usually reach the spawning areas;
(b) Bad conditions: they will be very vulnerable to the downstream fishery although less to fishing in the upper reaches. As a result, however, they are unlikely to reach the spawning grounds in time to reproduce. It is possible that these fish spawn in the lower reaches but the redds cannot be seen, due to the width of the river and the turbidity of the water).

These four scenarios encompass the extremes found during upstream migration. It is rare, though, that 'favourable' or 'unfavourable' conditions would persist throughout the system at any one time or throughout the year.

Spawning and return to the sea

The digging of redds usually starts at the beginning of November, the females almost always choosing sites immediately above riffles. We have attempted to count these redds since 1977, as an indication of the numbers of adults reaching the good spawning areas.

At this period, the salmon are very thin and often exhibit necrotic patches on the skin. The downstream migration of any survivors may be seen during the winter and the following spring. Only a very small proportion survive to return for a second time when about 1 per cent of adults sampled during the upstream run exhibit a spawning mark on their scales.

20.3 LIFE CYCLE STRATEGIES AND COMPARISON WITH OTHER FRENCH SALMON POPULATIONS

Scale reading shows that all possible combinations exist between a freshwater life of 1+ years (sometimes 2+) and a sea life of 2 or 3 summers, sometimes 4. There appears to be no invariable relationship between the two phases. Taking into account the duration of the adult upstream migration, it is apparent that there are four possible periods of time between generations:

a 4 year period (15 per cent of the adults)
a 5 year period (55 per cent of the adults)
a 6 year period (26 per cent of the adults)
a 7 year period (3 per cent of the adults)

These proportions are only averages and can vary greatly between years.

By comparison with other French rivers where the distance from the sea to the spawning grounds is relatively small; the 'mean' distance (half the sum of the distances to the first and second spawning areas respectively) is: 795 km for the Loire-Allier; 109 km for the Gave d'Oleron and the Aulne; less than 42 km for all other salmon rivers in France. The life cycle of Loire-Allier salmon is lengthy, although the freshwater stage is short (> 80 per cent 1+ smolts compared with 50-60 per cent in other river systems). This is responsible for two factors: first, the long period of growth in the sea (24 per cent 2 summers, 75 per cent 3 summers and occasionally (0.5 per cent) 4 summers) compared with other river systems where salmon spend 1 or 2 years in the sea (sometimes 3 in the Gave d'Oleron); and secondly, the very lengthy adult phase in freshwater: from 14 to 6 months compared to 10 to 2 months in other river systems.

The early estuarial entry appears to be a 'necessary evil', knowing the river distance to be travelled and the obstacles that must be surmounted. The earliest fish have the best chance of reaching the spawning areas. Slightly later fish run the risk of dying without spawning but in compensation, they will have enjoyed the benefit of a longer growing period (several months) in the sea and will be exposed to river mortality factors for a shorter time.

The long growth period in the sea can be explained by supposing: (a) that larger salmon can better overcome obstacles in the river and/or better survive a prolonged fast, and (b) that a high ratio of marine life/total life span is a good survival strategy.

Finally, the time of entry into the estuary is spread over a period of 6 - 8 months, which is probably an adaptive response to variable hydrological conditions in different years.

20.4 MANAGEMENT OF THE RIVER SYSTEM AND THE SALMON RESOURCE

Following the significant decrease in the population of salmon in the Loire catchment which has been evident since the beginning of the nineteenth century, considerable efforts have been made to reverse this tendency during the past 20 or 30 years. Without these efforts, it is probable that salmon would have disappeared by the present time.

Nevertheless, the results obtained have not come up to expectations. The level of the salmon population (at least, the numbers of adults returning to the river) has not noticeably increased. This failure is due, in part, to the growth of the Greenland fishery since 1960; but it is due also, in our opinion, to the general state of the river system.

Remedial measures to improve fish passage over dams have not been undertaken in the upper reaches of the system. On the Loire, where salmon are known only to professional fishermen, many weirs and dams have been built between 1960 and 1975, without any fish passage facilities having been planned. These facilities are now being planned but are still not under construction. On the lower reaches of the Allier itself (Départements de la Nièvre et du Cher), opposition by anglers, over three years, has delayed the provision of a fish pass at a weir (Pont-du-Guetin), where it will be necessary.

Net fishermen on the Loire blame the scarcity of salmon on the depredations of rod fishermen 'on the spawning grounds'. But the rod fishermen, each of whom catches 0.4 salmon per year on the Allier, fail to understand why net-fishing for salmon is still allowed on the Loire and why they are virtually the only people who contribute to the expenses of protecting the salmon (fish passes, bailiffs, restocking exercises, etc.). As for the estuary fishermen, no-one knows what they think nor what they take, in terms of fish. They are too far from the spawning grounds and the 'Domaine Maritime' administration has too much to do and too few resources, to worry about salmon.

Restocking is financed almost completely by rod fishermen, although they obtain barely half the captures in the home river and much less than half the total recaptures if one includes those off Greenland. Again, we find (Bomassi, 1985) that restocking with eggs of Scottish origin has given much lower survival rates than those obtained from Loire-Allier eggs. As a result, we are now trying to augment the numbers of eggs from native fish by rearing broodstock to maturity entirely in fresh water (Carmie and Jonard, 1985).

Sand and gravel extraction from the Allier has been forbidden since 1979 by most of the appropriate local authorities but is still allowed in some areas (lower reaches of the Allier and the Loire) even though the damage caused by such activities is considerable.

Certain riverside towns, for whom salmon quite evidently have no priority, are actually considering construction of dams across the Loire, to form marinas. One more problem for salmon!

Despite the progress that has been made in the Loire-Allier system in certain areas - reduced pollution, maintenance of dry-weather flows by large reservoirs (Naussac, Villerest) - there is still insufficient coordination of the regulations and the fishery protection service, especially as the latter relates to salmon held up below obstacles or in the control of catches in the estuary ('zone maritime') or in the 'mixed zone'.

The Loire-Allier system passes through (or borders) four Regional Boards of the CSP (primarily Poitiers and Clermont-Ferrand, secondarily Lyon and Montpelier) and more especially, eleven 'Departéments', each of whom have different ideas and priorities.

If we are serious about restoring the salmon population of such a river system, we must have an organisation which spans the geographical boundaries, responsible for coordinating management, for the dissemination of information and for the application of cohesive and effective regulations. It will be necessary too, to give serious attention to the allocation of the resource. In effect, the continual conflict between net fishermen at the bottom of the river and rod fishermen in the upper reaches is only nullifying any efforts at restoration, each group believing that any sacrifices made will only benefit the other. A restoration programme is in progress on the Dordogne at present and they risk encountering similar problems from the fact that four Departéments are concerned.

The Loire-Allier salmon must succeed in crossing 15 dams in one year, without feeding. How many years must we wait until a single 'Migratory Fish Authority for the Loire-Allier System' (also involving the Vienne-Creuse-Gartempe branch) is formed, capable of breaking down the barriers which separate the administrative departments and particularly those which separate 'Maritime' and 'River' administrations? An Authority which can make and enforce decisions? Whilst we wait, the Atlantic salmon will remain a threatened species in a river system of this size.

REFERENCES

Bachelier, R. (1963) L'histoire du Saumon en Loire, Bulletin Francais de Pisciculture, 211, 212, 213, 53 pp

Baril, D. and Gueneau, P. (1986) Radio-pistage de saumons adultes (Salmo salar) en Loire - 4ème DR du CSP Bulletin Français de Pisciculture, 302, 86-105

Bomassi, P. (1985) Comparaison des recaptures d'adultes après marquage de saumoneaux d'élevage selon l'origine des oeufs : souche Allier ou souche étrangere. Communication au Colloque franco-québécois à Bergerac (France), 28 May 1 June, 1985. Polyc., 11 pp

Bomassi, P. and Travade, F. (Personal communication) Projet de réimplantation du Saumon dans la partie supérieure de l'Allier : Expériences sur les possibilités de dévalaison des saumoneaux au barrage hydro-électrique de Poutès en 1983 et 1984

Carmie, H. and Jonard, L. (1985) Utilisation de saumons Atlantiques entièrement éléves en eau douce pour la production d'oeufs et de saumoneaux de repeuplement. Premiers résultats obtenus à Augerolles. Polyc. 21 pp

Clavel, P., Cuinat, R. Hamon, Y. and Romaneix, C. (1977) Effects des extractions de matériaux alluvionnaires sur l'environnement aquatique dans les cours supérieurs de la Loire et de l'Allier. Bulletin Français de Pisciculture, 268, 122-54

Cuinat, R. (1980) Le Saumon Loire-Allier : caractéristiques actuelles et perspectives. Symp. FAO, CECPI, Vichy Avril 80, dactyl. 18 pp

Cuinat, R. and Bomassi, P. (1985) Evolution de la situation pour le Saumon du bassin Loire-Allier, de 1979 a 1985. Polyc. 21 pp

Cuinat, R. (1981) Conséquences écologiques des extractions dans quelques grands cours d'eau d'Auvergne. Extrait d'Equipement Mécanique, Carrières et Matériaux, 190, 5 pp

Cuinat, R. (1986) Comportement migratoire du Saumon Loire-Allier. Problème des obstacles. Session Définition et contrôle de l'efficacité des passes à poissons - Société Hydrotechnique de France et Agence de Bassin Loire-Bretagne, Orléans 19 - 20 March, Polyc. 13 pp

Cuinat, R., Bomassi, P., Bousqet, B., Joberton, G. and Marty, A. (1980) Observations sur les juvéniles (smolts) de Saumon atlantique bloqués dans la prise d'eau d'une centrale nucléaire sur la Loire. FAO/CECPI, Vichy, Polyc. 17 pp

Prevost, E. (1985) Les populations de Saumon atlantique (Salmo salar L.) en France. Premières analyses biométriques et scalimétriques en liaison avec les caractéristiques des bassins versants. Mémoire de fin d'études à l'ENS A. de Rennes, 63 pp

Chapter Twenty One

THE RESTORATION OF THE JACQUES-CARTIER: A MAJOR CHALLENGE AND A COLLECTIVE PRIDE

Marcel Frenette[1], Pierre Dulude[2] and Michel Beaurivage[3]

1   Prof. University of Laval, Dept. of Civil Engineering and
     President of the Committee of Restoration of the
     Jacques-Cartier river (CRJC)
2   Biologist, Ministry of Leisure, Hunting and Fishing (MLCP)
3   Biologist, Committee of Restoration of the Jacques-Cartier
     river (CRJC)

> The salmon were always jumping in the river
> and their voices were singing in time to the
> boiling waters in a soft refrain:
>
>   Tow row row, will ye now
>   Take us while we're in the humor.
>
> (Account of a fishing trip on the Jacques-Cartier,
>   Tolfrey, 1814).

21.1   INTRODUCTION

The Jacques-Cartier river, a tributary of the Saint Lawrence, is situated in the heart of Quebec, some 50 km upstream from Quebec city and 200 km downstream of Montreal (Figure 21.1). Several peculiarities distinguish it, beyond the fact that it was considered formerly as 'one of the best salmon rivers in Canada' (Tolfrey, 1816) and that its history goes back to the second voyage of discovery in Canada, by the French navigator, Jacques-Cartier, in 1535, after whom it was named.

The first evidence of decline of salmon in the Jacques-Cartier appears in 1830 with excessive fishing and poaching; it became an established fact in 1913 with the completion of a barrage across the mouth of the river. This barrage, always in operation, situated at Donnacona served several purposes at the same time: supply of water to the paper industry, transport of wood and hydro-electric energy. A few years were then sufficient to result in the complete disappearance of this species; a few years compared to several regular cycles.

The official start of the project for the restoration of the Jacques-Cartier was in 1979 with the formation of the Committee for the Jacques-Cartier river (CRJC) and the

400

Figure 21.1: Situation of the Jacques-Cartier river basin

preparation of a joint project with the Ministry of Leisure, Hunting and Fishing (MLCP) seeking to define the problems of reintroduction of salmon in the Jacques-Cartier (MLCP-CRJ, 1979). In 1980, the CRJC undertook the evaluation of the general state of the river in relation to salmon in a study entitled: 'A biophysical study of the Jacques-Cartier river' (CRJC, 1980). In 1981, the MLCP proceeded to the first investigations while the CRJC completed the first socio-economic study (CRJC, 1981).

Meanwhile attention was drawn to the restoration from the results of a major undertaking several months beforehand by a group of anglers joined by several municipal riparian owners, a few scientists and various regional sports associations, economic and cultural. At the start, therefore, if it had not been for the perseverance of the anglers and of many people who met many difficulties and much scepticism, the project would have been in vain. The complex make-up of the 'restoration pioneers' constituted in itself the triggering-off of the project which like a chain-effect, has become the 'renaissance' of salmon in the Jacques-Cartier by being both a 'daily-occurrence' and becoming a collective pride.

21.2   A SHORT HISTORICAL REVIEW

According to the ethnologist, H. Germain (1984), at the beginning of the French regime, the Jacques-Cartier river served as a highway for Norse trappers for the fur trade, providing all with a source of subsistence through the practice of salmon fishing. It was particularly as a 'source of fish' that the Jacques-Cartier river was best-known, first for the Amerindians who occupied the territory and, following that, for the colonial inhabitants. It was also, moreover, where the brave Jacques-Cartier was located at Donnacona near the mouth of the river and served as a refuge for the French army after the battle of the Plains of Abraham in 1760, making Canada a British Colony.

The historian P.L. Martin (1980) emphasises, on the other hand, in his book History of Hunting in Quebec that: the wealth of salmon in the Jacques-Cartier is reported as early as 1768 in a place called 'the salmon fishery' and up to the Napoleonic War and the American War of Independence of 1812-1814, the colony received some officers on half-pay, mostly of British origin. The clan spirit, favoured by the organisation of the everyday life of armies and of the military camaraderie, is consequently the origin of the first practice of salmon-fishing in Canada. The Jacques-Cartier was the first place for the emergence of hunting and fishing clubs, of which 'The Jacques-Cartier River Fishing Club' was formed in 1877 to disappear in 1913 with the construction of the Donnacona barrage.

Some British officers like Frederick Tolfrey (1814) and J.P. Cockburn (1828) have upheld the times which bore witness to the merits of a fishing trip on the Jacques-Cartier, of the savage beauty of the river, and of feasts which one could have residing in these places. 'The stories of journeys' and illustrated albums of this period attracted a good number of strangers greedy for the experience, those, who could, saw as a consequence the success of the fishing diminish considerably towards 1830 (Gatien, 1955). To put an end to the fishing abuse, Joseph Knight Boswell reserved for himself, as early as 1840, some plots of land opposite the salmon falls, also constituting private fishing reserves. In order to benefit a fishing activity with special features, J.K. Boswell excavated, towards 1867, a canal more than 50 metres long to by-pass the Pont-Rouge fall (18 km upstream from the mouth of the river) and, towards 1871, planned the construction of three fish passes in a loop of the river, accessible with difficulty, called 'Déry Gorge', situated about one kilometre downstream from the canal.

Meanwhile, there was passed in Canada the promulgation of laws for the protection of the fauna. The salmon then became the object of attempts at conservation. Also the first fish hatchery in Canada, established by Richard Nettle, in Quebec in 1857, used trout and salmon eggs from the Jacques-Cartier.

In short, the Jacques-Cartier renowned as a celebrated fishing area was the object of several historic accounts in works describing the exceptional interest of this privileged fishing place and of residence among natural scenery which, again today, has conserved all its character.

21.3 THE OBJECTIVES OF THE CRJC

In 1979 at the time of the passing of its charter, the CRJC set several objectives which can be summarised in these words: Restore, Conserve and Value.

More specifically, the aims envisaged were:

(1) The restoration and conservation of the river and its environment.
(2) The increase in value of the salmon resource.
(3) The valuation of the banks.
(4) The provision of structures permitting the return of salmon and the use of the river for recreational tourists.
(5) The use of the Jacques-Cartier for teaching and research.
(6) The development of the culture and heritage of the river.
(7) The confirmation and strengthening of the development of the socio-economic aims.
(8) The control, the management and the taking into custody by the local people.

It is doubtful if such objectives could be achieved without the complicity of different levels of government (federal, provincial and municipal) in addition to a contribution from local associations and the regional population who are dedicated to the success of the scheme.

It is, therefore, time to make an attack on the waters of the Jacques-Cartier (pollution control etc.) and its shores (cleaning the banks, setting up reception centres, some view-points, camping sites and access areas); only then will the project take root to become an important development focus for the region, as well as a scheme for recreational tourism, socio-economic, cultural and heritage benefits.

21.4: HYDROGRAPHY OF THE RIVER

The Jacques-Cartier river is one of the important tributaries on the north bank of the Saint-Lawrence in the neighbourhood of the conurbation of Quebec. Its rather elongated catchment covers approximately 2515 km^2 (Figure 21.1).

Figure 21.2: Length profile and average monthly flow of the Jacques-Cartier river

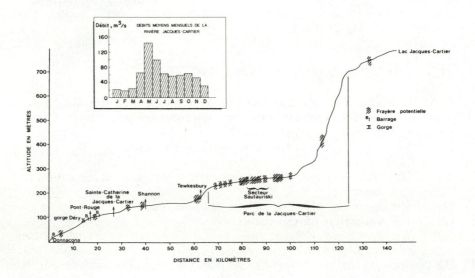

The source of the river is at 853 m in the Laurentian Park (Figure 21.2). It drains, over a distance of some 145 km, into the Laurentian Shield at the lower area of the Saint Lawrence. The high part of the Jacques-Cartier basin lies between Tewkesbury and Shannon, originating from Precambrian formations characterised by the presence of igneous rocks, which exert a preponderant effect on the water quality of this river. In contrast, the lower part of the drainage basin, with formations of more recent origin which are Ordovician, is characterised by the presence of sedimentary rocks which are associated with the lower area of the Saint Lawrence.

Half the catchment is covered with a homogeneous forest of conifers whereas in the southern half there is a mixed forest where there are some conifers but also some areas with non-forest trees.

The annual rainfall is between 1000 and 1100 mm i.e. a region of high precipitation which has an obvious effect on the river flow.

Figure 21.2 depicts the average monthly flow near the mouth of the Jacques-Cartier river. May, generally, is the month in which the highest flow occurs, which corresponds to the period of snow-melt. The time of large floods is spread over the period April to June. In July, August and September there is a diminution in flow which corresponds to summer drought. In October and November the flows may improve temporarily before they undergo a major drop which is prolonged throughout the winter; February is the most critical month.

The salmon, as a species, has consequently to adapt itself to variations in monthly flow of the order of 1 to 10. The period of high flow corresponds with the downstream smolt migration to the sea. In June, July and August the ascent commences of adult salmon for spawning which is from late October until November. During the winter period the eggs, deposited in the gravel by the spawners, incubate. The percolation of water through the river bed assures the development of the eggs. The eggs hatch in April and May, then the young live in the river for two, three or four years before migrating to sea at the time of the spring floods. The presence of several lakes at the head of the Jacques-Cartier catchment provides a natural regulation of daily flows. Moreover, the regulation of summer flows could be possible by local retention of the discharge from several of these lakes behind barrages eventually providing a major improvement in the management of the salmon population.

On the physiographic plan, the Jacques-Cartier moreover has a unique character, since it drains into the fresh water of the Saint Lawrence. This aspect at the same time confers on the Saint Lawrence an important start to the migration route since it gives to the river several standard indices of pollution, notably in the standard of the zone of salt and freshwater situated between the Île d'Orléans and the Île aux Coudres, which are more than 60 km downstream of the mouth of the Jacques-Cartier river (Figure 21.1).

To show that the Jacques-Cartier river can provide a new area for the reproduction and survival of Atlantic salmon it was also necessary to show that the Saint Lawrence itself did not constitute an obstacle to the migration of salmon, where there is a double challenge to migration.

21.5   THE SALMON RESOURCE - THE HEART OF THE RESTORATION

Restocking

In the salmon scheme, the restoration of the Jacques-Cartier is based on a massive restocking programme, using the spawners adapted to the river conditions in the north branch of the Saint Lawrence. At the start, within the limits of this programme, it was anticipated by the MLCP that restocking, to begin in 1981, would be with 100,000 alevins (unfed fry) and 30,000 1+ parr annually over a 7-year period.

The 20 May 1981 marked an important event in the restoration of the river when 5,500 smolts were released with the aim of seeing if the salmon could complete its life cycle in saltwater and return to its 'adopted' river in spite of the pollution in the main river (i.e. Saint Lawrence). Another batch of 5,325 post-smolts were also released into the river in September. At last, in June, the massive restocking programme began with the release of some 100,000 alevins (unfed fry).

In order that the Jacques-Cartier salmon should be recognised, notices were issued that the adipose fin had been clipped on all hatchery smolts and about 10 per cent of them had been tagged.

Over the course of the years, some problems in the hatchery unfortunately had not helped the restocking programme, notably during 1983-5. A total of 202,450 alevins, 83,932 1+ parr and 10,825 smolts were released between 1981 and 1985 (Table 21.1). The restocking programme will be continued until 1994 with priority being given to the use of spawners from the Jacques-Cartier. This will help to further the chances of developing the 'Jacques-Cartier strain' in order finally to attain as quickly as possible the production capacity of this river.

Table 21.1: Release of salmon carried out by the MLCP from 1981 to 1985 in the Jacques-Cartier river

Stage	Number of salmon released					
	1981	1982	1983	1984	1985	Total
Alevins	100,000	90,000	750	11,700	0	202,450
Parr		25,500	30,032	24,000	4,400	83,932
Smolts	10,825					10,825

The returns of salmon

Following the stocking of salmon smolts in 1981, four grilse were noted below the Donnacona dam in 1982, thus proving the feasibility of the project (Table 21.2 and Figure 21.3). In 1983

Table 21.2: Numbers of salmon returning to the Jacques-Cartier river since the MLCP restocking programme began in 1981

Category	Number of salmon					
	1982	1983	1984	1985	1986[a]	Total
Grilse	4	53	146	100	42	
Salmon		15	15	262	212	
Total	4	68	161	363	254	849

Note: a, to 29 July.

and 1984, when the first significant returns of salmon were expected, a 'false fish-pass' was installed on the lower edge of the dam and by this means, 68 and 161 salmon were recorded in the two years, respectively. These totals included fish taken by sport fishermen. During these two years, broodstock was transported to the rearing station in order to establish the 'Jacques-Cartier strain' whilst the majority of the grilse were released in the river above the Donnacona dam, for sport fishing.

Following this successful return of salmon to the river, priority was given to the construction of a fish pass at the Donnacona dam. After protracted negotiations with the Domtar company, which owned the pulp and paper mill, the fish pass was

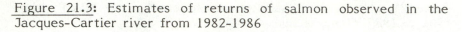

Figure 21.3: Estimates of returns of salmon observed in the Jacques-Cartier river from 1982-1986

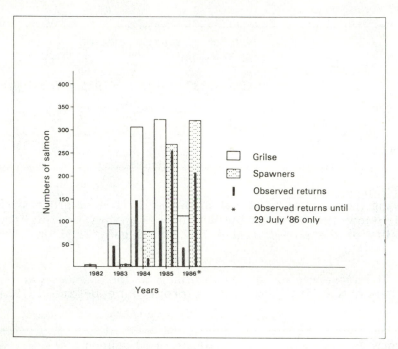

completed in 1985, thanks to financial aid from the Quebec government.

This fish pass has 17 concrete pools and overfalls, an auxiliary water feed-pipe to increase the amount of 'attraction' water and a deflecting wall lower down the river which forms an artificial holding pool. The installation also allows us to hold salmon after capture in order to transport them later to specific required areas, given the existence of other major river obstacles at Pont-Rouge.

The fish pass began operating on 12 July 1985 and by the next day, the first salmon was recorded, less than 18 hours later.

In 1985, when 361 salmon were recorded, most of the large salmon were transported to a tributary stream where all fishing was prohibited. The remainder were kept, either for artificial reproduction in the hatchery or were released in the river for sport fishing and natural reproduction. All these operations were carried out jointly by MLCP and CRJC.

1985 was an historic year for the Jacques-Cartier, in that for the first time since salmon disappeared from the river, salmon once again spawned naturally in the upper regions of the river, particularly in the area of the Jacques-Cartier National

Park.

The 1986 upstream run by 29 July, comprised 42 grilse and 212 salmon counted through the fish pass.

The fish pass has been fitted with a large observation window and this has proved to be a very popular tourist attraction. Indeed, almost 4,000 people visited the site in 1985, even though it was not officially open to the public. In 1986, more than 14,000 people have come to see the salmon although visitor reception facilities are still incomplete.

In brief, then, these returns have demonstrated the success of the restocking programme of the MLCP and have proved that the reintroduction of salmon to the Jacques-Cartier has already provided returns to the recreational and tourist programme.

Protection of salmon stocks and fishery organisation

As far as protection of salmon stocks is concerned, the important work of the voluntary conservators should be recognised. These are volunteers who actively support the efforts of the protection staff of the MLCP in maintaining a close watch on the river.

Since priority was given to the building-up of the salmon stock of the Jacques-Cartier, the sport fishing has not been organised along traditional lines. It is designed rather, to maintain the interest of the sport fishermen by allowing them to fish largely for grilse as well as a proportion of the larger salmon released for this purpose between Donnacona and Pont-Rouge.

However, even though sport fishing has been allowed along much of the course of the main river, certain restrictive measures have been imposed under the aegis of the regional director of the MLCP. These include: the forbidding of fishing in 300 m lengths below the dams at Donnacona and Pont-Rouge and restricting fishing to 'fly only' between Pont-Rouge and Donnacona. All forms of sport fishing are permitted in the remainder of the river, except for the R. Sautauriski, where broodstock salmon are released and the Epaule and Cachee rivers which are set aside for restocking with salmon juveniles. All the above three tributaries are situated in the Jacques-Cartier National Park.

The salmonid potential of the watershed

The biophysical potential of the river was re-evaluated in 1986 by CRJC and MLCP (CRJC, 1986), following a refinement in the photographic method of analysis. From this study of the productivity of the watershed, the numbers of spawners required and the numbers of fish available for rod-fishing was estimated at about 5,000 salmon (range 4,700 - 6,000) and at more than 200 holding pools for rod-fishing. The number of spawners required to assure this level of return was put at 2,800. Thus, once full reproductive capacity has been attained, a further 2,200 would be available for the river fishing.

Ongoing projects

Two important projects have been inaugurated jointly by MLCP and CRJC in order to plan the progression of the restoration scheme. The first consists of a telemetric study of the degree to which the rapids of the Dery Gorge at Pont-Rouge constitute an obstacle to migrating salmon, in combination with a study of the behaviour of salmon below the Bird dam at Pont-Rouge. Ten salmon were fitted with implanted radio transmitters and from tracking their behaviour, it was possible to plan what action is necessary for improving the upstream passage of fish.

The second project consists of a survey of potential sites for a kelt reconditioning installation. When this project is finalised, it should be possible to produce salmon alevins from the previous year's spawners, without affecting the current year's natural production, thus achieving the full potential productivity more rapidly.

21.6 THE RESTORATION SCHEME AND EDUCATION OF THE PUBLIC

The mandate of the restoration Committee of the Jacques-Cartier also included a public education programme involving this important part of our national heritage.

Since adult human behaviour is not easily influenced and any deep understanding requires a sustained in-depth approach, the CRJC has focused its attention on young people and in consultation with local expert youth counsellors, has prepared an information programme on the restoration of the Jacques-Cartier. In order to assure its use and usefulness, the programme has been integrated with that of the 2nd primary level in natural sciences. Relying more on observation and practical experience, we have suggested an all-round approach, establishing the bonds between the different components of the riverine environment, thus allowing a better understanding of the process of re-introduction of salmon.

A series of research activities are designed for young people to understand that the restoration of the Jacques-Cartier is an environmental project. The child at school is encouraged to ask questions about the role he or she can play and the extent to which everyone can contribute to the discovery of this wealth in our environment.

The programme, called 'Discover your River', was

introduced into nine schools in 1983 and is now being repeated for the fourth consecutive year. This has involved some twenty teachers and youth counsellors and about 350 children each year, i.e. about a thousand children in all since its inception.

We can confirm that young people have learnt a great deal about the aquatic environment, its animals, its plants and above all, its salmon. They have acquired new skills in both knowledge and practical ability as a result of their research and discoveries. In particular, they have developed a genuine feeling for the environment which will undoubtedly prove to be a source of joy and pleasure to them in the future.

21.7   VISITOR CENTRES - A VISIBLE PRESENCE OF THE CRJC IN THE ENVIRONMENT

The Corporation of the Jacques-Cartier river was given the task of supervising the management, the exploitation and the overall utilisation of both the bankside areas and the river itself. To this end, five Visitor Centres have been built along the length of the river, at Donnacona, Pont-Rouge, St Catherine, Shannon and Tewkesbury.

The design of the centres is very well adapted to the particular local environment and history. Built on an Anglo-Norman style, the centres resemble the old-style fishing lodges that were found along the river in the nineteenth century. As well as the building materials used, the decoration and colours employed also reflect the architecture of the period.

Each centre offers a range of tourist services for fishermen and a number of demonstrations on such themes as:

(1)   the life-cycle of the salmon
(2)   the achievements of the CRJC
(3)   leisure pursuits associated with the river
(4)   the history of fishing, human settlement and development in the Jacques-Cartier watershed.

As a result of these demonstrations, the Centres facilitate the appreciation of the riches of the natural environment whilst complementing this appreciation with an understanding of river management and the control of rod-fishing.

21.8   HUNTING AND FISHING ASSOCIATIONS - USEFUL ALLIES

With the advent of the CRJC, the regional sporting associations quickly became interested in involving themselves in the activities of the Corporation.

Initially however, it must be said that we met with a certain amount of scepticism and even apprehension, in respect of a possible conflict of interests between trout and salmon fishermen that the associations thought might arise from the restoration programme. The pleasure and pride associated with the capture of the 'king of freshwater fishes' has now given a new dimension to the relationship of the associations with the CRJC. This is why fishermen quickly became ardent advocates of the project in their area, as well as appreciating the 'spin-offs' to the local economy.

Today, the associations have assumed an even more important function, in that as well as helping with the growth of the CRJC, they have also involved themselves in the areas of protection, conservation and education. Many sportsmen have enrolled as 'Nature Conservation Volunteers' - recognised by the Government - with the aim of assuring the survival of the salmon by educating all river users.

21.9 REVENUE CONSIDERATIONS

From its inception, the CRJC has benefited from different grant programmes, directly and indirectly totalling almost $2 million. The Quebec government, for its part, has injected some $700,000 which has been utilised in the restocking exercises, the river works involved in the planning and construction of the Donnacona fish pass, river bank clearance, administrative support and sociocultural developments as well as a grant towards an ongoing overall plan for the management, exploitation and utilisation of the river and the riverside areas.

The Federal Government has given just over $1 million, used mostly in building the five Visitor Centres and recreational installations such as camping sites and ski trails, involving the creation of jobs. The townships along the river, ten in all, have together given almost $150,000 to the CRJC.

One should not forget an incalculable number of voluntary hours of labour - more than 40 people work voluntarily for the CRJC - as well as the contributions of our members and of various regional organisations. In addition, the river purification programme has involved an expenditure of almost $7 million, of which 70 per cent was provided by the provincial government.

This injection of funds has resulted in the creation of many hundreds of 'occasional' jobs (about 500 in five years), in five permanent jobs and a number of indirect benefits deriving from the tourist industry, purchase of materials and sporting and educational goods. Let us remember too, that the first exponents of the restoration programme - the sport fishermen - are now benefiting from the enjoyment of their favourite sport - salmon fishing.

21.10 SUMMARY

After seven years in existence, it is fair to say that the CRJC is developing in an impressive manner. From very simple beginnings, the reintroduction of salmon into the Jacques-Cartier by the MLCP is now part of a very considerable programme, involving a firmly based restoration programme.

The driving force of this restoration is the local population. From being largely opposed to the development, the population is now the foundation stone of the project. The motivations are:

(1)   <u>To conserve</u> - i.e. protect the river and its banks from inappropriate developments.
(2)   <u>To restore</u> - i.e. recreate the right conditions for the return of salmon, restore the original character of the river which it had lost over the years and improve the water quality and river banks.
(3)   <u>To derive economic benefit</u> - i.e. to manage the river as a function of the development of its salmon potential as well as its tourist, historic, heritage and cultural potential.

The restoration of the Jacques-Cartier was an innovative project from several points of view.

(1)   It was unique, involving the integration of a number of activities associated with a single watercourse.
(2)   It is a regional development laboratory where plans are both made and put into practice.
(3)   It is a political common ground, establishing political links which transcend the usual boundaries.
(4)   It is a project carrying hope for the future, with both human and financial investment in the future of a world too often preoccupied with the present.

To sum up, the Jacques-Cartier restoration project has stimulated feelings for 'our river' which are deeply embedded in the very heart of our local community.

Moreover, the project appeals to an often unconfessed willingness to help and results in a community feeling which has always sought expression, although not always successfully.

Finally, the spirit of unity implicit in the project has resulted in its becoming over the years, a melting pot of people and ideas, a meeting-place for people to discuss and exchange views whilst participating in a huge regional development where 'the salmon is at the heart of the action'.

REFERENCES

Cockburn, J.P. (1828) Oeuvre sur aquarelle: 'The Jacques-Cartier Bridge', Collection Archives publiques du Canada, C-12-634

CRJC (1980) Etude biophysique, rivière Jacques-Cartier, Bibliothèque CRJC, 189 pp

CRJC (1981) Etude socio-économique du bassin hydrographique de la Jacques-Cartier, Bibliothèque CRJC, 95 pp

CRJC (1986) Potentiel salmonicole du bassin de la rivière Jacques-Cartier et caractéristiques des fosses à saumon sur le cours principal, étude réalisée par la firme Gilles Shooner et Associés, biologistes, en collaboration avec le MLCP. Bibliothèque CRJC, 22 pp, plus tableaux et annexes.

Gatien, F. (1955) Histoire de Cap-Santé depuis la fondation de cette paroisse en 1769 jusqu'à 1955, Document interne, Archives de Cap-Santé, 176 pp

Germain, H. (1984) Aspects historiques de la pêche au saumon sur la rivière Jacques-Cartier. Les Carnets de Zoologie (Québec), 44, (1), 5-7

Martin, P.L. (1980) Histoire de la chasse au Québec, Editions du Boréal Express, Montréal, pp 56-80

MLCP-CRJC (1979) Avant-project de restauration du saumon dans la rivière Jacques-Cartier, Bibliothèque MLCP-CRJC, 64 pp

Tolfrey, F. (1816) Un aristocrate au Bas-Canada présenté par Paul-Louis Martin. Editions du Boréal Express, Montréal, 1979

Chapter Twenty Two

ATLANTIC SALMON RESTORATION IN THE CONNECTICUT
RIVER

R.A. Jones

Department of Environmental Protection, Bureau of Fisheries,
State of Connecticut

22.1   INTRODUCTION

The magnitude of the historic runs of Atlantic salmon in the
Connecticut River is not well documented. However, local
historical references and the nature of the 29,000 km^2 watershed
suggest that runs may have been the largest in North America. In
1798 the first dam to block fish migration in the Connecticut
River was completed in the vicinity of Turner's Falls,
Massachusetts, about 190 km from the mouth of the river. During
the same time period, known as the Industrial Revolution in the
United States, virtually every tributary of the Connecticut was
dammed to produce inexpensive power. By the 1820s Atlantic
salmon had disappeared from the river and the population of
American shad had been seriously depleted. In 1867 the Fish
Commissioners of the States of Connecticut, Massachusetts, New
Hampshire and Vermont, all orginally appointed to investigate the
problems of salmon and shad, met 'to cooperate in the restocking
of the Connecticut with salmon and shad'. The effort involved
the stocking of hundreds of thousands of juvenile salmon and
although over 800 adult salmon were known to have returned to
the river, this early salmon restoration scheme failed.
Unsuccessful fishways and an inability to control commercial
fishing were the major causes of the failure (Jones, 1978).
    Interest in salmon restoration seemed to re-awaken about
every quarter century since those early efforts. It was not until
the passage by the US Congress, in 1965, of the Anadromous Fish
Conservation Act that serious efforts were again initiated to
restore anadromous fish to the Connecticut River Basin. This Act
provided federal funding aid to state natural resource agencies
for such restoration projects. During 1966, the respective state
and federal biologists and administrators, recognising that the
funds available to the individual states were insufficient to
accomplish the needed effort, began discussions regarding

cooperative anadromous fisheries restoration in the river. Finally, in 1967, 100 years from the date of the historic meeting of their predecessors, the fish and wildlife agency administrators from Connecticut, Massachusetts, Vermont and New Hampshire joined with their counterparts in the federal government and agreed to support an anadromous fisheries programme for the Connecticut River Basin.

22.2 THE CONNECTICUT RIVER ANADROMOUS FISHERIES RESTORATION PROGRAMME

Programme administration

In order to provide overall direction to the restoration programme the state and federal administrators agreed to participate in a Connecticut River Fisheries Management Policy Committee. In a 'Statement of Intent' the Policy Committee set the framework for what turned out to be nearly 20 years of effort. In this statement goals were set, estimated benefits described and problem areas and research needs were defined. It was determined that the programme would return American shad to their historical spawning area in the river and attempt to restore Atlantic salmon to some portion of their historical range in the basin. Among its first official acts, this committee named a Technical Committee composed of senior state and federal biologists. The Technical Committee was designated to 'design and implement needed research programmes' and 'develop and recommend sound fishery management practices' (Divine, 1975).

During this period the Policy Committee with the assistance of the Technical Committee became a rather unique organisation in the management of regional fisheries programmes. As an organisation established on little more than a 'handshake', the Committee was able to bring together state and federal government agencies, private industry, special interest groups and average citizens in a cooperative effort with a common goal.

As efforts associated with shad restoration neared completion and public interest in the Atlantic salmon programme became more intense, it became clear that a more formal institutional arrangement was necessary for programme administration. Under the direction of the Policy Committee, legislation was drafted which, once enacted by all four basin states and ratified by the US Congress, would create an interstate agreement known as the Connecticut River Atlantic Salmon Commission. This legislation provided for membership to include a knowledgeable citizen from each of the basin states, appointed by the respective governors, as well as the members of the original Policy Committee. In addition to the duties assumed by the Policy Committee, the Commission was authorised to regulate the salmon fishery in the main stem of the Connecticut River and to issue a licence for participation in any Atlantic salmon fishery within the basin. On approval by the US Congress in 1983, the Connecticut River Atlantic Salmon Commission held its first meeting in January 1984 and took up the task of administering the salmon restoration programme in the Connecticut River Basin.

Fish passage

The basic restoration programme was directed primarily at providing fish passage facilities at the lower five main-stem dams: Holyoke Dam, in Massachusetts, at river km 140; Turners Falls Dam, also in Massachusetts, at river km 196; Vernon Dam, Bellows Falls Dam and Wilder Dam, all in Vermont and New Hampshire (the Connecticut River being the common boundary), at river km 228, 280 and 349 respectively. In addition, two dams on tributaries in Connecticut were selected for fishways, Rainbow Dam on the Farmington River which enters the Connecticut at river km 88; and Leesville Dam on the Salmon River which enters at river km 29 (Rizzo, 1970).

The process used to initiate the construction of these needed facilities involved principally the demonstration of programme viability and the level of public interest in the programme. All of the main stem dams were hydro-electric dams licensed under the Federal Power Act and licence language specified that fish passage facilities would be required when such was deemed to be in the best public interest. The Federal Power Commission, now known as the Federal Energy Regulatory Commission (FERC), was the licensing agency and was empowered to order the installation of fishways once programme viability and public interest were shown.

Since American shad restoration was a major project purpose and since this species was present in the river, the demonstration of programme viability was a relatively easy task. Major efforts involved the quantification of existing spawning and nursery habitat and the assessment of anticipated production in each area made available by succeeding fishway installation. Any cost/benefit analysis that was necessary was based on the restored shad population. Those benefits that might accrue from the attempt to restore Atlantic salmon were considered to be in excess of that needed for acceptability. Finally, through a combination of adversarial confrontation and cooperative negotiation, agreements were reached by all parties for a phased development of facilities at each dam. For each needed facility fishway design parameters were produced by the US Fish and Wildlife Service in cooperation with the Technical Committee and made available to the power companies as guidance in the development of final design.

These efforts to provide fish passage have resulted in the redesign and reconstruction, by the Holyoke Water Power Company, of a lift type facility, orginally built in 1955, at Holyoke, Massachusetts (completed in 1976); the construction by the Western Massachusetts Electric Company of a multi-fishway facility at Turners Falls, Massachusetts, completed in 1980; the construction of facilities by the New England Power Company at Vernon, Vermont (completed in 1981), at Bellows Falls, Vermont (completed in 1984) and at Wilder, Vermont (to be completed in 1987).

In Connecticut a fishway at the Rainbow Dam, on the Farmington River, was put into operation in 1976. Its construction was a cooperative effort by the Connecticut Department of Environmental Protection (DEP), the United States Fish and Wildlife Service, and the Stanley Works, owner of the dam, and is operated by the DEP. A fishway at the Connecticut state-owned Leesville Dam, on the Salmon River was completed by the DEP in late 1980 and put into operation in 1981.

New hydropower development

New hydropower development is an ongoing and continuing issue. During the early years of the restoration programme, a major pump storage facility was constructed at river km 203. This facility utilised the Turners Falls Dam impoundment as its lower reservoir and had the potential of seriously disorienting migrating anadromous fish. Subsequent studies indicated that disorientation did not occur and that entrainment was minimal. However, the power company, as a condition of its operating license, agreed to construct appropriate protective devices if such were shown to be necessary in the future.

With the advent of the world-wide oil crisis, a re-awakening of the historical interest in small hydropower development has occurred in the United States and particularly in New England. This matter is of continuing concern and every effort is being made to assure compatibility with fisheries interests wherever possible. Where compatibility is not possible, mitigation or denial of required licensing is sought through the appropriate state and federal regulatory agencies. Recent developments suggest that the Federal Energy Regulatory Commission may suspend action on all pending and new hydropower licence applications within the Connecticut River Basin until the Commission staff has completed a 'cumulative environmental impact' assessment of all existing hydro-electric facilities in the Basin.

Other issues

Although primary consideration was given to providing access to historical spawning grounds, other issues had to be addressed during programme development. Of these issues, water pollution, including thermal, primarily from nuclear-fuelled steam-electric stations, was the most significant. Continuing efforts on the part of state and federal agencies in recent years have caused great strides to be made in water pollution control. Early in the programme it was determined that water quality was not a limiting factor in the restoration of efforts. The development of nuclear power, however, posed unknown problems that required detailed study. A five-year study on the effect of heated water effluents on the behaviour of anadromous fish (primarily American shad) was conducted in conjunction with the construction and operation of a one megawatt nuclear plant on the Connecticut River at km 32. The results of this study indicated that the operation of this plant had an insignificant impact on migrating anadromous fish (Leggett, 1976). Due primarily to the efforts of fisheries interests, cooling towers were included in the construction of another nuclear facility at river km 230 in Vernon, Vermont.

22.3   THE ATLANTIC SALMON PROGRAMME

The salmon programme specifically involved the assessment of salmon habitat, the acquisition of salmon eggs and rearing of juveniles, the construction of smolt release facilities, the construction of new hatchery facilities, the conversion of existing trout hatcheries to salmon production, and the construction of adult salmon holding facilities.

The major problem identified in the salmon programme was the loss of the gene pool of Connecticut River salmon. Given that salmon inhabiting different river systems have a somewhat different genetic make-up, the 'Connecticut River' salmon was extirpated in the early 1800s. Since 1967 nearly six million juvenile salmon (fry, parr and smolts) have been introduced into the Connecticut River and its tributaries (Anon., 1982; Rideout and McLaughlin, 1985; Rideout and Sillas, 1985). Initially eggs were obtained from whatever sources were available, primarily Canadian, and reared in various existing state and federal hatcheries in the basin. Whenever possible, salmon were reared to smolt size and released in locations below the Holyoke Dam.

Starting in 1974, efforts were made to obtain salmon eggs from specific sources in an attempt to select genetic characteristics most suitable to the Connecticut River. A primary consideration was to use åt least one parent from strains as geographically close to the Connecticut River as possible. During

1976 approximately 29,000 smolts, the progeny from a cross between adults from the Penobscot River in Maine and a river system in Quebec, were released in the Farmington River. Also in 1976, the first Connecticut River salmon to survive until spawning, a male, was used to fertilise 41,000 eggs from Penobscot River females (Lance, 1977).

Although stocking efforts through 1975 (over 500,000 smolts and parr) had yielded insignificant returns (13 fish from 1967 to 1977), the 1976 releases produced the first major adult returns in the programme. In 1978, 83 salmon were known to return to the Connecticut River. Of these returns, 79 were taken from traps in the Rainbow and Holyoke fishways and transported to the Berkshire National Fish Hatchery in Massachusetts, to be held for spawning. The offspring from these fish were thought to be the beginning of a new gene pool of Connecticut River salmon.

The capture, transportation and holding of these wild, 4-6 kg fish under hatchery conditions and by personnel unfamiliar with the pecularities of Atlantic salmon was not without difficulty. Stress, bacterial infection, and parasite infestations all took their toll and only two of the 1978 returns survived until spawning. The experience, however, provided substantial benefit. Fish culturists, biologists and administrators all gained invaluable experience in the problems associated with the artificial culture of Atlantic salmon. The application of the knowledge gained has increased the survival of wild adults held to spawning to nearly 95 per cent in recent years.

Since the future of the programme rests with the progeny of returning fish, significant efforts have been undertaken to upgrade the ability to hold and rear Atlantic salmon. A new federal hatchery at Bethel, Vermont, dedicated to the production of salmon for the Connecticut River, went into production in 1978. The states of Connecticut and Massachusetts converted existing trout hatcheries into salmon production. Connecticut and the US Fish and Wildlife Service have constructed special facilities for the holding of salmon from capture until spawning. Major efforts have been undertaken to develop disease and parasite control measures and specialised fish culture techniques. Figure 22.1 shows the numbers of juvenile salmon released in the Connecticut River basin from 1977 to 1986.

The end result of the Connecticut River basin Atlantic salmon culture programme, the return of adults, has been as expected - poor. With the loss of the Connecticut River strain of Atlantic salmon in the early 1800s and with the loss of virtually all salmon stocks south of Maine, it has been recognised that the re-creation of this stock would be difficult at best. Results have improved with the ongoing effort. Returns during the early years of the Connecticut River restoration programme were disappoint-

Figure 22.1: Juvenile salmon releases 1977-1986

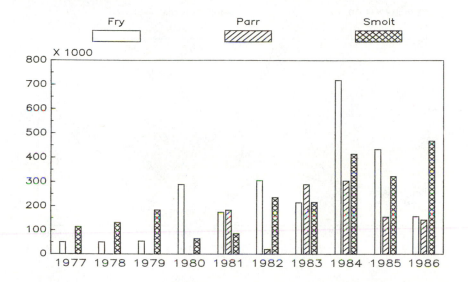

ing - 13 adults in the first 10 years. Returns during the second decade of the programme have been somewhat more encouraging but still disappointing. Figure 22.2 depicts the relationship between smolt stocking and the resulting return of two sea year adult during the last ten years of the programme. What is not shown is the relationship between the source of the stock and subsequent returns. Other factors such as size of smolts at release, disease problems at rearing stations, and climatological conditions at stocking time, as well as overall marine survival, are known to have significant impacts on return rates.

22.4   PROGRAMME COSTS

Although it is virtually impossible to quantify all the costs associated with the Connecticut River Atlantic Salmon Restoration Programme, certain isolated cost figures will indicate the magnitude of the financial investment that has been made. The reported cost of the construction of the multi-fishway facility at Turners Falls, Massachusetts was 13.5 million dollars; the fishway at Vernon, Vermont approximately nine million dollars; and the redevelopment of the fishlift at Holyoke, Massachusetts 3.5 million dollars. Connecticut has spent over

Figure 22.2: Smolt release vs. two sea year adult returns

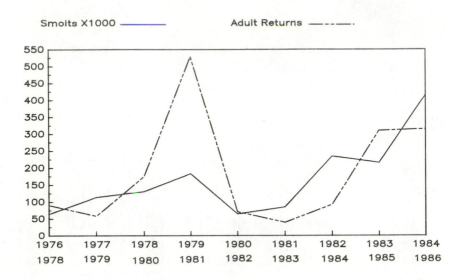

Smolts X1000 ————   Adult Returns —·——·—·—.

$100,000 on the conversion and redevelopment of an existing trout hatchery to a salmon production facility. Connecticut's present operation and maintenance costs for two fish passage facilities and our adult holding facility exceed $200,000 annually.

22.5 THE FUTURE

Present salmon production facilities in the basin have a capability of handling up to 1,000 adults and can produce 3.5 million eggs, two million fry and 500,000 smolts annually. It is anticipated that all returning fish will be retained for hatchery purposes until these figures are reached. Based on habitat assessments, it is estimated that 5,570 adult salmon will be needed to utilise the available spawning area in the Connecticut River system (Stolte, 1982). When the total of 6,570 returning adults is reached and a number of additional fish are available for recreational fishing, the goals of the programme will have been fulfilled.

Figure 22.3 shows two mathematical projections based on existing production capabilities and reasonably conservative freshwater survival rates but with different (and in each case increasing) marine survival rates. It is believed that as additional generations of 'Connecticut River salmon' are produced and as the numbers of wild smolts increase return rates will improve.

422

Figure 22.3: Salmon return projections at differing marine survival rates

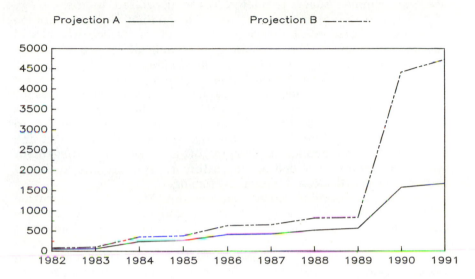

Projection A is considered a minimal return and is calculated on the basis of an average marine survival of 0.4 per cent. Projection B, which can be judged to meet the programme goal, is based on a 1.4 per cent marine survival rate. Predicting the year during which the programme goals are likely to be reached is difficult at best and, from a public relations standpoint, potentially dangerous. It is noteworthy that both projections predict a return in 1985 and 1986 somewhat above that which actually occurred - 310 and 315 respectively.

Yet to be addressed in any detail are the management issues associated with the development of a recreational fishery for Atlantic salmon in southern New England. The Connecticut River Atlantic Salmon Commission has promulgated regulations which prohibit the taking of Atlantic salmon in the main stem of the river. These regulations provide that the Commission may declare an open season if, in its judgement, adequate numbers of salmon are available. During any such open season, salmon fishing shall be by fly fishing only and the Commission shall determine the daily and season catch limits. The states within the basin have also promulgated similar regulations for their respective tributaries. Provisions for a Connecticut River Basin Atlantic Salmon licence and tags are included in the Commission's regulations. Under the provisions of the enabling agreement legislation, anyone wishing to take or possess Atlantic salmon anywhere in the basin must have obtained such a licence.

Clearly the measures now in place are insufficient for the

future. The imposition of a developing Atlantic salmon fishery upon southern New England's existing mixed recreational fisheries presents complex social problems, often unrelated to the biology of the species involved and not found in historic Atlantic salmon fisheries. Coupled with the potential level of effort which will accrue to a new fishery within easy reach of a population of more than 20 million (Massachusetts, Connecticut, Rhode Island and southeastern New York), these problems must be addressed well before they become a reality (Jones, 1983). The states of Connecticut and Vermont have begun the public process associated with the development of a rational, acceptable future management programme. To date this activity has involved the development and public discussions of a range of management options which are intended appropriately to allocate the available resource and still allow for quality fishing.

The Connecticut River Anadromous Fisheries Restoration Programme has had a long history of cooperative effort. State and federal resource and regulatory agencies, power companies and other private industries, academic institutions, commercial fishermen, citizen groups and many private individuals have worked closely together for the restoration of Atlantic salmon in New England's major river system. The return of nearly 1,700 salmon in the past ten years is only the beginning. With the confidence gained from these returns and the continued cooperation of all concerned, we have the ingredients for success.

22.5 SUMMARY

The Atlantic salmon was exterminated from the Connecticut River in the early 1800s and although historical documentation as to the size of the runs is lacking, it has been suggested that the Connecticut River run may have been the largest in North America. The first restoration effort since the 1870s began in 1966 when the Connecticut River basin states joined with federal agencies to implement a cooperative Connecticut River Anadromous Fisheries Restoration Programme. Early administration of the programme was carried on by a Policy and Technical Committee made up of state and federal fisheries administrators and biologists. Subsequently, legislation created the Connecticut River Atlantic Salmon Commission, which now manages the programme. The basic goal of the programme is to restore American shad to their historical spawning grounds and to attempt to restore Atlantic salmon to some portion of their historical spawning area.

The restoration programme involved the provision for fish passage facilities at main stem dams and selected tributaries, the assessment of available spawning and nursery habitat, and the acquisition, rearing and release of juvenile Atlantic salmon. A major effort involved the construction of new fish rearing and holding facilities and the conversion of existing trout hatcheries to salmon hatcheries. Nearly six million juvenile salmon (fry, parr and smolts) have been introduced since the programme inception. Returns of adults has been poor as was to be expected - a total of less than 1,700 in the past 10 years. The programme goals of full utilisation of available spawning area, adequate brood stock to provide supplemental stocking and a surplus for a recreational harvest are expected to be reached with the availability of additional generations of 'Connecticut River salmon', with increased knowledge, and with the continued cooperation of all concerned.

REFERENCES

Anon. (1982) Annual Progress Report. The New England Atlantic Salmon Program. US Fish and Wildlife Service, 9 pp

Divine, G.E. (1975) Connecticut River Fishery Program - Status Report. US Fish and Wildlife Service, 9 pp

Jones, R.A. (1978) Success with salmon. In: Oceans, 11 (5), 59-60, Oceanic Society

Jones, R.A. (1983) Options for the recreational harvest of Atlantic salmon in Connecticut - A Public Discussion Paper. 15 pp, Connecticut Department of Environmental Protection

Lance, R.I. (1977) Connecticut River Anadromous Fish Restoration Program - Biennial Report. US Fish and Wildlife Service, 38 pp

Leggett, W.C. (1976) The American shad (Alosa sapidissima), with special reference to its migration and populaton dynamics in the Connecticut River, In: D. Merriman and L.M. Thorpe (eds.) The Connecticut River ecological study: the impact of a nuclear power plant. American Fisheries Society Monograph 1, pp. 169-225

Rideout, S. and McLaughlin, E.A. (1985) Progress Report - Connecticut River Anadromous Fish Restoration Program - January 1, 1977 - December 31, 1982. US Fish and Wildlife Services, 65 pp

Rideout, S. and Sillas, A.U. (1985) Progress Report - Connecticut River Anadromous Fish Restoration Program - January 1, 1983 - December 31, 1984. US Fish and Wildlife Service, 21 pp

Rizzo, B. (1970) Fish passage facilities design parameters for Connecticut River dams. US Fish and Wildlife Service, 54 pp

Stolte, L.W. (1982) A strategic plan for the restoration of Atlantic salmon in the Connecticut River basin. US Fish and Wildlife Service, 82 pp

Chapter Twenty Three

THE ANGLER'S POINT OF VIEW

G.O. Edwards MBE

Awdurdod Dwr Cymru/Welsh Water Authority

23.1 INTRODUCTION

I would stress from the outset that I am a wholly dedicated angler. While I enjoy all branches of the sport (coarse, sea and game) and will fish with any legal method (spinner, bait or fly), my preference is fly fishing for salmon and sea trout. Although perhaps better known as a rugby player than an angler, my first love has always been angling. I am on record as stating that if I had to choose between playing rugby or going fishing I would choose to go fishing. That still applies today and I consider it fortunate that my sporting career has provided opportunities to fish in many parts of the world and on many hallowed waters.

23.2 BACKGROUND

In 1982, I was invited by the Secretary of State for Wales to serve as the appointed member representing 'fisheries interests' on The Board of the Welsh Water Authority. Ultimately, after careful consideration, I gladly accepted the appointment. My main reason for doing so was because it gave me an opportunity to give something back to the sport which meant so much to me. I could but only hope that my contribution might, in some small way, allow my children to experience the same pleasures from going fishing that I had enjoyed. One of my first major tasks as 'fisheries member' was to chair a Special Working Group convened by the Board to investigate the widespread and increasing concern expressed by anglers, fishery owners, commercial fishermen and tourist bodies that the quality of salmon fishing (and sea trout fishing had declined to such an extent that the future of many fisheries was in jeopardy. Quite apart from the fact that the Welsh Water Authority (one of ten regional water authorities in England and Wales created in 1974) was the statutory body responsible for the conservation and regulation of the salmon and sea trout fisheries and had a duty to 'maintain, improve and develop' such fisheries, the tourist industry is crucially important to the economy of Wales and any decline - whether real or apparent - in the quality of the fisheries was also cause for concern in this wider context.

The Working Group took written and oral evidence from respondents representing a wide range of interests. It also spent a great deal of time and effort examining the historical record of catch statistics and, when available, redd counts for each of the 40 or so salmon and sea trout rivers in Wales - along with any available data from electronic fish counters and parr surveys for certain systems. The Group's report (Harris, 1985) concluded that there had been a demonstrable and marked decline in the quality of the salmon fishing in terms of the length of the 'worthwhile' fishing seasons since fewer fish were now caught and the season had got progressively later with a marked reduction in the number of early-running fish. There was clearly a crisis of confidence in the fisheries and this was reflected by a decrease in the number of anglers obtaining licences and permits: with serious implications in relation to angling tourism in many areas.

What was less clear was whether or not there had been a decline in the quality of the fisheries. In the almost total absence of any other information the answer to the question 'Have stocks declined?' depended upon an interpretation of catch statistics from the commercial and rod fisheries and the Group soon became painfully aware that the historical data was incomplete, inaccurate and inconsistent for most fisheries and that the circumstantial and anecdotal evidence provided by anglers, when viewed as a whole, was of more practical value than the purported scientific evidence. Thus statements like 'Fish are visibly absent', and 'Salmon parr are no longer a nuisance' assumed major importance when attempting to make any general conclusions. Fortunately some good information on catches and effort (number of anglers) was available for a few fisheries where it was evident that despite an increase in angling effort the total catch over recent years had declined. This is one of the classic symptoms of a declining fishery. The Group concluded on the basis of all the evidence (technical, circumstantial and anecdotal) that stocks had declined overall to a level that was now critical on some systems.

My involvement with this Working Group was in many respects one of my most challenging, enlightening and, ultimately, sobering experiences. In short, as an angler, I learnt a great deal. Most anglers have strongly held views about exactly what is wrong with our salmon fisheries and precisely how to go about remedying the problem. I was no different in that respect (less netting, more restocking, import eggs from early run Scottish fish, eradicate predators and competitors, ban this, prevent that etc.). I now realise that the management of our salmon and sea trout fisheries is an exceedingly complex and difficult business with enormous constraints and obstacles to progress. The remainder of this paper deals with some of the problems, constraints and obstacles identified by the Working Group as significantly affecting the future wellbeing of salmon and sea trout fisheries and which, in my view, are of paramount importance for the future.

I can here highlight the main issues only. These I have grouped under broad headings and would mention that my comments and observations refer only to Wales and England, any wider implication is accidental rather than incidental.

23.3   TECHNICAL AND SCIENTIFIC CONSTRAINTS

Despite the fact that various statutory bodies in England and Wales have been charged by Parliament with the management of the salmon and sea trout fisheries of England and Wales for almost 130 years, it is depressingly clear that we still do not know precisely what it is that we are managing in relation to the number of salmon and sea trout and the composition and timing of the runs in any single river in England and Wales. Thus we cannot say with any certainty whether stocks have increased, decreased or remained relatively stable over any period. Nor can we say whether spawning escapement is at, above or below optimal replacement levels. The sole basis for determining the status and wellbeing of the resource and, thus, for defining and reviewing management policies has been based on the historical record of declared catches. However, it is apparent that, with very few exceptions, the long-term catch record for England and Wales is not worth the paper on which it is written! It is only since 1976 that the catch record for Wales has been prepared on a standard comprehensive and consistent basis that allows year to year comparisons to be made with any real confidence for the rod fishery. But, irrespective of that, it is clearly premature to interpret the catch record in terms of stock levels until the essential research on exploitation rates has been undertaken and the assumption that stocks and catches are linked in any way that is meaningful has been validated. The parlous state of our knowledge about the abundance and key characteristics of the resource in England and Wales is an indictment.

Despite the fact that a vast amount of research has been undertaken into the Atlantic salmon and its related fisheries, there are still too many important gaps in our knowledge that limit our ability to manage the resource efficiently and effectively. (This applies to sea trout even more so.) It would seem that not only do we not know what it is that we are managing but also that, even if we did, we would not know how to manage it properly!

I refer here not to our lack of understanding about the impact of new or recently identified problems - such as acidification or the effect of high seas and interceptory fisheries on regional stocks - but to some of the fundamental and basic principles of day-to-day fisheries science - such as 'How many salmon are needed to maintain stocks?' 'Does like breed like?' 'What is a sea trout?' and, of paramount importance 'What is the relationship between stock and recruitment?' In addition to these and other 'strategic' topics we still need answers to important 'tactical' questions concerning, for example, the maintenance of genetic integrity, the 'best' stage at which to restock, and the means of restoring salmon habitats.

The history of fisheries management in England and Wales would suggest that we have been guilty of making too much of too little: of widely over-extrapolating from work carried out elsewhere and in a completely different context. 'Rules-of-thumb' are very useful, but the embarrassing failure of several important fishing protection schemes intended to mitigate damage to the resource in England and Wales (and, I anticipate, elsewhere also) indicates that there are considerable risks in applying precepts developed in Canada and Scandinavia to management problems on British rivers without proper qualification.

The duplication of research is wasteful of resources but parallel research, which is something very different, would seem to be essential in relation to several key topics because of the differences that seem to exist between (and sometimes within) different river systems. My impression is that rather too much duplication and not enough parallel research has occurred. In this, and other, contexts there would seem to be four main problems:

(1)   Research is badly co-ordinated nationally;
(2)   Much research is opportunistic and, inevitably linked to the perception and inclination of the originator;
(3)   The generators of information are frequently not the users of the information;
(4)   There is no prioritised 'shopping list' of research requirements that has official status.

Many of the important gaps that exist in our knowledge continue to do so because the essential research is too difficult, costly and uncertain in the long-term to be (respectively) profitable, contemplated and attractive. I fully appreciate that there are enormous difficulties in working within the aquatic environment and of researching animals so variable, engimatic and political as salmon and sea trout, <u>but</u> it has been obvious for decades that many of the highest priority needs in respect of research and monitoring can be undertaken only with fixed traps and counting stations on our rivers. I wholly endorse the view expressed by many managers and scientists that our future ability to manage the resource and provide for its future will continue to be based on speculation, intuition and guesswork unless we finally 'grasp the nettle' and construct the facilities necessary to carry out the essential programmes of research and monitoring. Clearly it would be prohibitively costly to contemplate the construction and servicing of upstream and downstream traps on more than just a few carefully selected river systems in England and Wales; <u>but</u> surely something is better than nothing? I consider it vital that we at least make a start to evaluate properly the status and wellbeing of the resource and to determine the relationship between stock-and-catch and between stock-and-recruitment on at least one river in England and Wales (preferably Wales).

23.4 LEGISLATIVE CONSTRAINTS

The main statute relating to the management of salmon and sea trout fisheries in England and Wales is the <u>Salmon and Freshwater Fisheries Act</u> 1975. It is important to note that this is not a recent piece of modern legislation but merely a consolidation of previous legislation going back to the 1923 Act which, in turn, was largely a consolidation of earlier legislation going back to the 1860s. Thus, the current legislation was framed in the last century when problems and perspectives were very different from those of today. The 1975 Act is widely held by front-line managers to be a hopelessly outdated piece of legislation which is deficient in many important respects and inadequate in relation to modern requirements. This can be shown by example.

The 1975 Act exists solely to conserve the fisheries and to provide for their regulation. That it does neither is evident by the fact that while the number of licences available for the commercial fishery can, after a protracted and onerous procedure, be limited for up to ten years by the provisions of a local Net Limitation Order, it is impossible to limit either the number of licences available to anglers or, otherwise, to restrict the catch by the rod and net fishermen; except retrospectively, imprecisely and with inordinate difficulty.

The principal deficiency of the outmoded legislation is that it specifically excludes any allocation of the harvest between competing interests - namely anglers and commercial fishermen.

Thus the Regional Water Authorities must maintain, improve and develop the salmon fisheries regardless of whether or not they are exploited at all and regardless of any actual or potential benefits that may accrue to society by their exploitation. This is clearly a nonsense and, indeed, a facet of the legislation which may be counterproductive. At all times the interests of the resource must be paramount, but bearing in mind that the future of the salmon depends on more than just the sectional and minority interests of anglers and commercial fishermen, it is of greatest importance to focus the social and economic benefits from the exploitation of any surplus of stock on the community as a whole. Studies undertaken in Wales have established that the socio-economic value of the sport fishery to the community is far in excess of that from the commercial fishery. But, it is impossible to reduce the share of the harvest taken by the commercial fishery to benefit the rod fishery even though such an action would increase the benefits derived by the community and enhance the value of the resource.

While it is possible to reduce the number of licences issued in the commercial fishery by applying for a Net Limitation Order, any such proposals can only be promoted under the enabling legislation on the basis of the need to conserve stocks. However, even though the NLO may reduce the number of licences for a particular local area, the 1975 Act requires that any person who held a licence prior to the reduction should continue to be issued with a licence in the future! Only when the netsmen die or relinquish their licences voluntarily can the number of licences be reduced to the level fixed by the Order even though the reduction was based on the immediate need to reduce exploitation to conserve the resource! It would seem therefore, that the 1975 Act gives greater importance to protecting the interests of commercial fishermen than to protecting the interests of the fish. This, again, is a nonsense.

Another serious deficiency of the legislation is that it provides an enormous loophole which enables the public right of fishing for sea fish to be used as a pretext for the illegal capture of salmon and sea trout. The conservation and regulation of sea fisheries in England and Wales is not a function of the water authorities and is covered by separate legislation to that relating to salmon and sea trout. This dichotomy in the legislation makes it impossible to regulate sea fishing for the purpose of protecting salmon - no matter how great the need.

The Salmon and Freshwater Fisheries Act 1975 is deficient in many other respects also - it allows for no direct control over the amount of water abstracted into fish farms or the screening of inlet and outlet streams to prevent the diversion of smolts, kelts and adults into the farm and there is no mandatory system of 'referrals' in respect of actions likely to be potentially damaging to the aquatic environment and to fisheries (e.g. afforestation).

Proposals for a major review and update of the fisheries legislation affecting wild stocks have received, at best, a lukewarm response from the Ministry of Agriculture, Fisheries and Food. Although long-awaited proposals for new legislation to combat illegal fishing are currently before Parliament, these cover just one aspect of a wider need.

23.5 PERSPECTIVES AND ATTITUDES

The traditional basis for the management of salmon and sea trout fisheries in England and Wales is in relation to their importance as a source of food and, incidentally, in providing employment and income in the commercial fishery. While exploitation of the fisheries by commercial means was important in the past it is relatively unimportant today. The total catch of wild salmon makes an insignificant contribution to the nation's fish food requirements and its importance in this respect has been made even more irrelevant by the dramatic and continuing increase in the output of farmed salmon. The decline in the importance of the commercial fishery over the last century has been such that it would now seem that (with a few very local exceptions) it provides neither an important source of employment nor a material contribution to the income of the fishermen. The decline in the importance of the commercial fishery has been offset by a remarkable increase in the demand for angling to the extent that the social and economic value of the sport fishery now exceeds that of the commercial fishery by a substantial amount.

It is to be recognised that the concepts applied to the management of food fisheries are inappropriate in relation to the exploitation of stocks where there is competition between the sport and the commercial fisheries and that a more flexible approach is needed which optimises the economic benefits to the community as a whole. Sadly, the Ministry of Agriculture, Fisheries and Food, which controls general policy, has historically been concerned with food production rather than optimising the economic yield of the fisheries and still seems reluctant to recognise the greater value of the recreational fishery.

If Government and society fail to recognise the value and importance of our salmon and sea trout fisheries it is, perhaps, because the social and economic benefits have not been adequately assessed. There are surprisingly few socio-economic evaluations based on fisheries in the British Isles and there is an urgent need to expand the limited local information (which is mainly from Wales) by a wider regional and national study embracing both the rod and commercial fisheries.

In addition to changing the perspective and attitudes of Government it is also necessary to change the awareness of anglers in relation to many key issues.

The dramatic increase in angling effort over the last few decades must ultimately result in a decline in the catch of each participant if overfishing is to be avoided. This basic truth is not properly appreciated by anglers and there are signs that the future wellbeing of the resource in many rivers requires the imposition of control measures to limit the level and rate of angler exploitation more effectively than hitherto. It is difficult to see how the introduction of bag limits can be avoided for much longer if the overall angling effort continues to increase; and it is evident that there is a need for a revival of the sporting ethic that distinguishes 'anglers' from 'fishermen'. Far too many anglers seek to profit from their fishing and this inevitably begs the question as to whether anglers should be prohibited from selling their catch and whether we must look to the introduction of catch-and-release as a means of providing for increasing recreational pressure while maintaining a defined level of exploitation.

One of the greatest needs is for the angling community to adopt a broader and less parochial perspective and to act more cohesively and purposefully in its approach to resolving some of the fundamental issues affecting the fisheries. That a well organised, responsible and determined angling lobby can achieve much is shown by the difficulty now being experienced by the Government in the passage through Parliament of what has been described as its inept and timorous 'Salmon Bill' which proposes new legislation to combat illegal fishing. It can only be hoped that the continued efforts of the anglers in concert with the strong representations of the water authorities and others will result in a new Act which is far better than the Bill on which it is based.

In the absence of adequate legislation to protect fisheries from the deleterious effects of a further deterioration in the aquatic environment, it is often necessary to fall back on the individual fishery owner's right under Common Law to claim compensation if the fishery is damaged by the actions of another. In many respects Common Law is a very powerful deterrent. The threat of an action in the Civil Courts has done much to protect and preserve the resource in England and Wales in the past. But litigation is very costly and very few individual fishery owners are in a position to bring an action for damages. Many fishery owners seem to be unaware of their common law rights and there would be obvious merit in the creation of a national 'fighting fund' to allow greater use for the civil courts to oppose environmentally damaging schemes to a much greater extent than has occurred in the past. There is such a fund to fight polluters and provide for compensatory restocking following a fishkill incident, but there is a need to widen the use of common law rights in several other contexts. I find it surprising that the national bodies representing angling interests have yet to get their act together in this respect.

23.6   COMMENT AND CONCLUSIONS

It is a mistake in any paper of this kind to fire off in all directions and lose sight of the target. It is all too easy to take a pessimistic view about the future of our salmon and sea trout fisheries as they are under considerable pressure and these pressures will inevitably increase unless we take purposeful and proper action to tackle the real problems. While the greatest threat is complacency and inaction, it is equally important to avoid hysteria and over-reaction. The tendency, prevalent among the angling community, to look for a single cause to explain any perceived decline in the fisheries is to be deplored as it diverts attention and essentially limited resources away from tackling the fundamental and basic problems. It has been said that fisheries management is as much about managing people as it is about managing fish. I now realise that this is all too true and would only add that, in England and Wales, it is also about influencing the perspective and attitude of Government.

The future of the salmon and sea trout fisheries depends upon identifying the problems affecting the stocks and catches, ranking them in order of priority and tackling them accordingly. Thus in Wales the main pressures can be ranked as: (1) illegal fishing, (2) farm pollution, (3) water abstraction, (4) acidification, (5) forestry, (6) industrial pollution - with river regulation, navigation, competition by coarse fish and obstructions being important 'priority' problems in a more local, river-specific context. Different regions and different countries will have different lists. However, while each individual pressure affecting the abundance and composition of the resource must be tackled in due course, it is my view that the main thrust of future endeavour must be to create an environment which makes progress possible. That many of the problems long since recognised as being significant still persist as important constraints limiting the abundance of the resource and our ability to manage it properly is, perhaps, because we have neglected the socio-political environment in which we operate. Thus in order to make progress in providing a future for the resource we must tackle a range of priority issues.

At a technical level it is of paramount importance to implement an effective programme of routine monitoring to determine the status and composition of our stocks. It is also equally important to fill the gaps in our knowledge that limit our ability to manage the fisheries effectively.

There is a need to identify a shopping-list of key research requirements and for more effective co-ordination of the national research effort. In these respects it is difficult to see how the construction of fixed traps on a suitable number of carefully selected 'index' rivers can be avoided for much longer even though the cost of construction and operation would be very costly.

At a non-technical level high priority must be given to creating a more favourable political environment. In this respect further socio-economic studies to establish the value and importance of the fisheries in a local, regional and national context are considered essential in order to change the attitudes and perspectives of society and, particularly, Government towards the future wellbeing of the resource.

In order to increase public interest in the fisheries it is essential to direct the economic benefits from the fisheries on the community as a whole. The historical 'food' orientated approach to the management of the wild resource is no longer necessary or desirable in that the economic value of the sport fishery to society far exceeds that of the commercial fishery and has far greater potential for enhancement.

The lack of any clearly defined National Policy for the management of salmon and sea trout fisheries in England and Wales must be remedied and this should give priority to optimisation of the economic value of the resource. That the conservation of the resource must at all times take priority over the interests of its consumers is to be accepted, but there must be provision within the legislation to provide for a greater allocation of the harvest towards the recreational fishery where it is clear that this is necessary to maintain and increase community benefits.

Angler attitudes and perspectives must change also. Increasing recreational pressure increases economic values but imposes pressures on the resource which, at a certain point, must lead to controls to prevent over-exploitation. Bag limits, catch-and-release and a ban on the sale of angler-caught fish are, inevitably, things of the future. The angling community must become better organised, more cohesive and, above all, more aware of the real issues and needs.

The future of the salmon and sea trout in England and Wales depends very much on being able to apply the correct information to the right problems in a favourable environment. Without that favourable environment, which requires the support of the angling community, society and the Government, the future management of the resource will be the subject of slow progress and very limited achievement.

The views expressed are those of the author and may not reflect the opinions of the Authority.

REFERENCE

Harris, G.S. (ed.) (1985) Working Group Report on Welsh salmon and sea trout fisheries. Welsh Water Authority, 152 pp

Chapter Twenty Four

EXPLOITATION AND MIGRATION OF SALMON ON THE HIGH SEAS, IN RELATION TO GREENLAND

Jens Møller Jensen

Greenland Fisheries and Environment Research Institute, Tagensvej 135, DK-2200 Copenhagen N, Denmark

24.1 INTRODUCTION

The presence of Atlantic salmon (<u>Salmo</u> <u>salar</u> L.) off West Greenland was first mentioned by Fabricius (1780), who, however did not see any himself. He recorded it as rare, in two localities in the district of Qaqortoq and in a bay near Nuuk (Figure 24.1). The latter bay is probably Kapisillit. The Kapisillit river is the only river in Greenland where salmon spawn annually. From time to time salmon have been observed in other rivers and on some occasions spawning.

Jensen (1948) also records salmon at Kapisillit (Nuuk district) and includes the district of Sisimiut. No spawning rivers are known in the last-mentioned locality. He suggested that salmon from areas other than Greenland were present in Greenlandic waters during the 1930s.

From the beginning of the 1950s the captures of salmon on the coast indicated that salmon were distributed during the autumn from Cape Farewell in the south to Disko Island in the north.

On 5 July, 1965 the salmon fishery at Greenland appeared for the first time on the scene of international politics. On that day a question was posed in the House of Lords to the British Government concerning the influence of the high catches of salmon at West Greenland on the fisheries and stocks in home waters (Netboy, 1968). Since then the high seas fishery for salmon at Greenland has been discussed by various international bodies.

Figure 24.1: Area map of West Greenland showing NAFO statistical divisions

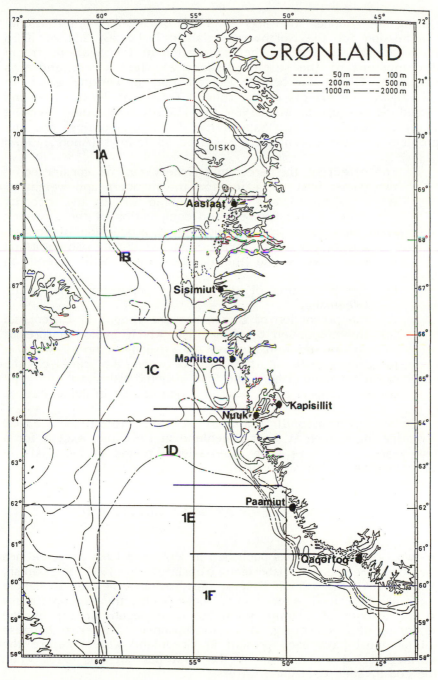

24.2 THE INTERNATIONAL SALMON INVESTIGATIONS AT WEST GREENLAND

During the Annual Meetings in 1965 of ICNAF (International Commission for the Northwest Atlantic Fisheries) and of ICES (International Council for the Exploration of the Sea), the rapid development of the salmon fishery at West Greenland was discussed. These two bodies decided that a working group to investigate the effects of the salmon fishery at West Greenland should be set up. In May 1966 the ICES/ICNAF Joint Working Party on North Atlantic Salmon had its first meeting, and the problem at West Greenland was stated as follows (Anon., 1967):

> The effect of the Greenland fishery can be considered in two parts; first the effect on the numbers and weight of fish returning to, and caught in home waters, and secondly the effect on the numbers and composition of the spawning stock and hence on the subsequent production of smolts.
> The effect on the numbers and weight of fish returning to home waters and the catches there will depend on
>
> (a) the proportion of the original population that visits Greenland;
> (b) the proportion of those that are caught in Greenland;
> (c) the proportion of those fish which avoid capture at Greenland and survive to return to home waters;
> (d) the growth of the fish between the time of the Greenland fishery and that in home fisheries;
> (e) the proportion of returning fish caught in home waters.

An international team of salmon biologists started investigating salmon at West Greenland in 1965 and, except for a few years, there have been international teams of biologists in Greenland each year since then.

24.3 THE SALMON FISHERY AT WEST GREENLAND

Description of the Fishery

In two Greenlandic books published in 1922 and 1923 as guidelines for the Greenlandic fishermen (reprinted in 1964 in Danish, Anon., 1964), the salmon fishery, the gear and the way to handle these are described. According to these books, the salmon fishery in the district of Sisimiut seems to have begun already by the beginning of this century. This information is of interest because salmon must have been present in Greenland waters before the warmer period which began in the first part of the 1920s when the cod invaded Greenland waters.

There is information about catches of salmon in the area from Sisimiut to Qaqortoq on the west coast of Greenland from the 1930s (Nielsen, 1961). The salmon were used for local consumption, fresh or salted, as there were no freezing facilities.

The present salmon fishery started in the mid 1950s in the district of Sukkertoppen, but by 1961 it was distributed along the west coast. It takes place in the autumn, between August and November.

Before 1965 only Greenlandic vessels took part in the fishery, and it was an inshore fishery carried out by small boats using fixed gill nets. During the period 1965 to 1975 vessels from Denmark, the Faroes and Norway participated in the fishery, and these vessels introduced the drift nets to Greenland. The material used for the first drift nets was multifilament nylon, and fishing took place during darkness. Later, Danish vessels began to fish with drift nets of monofilament nylon, which could catch salmon both during daylight and darkness, thereby increasing the efficiency of the fishery. Greenlandic vessels also used drift nets in the 1960s.

From 1976 only Greenlandic vessels have been allowed to fish for salmon at Greenland, most of them using drift nets, even small boats of 16 feet, but fixed gill nets are still in use. The vessels using this gear are small cutters and boats powered by outboard motors.

The Fishery from 1960 to 1985

From 1960 to 1964 total landings increased rapidly from 50 to 1,539 tonnes (Table 24.1 and Figure 24.2) and from 1965 to 1968 fluctuated around 1,200 tonnes. From 1969 to 1975 the landings ranged between 1,917 and 2,689 tonnes, and from 1976 to 1982 they fluctuated around 1,200 tonnes. In 1983 and in 1984 the landings decreased dramatically to 300 tonnes but increased again in 1985 to 864 tonnes (Anon., 1986).

Between 1964 and 1982 the catches by Greenland vessels fluctuated around 1,200 to 1,300 tonnes but from 1976 these catches were limited by the quotas or TACs. However, in 1983 and 1984 the catches were well below the respective TACs.

Distribution of the Fishery

The fishing area is from Disko Island in the north to Cape Farewell in the south, both inshore and offshore, up to 40 nautical miles from the baseline. The spatial distribution of catches taken by Greenland vessels is given in Table 24.2 and Figure 24.1 shows the NAFO statistical divisions used.

Most of the Greenlanders' salmon catch is taken in NAFO Divisions 1B to 1E, and particularly in 1B and 1C (Table 24.2).

Table 24.1: Nominal catches at West Greenland 1960 to 1985.
(tonnes, round fresh fish)

Year	Nor-way	Faroes	Swe-den	Den-mark	Total excl Green-land	Green-land	Total	TAC
1960	-	-	-	-	-	60	60	-
1961	-	-	-	-	-	127	127	-
1962	-	-	-	-	-	244	244	-
1963	-	-	-	-	-	466	466	-
1964	-	-	-	-	-	1,539	1,539	-
1965	+	36	-	-	36	825	861	-
1966	32	87	-	-	119	1,251	1,370	-
1967	78	155	-	85	318	1,283	1,601	-
1968	138	134	4	272	548	579	1,127	-
1969	250	215	30	355	850	1,360	2,210	-
1970	270	259	8	358	895	1,244	2,139	-
1971	340	255	-	645	1,240	1,449	2,689	-
1972	158	144	-	401	703	1,410	2,113	1,100
1973	200	171	-	385	756	1,585	2,341	1,100
1974	140	110	-	505	755	1,162	1,917	1,191
1975	217	260	-	382	859	1,171	2,030	1,191
1976	-	-	-	-	-	1,175	1,175	1,191
1977	-	-	-	-	-	1,420	1,420	1,191
1978	-	-	-	-	-	984	984	1,191
1979	-	-	-	-	-	1,395	1,395	1,191
1980	-	-	-	-	-	1,194	1,194	1,191
1981	-	-	-	-	-	1,264	1,264	1,265
1982	-	-	-	-	-	1,077	1,077	1,253
1983	-	-	-	-	-	310	310	1,191
1984	-	-	-	-	-	297	297	870
1985	-	-	-	-	-	864	864	852

One of the reasons for the lower catches in 1F is drift ice, which is very often present at the beginning of the salmon season. The salmon fishing in 1A usually takes place inside Disko Bay, which could explain the lower catches taken in this division. The reason for high catches in 1B to 1E seems to be a combination of the high number of fishermen and the presence of a feeding area found just outside the ports.

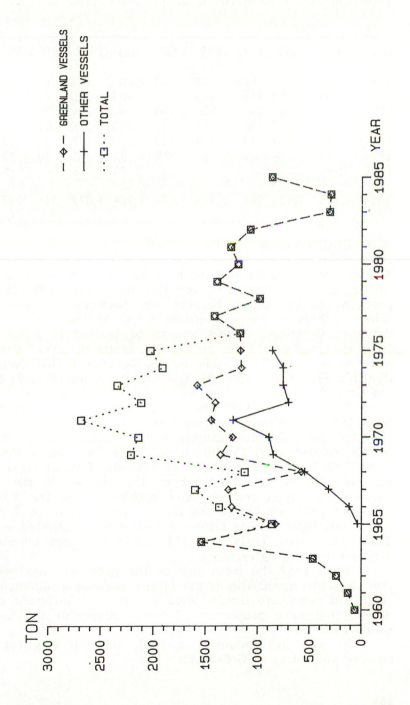

Figure 24.2: Nominal catches at West Greenland 1960-1985 (tonnes, round fresh fish)

Table 24.2: Nominal catches (tonnes) of salmon by Greenland vessels in 1975-1985 by NAFO divisions according to place landed

Div.	1976	1977	1978	1979	1980	1981	1982	1983	1984	1985
1A	171	201	81	120	52	105	111	14	33	85
1B	299	393	349	343	275	403	330	77	116	124
1C	262	336	245	524	404	348	239	93	64	198
1D	218	207	186	213	231	203	136	41	4	207
1E	182	237	113	164	158	153	167	55	43	147
1F	43	46	10	31	74	32	76	30	32	103
1NK	-	-	-	-	-	20	18	-	5	-
Total	1,175	1,420	984	1,395	1,194	1,264	1,077	310	297	864

24.4   REGULATION OF THE FISHERY

At the annual meeting of ICNAF in June 1972 a recommendation was agreed to phase out over the four years 1972-1975 the participation of Danish, Faroese and Norwegian vessels in the salmon fishery in the Convention Area. At the same time the catch by Greenland vessels was to be limited to 1,100 tonnes from 1972 and catches in territorial waters off West Greenland were to be included in this figure. In 1974 ICNAF noted the Danish adjustment in the TAC from 1,100 tonnes to 1,191 tonnes, the Greenlandic mean annual catch for the period 1964 to 1971.

From 1972 to 1978 the TACs were negotiated in ICNAF, from 1979 to 1983 between the EEC and other parties and from 1984 in the new international body, NASCO (North Atlantic Salmon Conservation Organisation). Table 24.1 shows the TACs from 1972 to 1985. In 1981 and in 1982 the TAC of 1,191 tonnes was adjusted to take account of the growth of the salmon between 10 August and the later opening dates. The TACs for 1986 and 1987 are 850 tonnes for an opening date of 1 August, but if this fishery opens later the TAC can be adjusted upwards. Table 24.3 gives TACs of 1,191 and 850 tonnes adjusted for different opening dates.

From 1982 the mesh size of the nets used has been 140 mm stretched mesh. This is the target, and not a minimum mesh size. This mesh size should catch - even with different opening dates - the same proportion of North American and European salmon, i.e. 1:1.

In order to administer a TAC, the following regulatory scheme was set up in Greenland:

Table 24.3: The 1,191 and 850 tonnes TACs adjusted to different opening dates

Opening date		Mesh size 140 mm
1 August		850
10 August	1,191	890
15 August	1,214	909
20 August	1,235	924
25 August	1,253	938
1 September	1,271	952
5 September	1,285	957

(1) No commercial fishing is allowed without a licence issued by the Greenland authorities.
(2) Fish-processing plants are not allowed to buy salmon except from licensed fishermen and all plants are obliged to report their accumulated catch, daily.
(3) Fishing has to stop immediately when ordered by the authorities. The order is cabled to all fish plants and announced over the radio.

In order to give all licensed fishermen along the West Greenland coast a chance to fish salmon, the TAC is divided into two quotas, the so-called 'free-quota', which all boats and fishermen can fish, and 'the local small boat-quota'. The latter quota is allocated on a district basis and only to boats smaller than 30 feet.

24.5 THE SALMON PRESENT AT WEST GREENLAND DURING THE AUTUMN

Salmon originating in other countries are present at Greenland from June to September and numbers seem to reach a maximum from August to October.

Country of origin
The results obtained from salmon scale analysis indicated the occurrence of fish derived from other countries during the autumn (Nielsen, 1961). In 1956, the first salmon tagged outside Greenland was recaptured in Greenland; it was tagged as a kelt in Scotland (Menzies and Shearer, 1957). The first recapture from North America occurred in 1961. This fish had been tagged as a smolt in Canada. Since that time several thousand recaptures

have been reported in Greenland waters. In order to estimate the exploitation rate in the West Greenland salmon fishery and to demonstrate that the salmon which visit Greenland return to their country of origin, it was necessary to tag fish in Greenland. After several trials with different gear, it was found that only drift nets were able to catch enough salmon suitable for tagging, although the quality of fish caught on longline and in trap nets was better neither gear could catch sufficiently large numbers (Table 24.4).

Table 24.4: The number of salmon tagged at West Greenland from 1965 to 1972. The number tagged and recaptured are given per tagging gear. Figures in parentheses are percentage of number tagged. TN = trap nets, LLD = drift longline, GNS = set gill nets, GND = drift gill nets, GRL = Greenland, NA = North America, EU = Europe

Period	Gear	Number tagged	GRL	Number recaptured NA	EU	Total
65-68	TN	72	4(5.6)	-	-	4(5.6)
68-70	LLD	70	1(1.4)	-	2(2.9)	3(4.3)
65-71	GNS	1,393	38(2.7)	4(0.3)	8(0.6)	50(3.6)
69-71	GND	758	25(3.3)	11(1.5)	12(1.6)	48(6.3)
72	GND	2,364	164(6.9)	12(0.5)	44(1.9)	220(9.3)
Total		4,657	232(4.9)	27(0.6)	66(1.4)	325(6.9)

The results from tagging experiments both at Greenland and in home countries have demonstrated that some salmon from nearly all salmon producing countries around the North Atlantic migrate to Greenland and some of these, if not caught, will return to home waters. The countries contributing to the Greenland salmon catch are: USA, Canada, Iceland, Norway, Sweden, Denmark, Finland, Scotland, England, Ireland, France and Spain. There are also probably salmon in Greenland waters from the USSR and Portugal. The main contributors to the stock at Greenland are Canada and Scotland (Møller Jensen, 1980).

Another result from the tagging experiments is that not all rivers producing multi-sea-winter fish (MSW) contribute salmon to the same degree in the Greenland catch even if they are situated in the same area. Similarly, the contribution made by tributaries in the same river system may be different.

Sea and river age composition, sex ratio and continent of origin
The age composition of salmon caught by research vessels and commercial vessels shows that only salmon which have already spent at least one winter in the sea are present (Table 24.5).

Table 24.5: Sea age composition from research vessel and commercial catches of salmon at West Greenland

| | Sea age composition (%) | | |
Year	1 SW	MSW	PS
1969-80	93-99	1-6	1-2
1981	97.0	2.5	0.5
1982	93.6	6.0	0.4
1983	90.5	8.1	1.4
1984	87.6	11.6	0.8
1985	93.8	5.9	0.3

1 SW = One-sea-winter fish.
MSW = Multi-sea-winter fish.
PS = Post spawners

These salmon would be multi-sea-winter fish or 'salmon' when caught in home waters. Therefore the fishery at West Greenland has an effect only on the 'salmon stock', and not on the 'grilse stock'.

In all years, except those (1983 and 1984) with low catches, the grilse or one-sea-winter (1-SW) component ranged between 93 and 99 per cent (Table 24.5). During 1983 and 1984 the MSW fish component was above the average. An explanation for the higher proportion of MSW fish in those years could be that 1-SW and MSW fish have different migration patterns and arrive at Greenland from different areas of the sea. The result of the tagging experiment in 1972 showed that some salmon overwinter in the sea and return to Greenland in the following year (Møller Jensen, 1980).

From analysis of the pattern on the scales it is now possible to identify North American salmon and European salmon (Lear, 1972; Lear and Sandemann, 1982, Anon., 1986). Table 24.6 summarises the proportions by continent of origin in the commercial catches from 1978 to 1985. The average proportion is close to 1:1. Differences in the annual proportions both between and within the two groups could be caused by changes in their migration patterns, the size of the

Table 24.6: Continent of origin (%) of North American (NA) and European (EU) salmon, from commercial samples, 1978 to 1985

Year	NA	(95% CL)	EU	(95% CL)
1978	52	(57,47)	48	(53,43)
1979	50	(52,48)	50	(52,48)
1980	48	(51,45)	52	(55,49)
1981	59	(61,58)	41	(42,39)
1982	62	(64,60)	38	(40,36)
1983	40	(41,38)	60	(62,59)
1984	50	(53,47)	50	(53,47)
1985	50	(53,46)	50	(54,47)

spawning stocks, the post-smolt mortality, and possibly also sampling bias.

The overall sex ratio (F:M) at West Greenland estimated from catches by research vessels varies between 2.8:1 and 4.1:1 with a mean around 3:1 (Anon., 1982; Munro and Swain, 1980).

Table 24.7 shows the smolt age composition of the salmon

Table 24.7: Smolt age composition of salmon sampled in 1982 to 1985 at West Greenland

Year	Smolt age composition (%)						7 and more years
	1	2	3	4	5	6	
North American							
1982	1.4	37.8	38.3	16.2	5.6	0.5	0.2
1983	3.1	46.9	33.2	12.4	3.4	0.8	0.1
1984	4.9	51.7	28.8	9.0	4.6	0.9	0.2
1985	5.1	41.0	35.7	12.1	4.9	1.1	0.1
European							
1982	15.7	56.1	23.2	4.3	0.8	-	-
1983	34.7	50.3	12.2	2.3	0.3	0.1	0.1
1984	22.8	56.9	15.1	4.2	0.9	0.2	-
1985	20.2	61.6	14.9	2.7	0.6	-	-

caught by commercial vessels in 1982-1985 by continent of origin, and Table 24.8 gives the mean lengths and weights of these fish (Reddin, personal communication). North American salmon have a higher smolt age than European salmon. This was to be expected due to the differences in the water temperature of the rivers. Generally, the lower the river temperature the older the smolts, and vice versa. North American salmon were shorter and lighter than European salmon. These differences were less in the later years, which makes separation of these two groups on the basis of scale patterns more difficult.

Table 24.8: Mean fork length (l,cm) and weight (wt,kg) of North American and European salmon, caught by commercial vessels at West Greenland 1982 to 1985

Year	Sea-age class							
	1		2		PS		Total	
	l	wt	l	wt	l	wt	l	wt
	North American							
1982	62.7	2.79	78.4	5.59	71.4	3.96	63.5	2.92
1983	61.5	2.54	81.1	5.79	68.2	3.37	64.4	3.02
1984	62.3	2.64	80.7	5.84	69.8	3.62	65.5	3.20
1985	61.2	2.50	78.9	5.42	79.1	5.20	62.6	2.72
	European							
1982	66.2	3.21	77.8	5.59	80.9	5.56	67.2	3.43
1983	65.4	3.01	81.5	5.86	70.5	3.55	66.2	3.14
1984	63.9	2.84	80.0	5.77	779.5	5.78	65.0	3.03
1985	64.3	2.89	78.6	5.45	77.0	4.97	65.0	3.01

The influences of environmental conditions on the migration of salmon to west Greenland

In 1983 and 1984 the landings of salmon at Greenland were the lowest (310 and 297 tonnes) recorded since 1962, when the fishery at West Greenland had just started. During these two years the landings declined relatively more than in any other fishery in the North Atlantic area, even allowing for home water regulations.

In the 1984 and 1985 North Atlantic Salmon Working Group reports, it was stated that at least three factors could have caused the low catches.

(1)   Adverse environmental factors.

(2)   Lower than normal sea survival rate of the relevant smolt classes and low stock abundance.
(3)   Reduced fishing effort at Greenland.

During 1983 the cod stock at West Greenland nearly disappeared from the fishing area. The situation could not be explained either by overfishing or the normal spawning migration from West to East Greenland and to Iceland. Other factors seem to have been important, and of these the most likely are the environmental conditions. Only these factors, or the effect of these, would explain or partly explain the sudden decline in stock abundance of both salmon and cod at West Greenland during the same period.

During 1980 and 1981 the air temperature at Nuuk fluctuated around normal, i.e. the 30-years monthly mean. The period from February 1982 to November 1984 was characterised by strong negative monthly temperature anomalies, and particularly the winter months in 1983 and 1984, which were extremely cold. The reason for the cold conditions was the displacement of the Arctic Canadian cold air mass to the Davis Strait area with its centre situated near Aasiaat at West Greenland, where the temperatures during the winters 1983 and 1984 were 12°C and 14°C below the respective mean (Rosenørn, Fabricius, Buch and Horsted, 1984). In 1985 the temperature almost returned to normal, except in the winter months of 1984-85 and 1985-86 when high positive anomalies were observed.

A positive correlation in June between the sea surface temperature on Fylla Bank off Nuuk (the upper 40 metres) and the air temperature in Nuuk has been demonstrated several times (Buch, 1986). Therefore, there is reason to believe that the sea surface temperature from February 1982 to November 1984 was affected by the cold air temperature. The sea surface temperatures in June 1983 and in 1984 were low and the same low temperatures were recorded in June 1969 and 1970 but the salmon catches in these years were not abnormally low. Therefore, it is not the sea surface temperature at the beginning of the salmon season which prevents the salmon from migrating to Greenland.

After the two cold winters of 1982-83 and 1983-84 low catches of salmon were taken in 1983 and 1984, but after the winter of 1984-85, which was warmer than normal, the catch in 1985 increased to 864 tonnes. This catch was limited only by the TAC of 852 tonnes. This indicates that environmental conditions such as cold and warm winters at Greenland could be important factors for the occurrence of salmon at West Greenland.

An analysis was made of variation in air and water temperatures and ice coverage based on monthly charts from the meteorological office (Anon., 1985). The results showed that 1983 did not differ significantly from the period around 1970 when low sea surface temperatures had been observed in Greenland but showed that 1984 was significantly colder. Fourteen different environmental variables were analysed. The four environmental variables further studied together with two sets of biological data are given below.

(1)   The northward extent in January of the 4°C water isotherm west of 45°C.
(2)   The difference between the area (km^2) covered by 4°C warm water (or warmer) in January and that in March north of 50°N and west of 40°W.
(3)   The northward extent in August of the 4°C water isotherm west of 45°W.
(4)   The extent (percentage) of West Greenland coast covered by the 4°C water isotherm north of 60°N and 44°W.
(5)   The catch of MSW salmon in Canada and Scotland in the year following the West Greenland fishery.
(6)   The abundance of salmon in Greenland from 1970 to 1984, based on qualitative statements of the success of the Greenlanders' fishery, classified from 1-5.

The annual values of these variables ((1)-(5) per cent), from 1973 to 1983 excluding 1977, were ranked in increasing order. Regression analysis of the stock abundance index at West Greenland on the five ranked variables showed a weak correlation. The warming of the water during spring (variable (2)) seems to be the most important one, which also seems to be in agreement with the shift between warm and cold winters. However, since the correlations are weak, factors other than the environment seem to influence stock abundance at West Greenland.

In addition to the abundance of salmon in home countries and subsequent marine mortality of the smolts, the environmental conditions in Greenland waters and in the surrounding area could affect the occurrence of salmon at West Greenland and also the pattern of the yearly feeding migration in the North West Atlantic area.

24.6   THE EFFECT OF EXPLOITATION OF SALMON AT WEST GREENLAND

Even if environmental conditions are disregarded it is not yet possible to assess the size of the population and the exploitation rate at West Greenland in such a way that these figures can be used to project the annual catch when a TAC is set. The two most important reasons are that the fishery at West Greenland depends almost entirely on only 1-SW salmon and it is not yet possible to estimate recruitment in relation to the fishery. In reality only one of the original questions posed so far has been answered (see Section 24.2), the growth of the fish between the time of occurrence at Greenland and in home water.

The exploitation rate in 1972

In 1972 ICES and ICNAF organised an international tagging experiment at West Greenland, which involved five research vessels and eight commercial vessels with observers on board. From 1 August to 21 October, 2,364 salmon were tagged, and of these 164 were recaptured at Greenland (156 in 1972, 7 in 1973 and 1 in 1974), 12 were recaptured in Canada (11 in 173 and 1 in 1974) and 44 were reported from Europe (38 in 1973 and 6 in 1974). The low rate of recapture is probably the result of a high tagging mortality (Table 24.4). The difference in the rate of recapture in Europe and North America may be explained by differences in the reporting rate and/or in the exploitation rate (Møller Jensen, 1980).

The estimated rate of exploitation for the salmon present at West Greenland was obtained from mark-recapture data utilising the model proposed by Ricker (1975), suitably modified. This showed that in 1972 the fishing removed 33 per cent of the fish present at the beginning of the season and based on a catch of 585,000 fish the estimated number present was 1.75 million (Andersen, Horsted and Møller Jensen, 1980).

The losses to the home water stock

The loss to home water stocks from catching salmon at West Greenland has been estimated using data from several years. The latest assessment was made in 1980, using the following parameters:

(1)   The proportion of fish of North American origin (NA) and the proportion of fish of European origin (EU) in the landings.

(2)   The catch of fish of NA and EU origin broken down into smolt year classes.

(3)   WG = the mean weight of each smolt year class of NA and EU salmon at Greenland.

(4)   WH = the mean weight of each smolt year class in home waters after returning from Greenland.

(5)   N = the non-catch fishing mortality, which is the mortality generated by fishing but not recorded as catch. The upper and lower values used are 0.1 and 0.3.

(6)   S = the survival rate of salmon between Greenland and home waters. The values used for NA and EU salmon are 0.85 and 0.90 and 0.90 and 0.95, respectively.

For each group of fish present at Greenland the short-term losses on home water stocks can be estimated from:

$$\text{Losses} = WH/WG \times S \times 1/(1 - N)$$

The estimated loss to their home water stocks for each tonne of NA salmon caught at West Greenland ranges between 1.47 to 2.00 tonnes. The corresponding values for European home water stocks are 1.29 and 1.75 tonnes. The total landings at Greenland in 1985 were 864 tonnes of which 415 tonnes were estimated to be North American salmon and 449 tonnes were estimated to be European salmon. Therefore 610 to 830 tonnes were estimated to have been lost to North American stocks and 579 to 786 tonnes to European stocks.

USA salmon caught at West Greenland

In 1986 the Working Group attempted to estimate the number of USA salmon caught at West Greenland for the first time. It noted that nearly all salmon from 1-year smolts of North American origin are produced in hatcheries and nearly all USA origin salmon are of hatchery origin. Thus, it is possible to estimate the harvest of USA origin salmon at Greenland from observations on the number of North American salmon in the Greenland fishery and the relative proportions of 1-year smolts from Canada and USA hatcheries among them.

The result from the above analysis was, that the number of USA salmon caught in the West Greenland fishery in 1984 and in 1985 was 2,600 (2.9 per cent) and 8,090 (2.7 per cent) salmon of 90,000 and 300,500 salmon landed in the two years.

24.7   SUMMARY

(1)   This paper describes the occurrence, migration and exploitation of salmon at West Greenland.

(2)   The occurrence of salmon at Greenland was first recorded by Fabricius in 1780. In the 1950s it was shown that salmon occurring during the autumn were derived from rivers outside Greenland.

(3)   In 1966 the ICES/ICNAF Joint Working Group on North Atlantic Salmon began its work. The main questions posed were:

(a)   The proportion of home water stocks which visit Greenland.

(b)   The proportion which are caught at Greenland.

(c)   The proportion which avoid capture at Greenland and survive to return to home waters.

(d) The growth of the fish between leaving Greenland and being captured in home fisheries.

(e) Exploitation rates in home waters.

(4) International teams of salmon biologists have visited Greenland every year since 1965.

(5) The landings increased from 60 to 2,210 tonnes between 1960 and 1969. From 1970 to 1975 they fluctuated between 2,000 and 2,600 tonnes. From 1976 to 1982 the catch level was relatively constant at around 1,200 tonnes. 310 and 297 tonnes were taken in 1983 and 1984, respectively, and 864 tonnes were landed in 1985. The main bulk of the fish are caught in NAFO Divisions 1B to 1E.

(6) Since 1972 the salmon fishery at West Greenland has been regulated by quota and from 1976 only Greenlandic vessels have been allowed to fish for salmon. The TAC from 1972 to 1973 and from 1974 to 1983 was set at 1,100 and 1,191 tonnes, respectively. In some years it was adjusted upwards to allow for a later opening date than 10 August. In 1984 and 1985 the TAC was reduced to 870 and 852 tonnes, respectively, and for 1986 and 1987 has been set at 850 tonnes for an opening date of 1 August. This TAC can be adjusted according to the opening date.

(7) Tagging experiments at Greenland and home waters have shown that salmon from Canada and Scotland make the major contribution to the fishery. However, salmon from the USA, England, Ireland, Iceland, Norway, Sweden, Denmark, Finland, France, Spain and possibly Portugal and the USSR make some contribution to the stocks of salmon at Greenland. Almost equal proportions of salmon of North American and of European origin are present in the catch.

(8) Only salmon which have already spent at least one winter in the sea (except for fish of native origin) are caught at Greenland, and between 93 and 97 per cent of these have spent no more than one winter in the sea. Thus the Greenland fishery has no effect on grilse stocks, as all the salmon caught would have been MSW fish by the time they had returned to home waters.

(9) The sex ratio of salmon at Greenland is 3:1 in favour of females. North American salmon are shorter and lighter than European salmon of the same age.

(10) Factors other than the abundance of salmon in home waters and the natural mortality of smolts in the sea can influence the number of salmon at West Greenland. Environmental conditions including cold winters in West Greenland or in combination with the speed of the warming of the surface water north of 50°N and west of 40°W from January to March also seem to influence the emigration of

salmon to West Greenland and, as a result, the numbers present in any one year.

(11) The only available estimate of the exploitation rate at West Greenland refers to 1972. In that year it was estimated that 33 per cent of the salmon present at the beginning of the season were caught.

(12) The estimated losses per tonne of salmon caught at Greenland are 1.47 to 2.00 tonnes and 1.29 to 1.75 tonnes for North American and European stocks respectively.

(13) Provisional estimates of the number of salmon of USA origin caught at Greenland in 1984 and 1985 are 2,600 and 8,090 fish respectively, or 2.9 per cent and 2.7 per cent of the number of salmon landed in each of these years.

REFERENCES

Andersen, K.P., Horsted, Sv. Aa. and Møller Jensen, J. (1980) Analysis of recaptures from the West Greenland Tagging Experiment. ICES, Rapports et Procés-Verbaux Reunion Conseil International Exploration de la Mer, 176, 136-41

Anon. (1964) Laerebog i fangst for Syd - og Nordgrønland. Den kongelige grønlandske Handel, København 1964

Anon. (1967) Report of the ICES/ICNAF Joint Working Party on North Atlantic Salmon, 1966. ICES, Cooperative Research Report Services A, No. 8

Anon. (1981) Report of the ICES Working Group on North Atlantic Salmon, 1979 and 1980. ICES, Cooperative Research Report No. 104

Anon. (1982) Report of ICES Working Group on North Atlantic Salmon. CM 1982/Assess: 19

Anon. (1984) Report of ICES Working Group on North Atlantic Salmon. CM 1984/Assess: 16

Anon. (1985) Report of ICES Working Group on North Atlantic Salmon. CM 1985/Assess: 11

Anon. (1986) Report of ICES Working Group on North Atlantic Salmon. CM 1986/Assess: 17

Buch, E. (1986) A review of the hydrographic conditions off West Greenland in 1980-85. NAFO SCR. DOC. 86/48

Fabricius, O. (1780) Fauna Groenlandica, Hafniae et Lipsiae

Jensen, A.S. (1948) Contributions to the Ichthyofauna of Greenland, 8-24, Spolia Zoologica Musei Hauniensis, 9, 65-78

Lear, W.H. (1972) Scale characteristics of Atlantic Salmon from various areas in the North Atlantic. International Council for the Exploration of the Sea, CM 1972/M:10

Lear, W.H. and Sandemann, E.J. (1980) Use of scale characters and discriminant functions for identifying origin of Atlantic Salmon. ICES, Rapports et Proces-Verbaux Reunion Conseil Internationale Exploration de la Mer, 176, 68-75

Menzies, W.J.M. and Shearer, W.M. (1957) Long-distance migration of salmon. Nature, 179, 790

Munro, W.R. and Swain, A. (1980) Age, weight and length distribution and sex-ratio of salmon caught off West Greenland. Rapports et Procés-Verbaux Reunion Conseil Internationale Exploration de la Mer, 176, 43-54

Møller Jensen, J. (1980) Recaptures from the International Tagging Experiment at West Greenland. ICES, Rapports et Procés-Verbaux Reunion Conseil Internationale Exploration de la Mer, 176, 122-35

Netboy, A. (1968) The Atlantic Salmon. Faber and Faber, London

Nielsen, J. (1961) Contribution to the biology of the salmonidae in Greenland, I-IV. Meddelelser om Grønland, 159, no. 8

Ricker, W.E. (1975) Computation and interpretation of biological statistics of fish populations. <u>Bulletin of the Fisheries Research Board of Canada</u>, <u>191</u>

Rosenørn, S., Fabricius, J.S., Buch, E. and Horsted, Sv. Aa. (1984) <u>Isvintre ved Vestgrønland</u>. Firskning/Tusaut i Grønland. Vol. 2, pp 2-19

Chapter Twenty Five

EXPLOITATION AND MIGRATION OF SALMON IN FAROESE
WATERS

Stein Hjalti i Jákupsstovu

Fiskirannsóknarstovan, Debesartrød 100 Tórshavn, Faroe Islands

INTRODUCTION

In the older literature on the Faroese fish fauna (Svabo, 1782;
Landt, 1800), both Atlantic salmon (Salmo salar) and sea trout
(Salmo trutta) are mentioned as present. In more recent works
(Joensen and Tåning, 1970), however, salmon is not considered a
native species.

In the period 1947-51 salmon fry of Icelandic origin were
released into several Faroese rivers. Since then salmon have
spawned regularly in some of these (Joensen and Tåning, 1970).

The first verified record of salmon caught in the sea
around Faroes was made in 1958, when a multi-sea-winter salmon
was fished north of the Islands at 130 m depth. In the years
1965-67 seven specimens were caught on the Faroe Banks, and
since 1967 exploratory and commercial fisheries for salmon in the
sea around Faroes have been annual events (Table 25.1).

Until 1977 the total annual landings of salmon by Faroese
vessels was at a relatively low level (less than 40 tonnes). In the
first years following the general introduction of economic zones,
however, the salmon fishery increased rapidly. In the most recent
years the fishery has decreased again due to quota agreements.

The present fishery at Faroes is a licensed fishery with a
restricted number of vessels (26 in the 1984/85 season) allowed to
participate. Each vessel is allocated an individual quota. The
average vessel is 82 feet long and is propelled by a 430 hp
engine. The range, however, is quite wide; 48 to 118 feet and 220
to 12,00 hp. One of the smaller vessels is shown in Figure 25.1.
Each vessel has a crew of 5 to 8 men.

The fishing method is by floating longline and is to a large
extent a replicate of the longline fishery for salmon in the Baltic
as it is performed by Danish fishermen. A schematic drawing of
the line set is shown in Figure 25.2. A more comprehensive
description of the fishery and the fishing methods is given by

458

Table 25.1: Reported nominal catches by Faroese vessels in the Faroese area longline fishery 1968-85. (Converted from gutted weight with a factor 1.11)

Year	Catch
1968	5[a]
1969	7
1970	12[a]
1971	0
1972	9
1973	28
1974	20
1975	28
1976	40
1977	40
1978	37
1979	119
1980	568
1981	1025[a]
1982	960
1983	753
1984	697
1985[b]	672

Notes: a, A small part of the catch taken more than 200 miles from the Faroese base line.
b, Provisional

Source: Anon. (1986a)

Mills and Smart (1982)
Initially the salmon caught were iced in boxes and the trips then lasted no longer than 9 days from port to port. In later years most vessels have installed freezing equipment to freeze and store the fish caught, and the length of the trips is now limited only by provision capacities.

The season has in later years with some small variations lasted from mid October to the end of May.

The Faroese fishery has raised a number of important questions. The countries having rivers in which the salmon fished originate as smolts are concerned about their relative contribution to the fishery and the corresponding loss to home water stock and catches. Similarly the Faroese government has been interested in obtaining knowledge on the standing stock of

Figure 25.1: A Faroese salmon longliner

Figure 25.2: Schematic drawing of a salmon longline

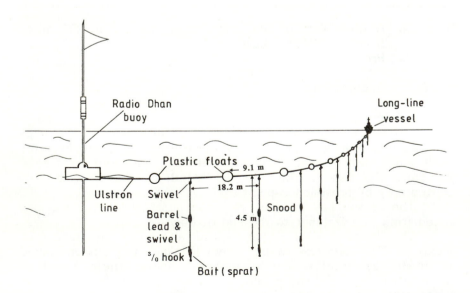

salmon in Faroese waters at any time throughout the year, the food consumed and the resulting growth of salmon.

ICES through the Atlantic Salmon Working Group has addressed these and other questions raised by home and host countries and by NASCO, but only with limited success. The Faroese fishery, however, has initiated a greater effort into salmon research in Faroese waters on feeding salmon and some new information has been gained.

I will in this paper present some of the research done and the results obtained.

MATERIAL

Biological samples

In 1968 research surveys and exploratory fisheries for salmon were initiated at Faroes. Since then the catches and landings of salmon have been sampled for length (fork and total), weight (gutted and ungutted), sex, maturity and age composition (river and sea) from scales (Struthers, 1981). Much of this work has been done in cooperation with the Freshwater Fisheries Laboratory of Scotland, which has also aged all scale samples. In 1981 ICES established a subgroup of the Working Group on North Atlantic Salmon to study the salmon fishery at Faroes. Under the auspices of this group a number of international observers have participated on commercial trips, collecting biological samples of various kinds in addition to collecting statistics on the line settings and haulbacks. The methodology for this work and the results have been discussed in detail by the Study Group (Anon. 1982, 1983, 1984a, 1986a). In addition, since 1982, a market sampling scheme has been established in which a Faroese observer has been hired for each season to sample the landings. This scheme is financed jointly by Faroes, Iceland, Ireland, Norway and the United Kingdom.

Tagging

Tagging at sea.

From 1969 to 1976, during the research trips with the Faroese research vessel Jens Christian Svabo, a total of 1,946 salmon were tagged off the Faroes in order to acquire information on their subsequent migrations, and in particular to determine their country of origin. In 1986 during a research survey with a Faroese salmon longliner, 100 salmon were tagged, again in order to study the feasibility of undertaking a new tagging programme at sea.

Smolt tagging.
During the years many countries have tagged a very large number of smolts both with external tags (Norway, Scotland, Sweden and UK) and micro tags (Iceland, Ireland and UK).

At Faroes, since 1980, a special effort has been made to increase the return rates of external tags by the fishermen. According to the licence regulations the fishermen are obliged to return all tags found to the Fisheries Laboratory at Faroes together with the necessary information. The tag return premium has been increased to 30 Danish Kr. per tag and each vessel is provided with an ample supply of pre-paid tag return envelopes. Based on the number of tags recovered during observed fishing trips the tag return rate of external tags is estimated to be in the order of 75 per cent.

During the market sampling a number of landings are monitored for adipose fin clipped salmon, which are all subsequently scanned for microtags. In the 1984/85 season approximately 27 per cent of the total landings were monitored for this purpose.

Fisheries statistics

Logbooks.
As mentioned previously, the Faroese longliners fishing for salmon are licensed, and in order to keep the licence a number of requirements have to be met. One of the most important is that the vessels have to keep daily logbooks noting positions fished, number of hooks set and the resulting catch. From these data it is possible to analyse the distribution and success of the fishery both in time and space.

Landing-sheets.
All salmon landings are monitored by governmental quality inspectors, and at the same time the salmon landed are grouped into seven weight categories. This information is noted on special landing sheets giving the number and weight in each category landed. From the landing sheets and age/weight keys it is possible to obtain very accurate estimates of the catch in numbers by sea-age class.

Research surveys
In 1985 and 1986, using a hired salmon longliner the Faroese fisheries laboratory conducted research surveys for Atlantic salmon within the Faroese fishing zone. The methodology and the results of the 1985 survey is described in Jákupsstovu, Jørgensen,

Mouritsen and Nicolajsen (1985). During the surveys a number of fishing stations were worked and the catch sampled for age, length, weight and stomach contents. For each set the positions for setting and haulback was noted and in 1985 also the surface temperatures at the same positions. Most haulbacks were monitored for lost fish and discards. During the survey in 1985 two salmon were tagged successfully with acoustic tags and tracked for 7 and 19 hours respectively. In 1986 two more salmon were tagged successfully and tracked for 5.5 and 4 days respectively.

RESULTS

Distribution of the fishery

In the period 1969-1979 the exploratory salmon fishery was conducted close to the Faroe Plateau (Figure 25.3) with no fishing north of 63°30'N. Furthermore, most fishing took place in one rectangle NE of the plateau. During the period 1978-82 the vessels explored a much wider area in the Norwegian Sea and for example in the 1981/82 season Faroese vessels fished salmon from 62°N to 71°30'N (Figure 25.4). The main area fished, however, lay between 64°N and 68°N. Following the NASCO Convention the Faroese salmon fishery has been restricted to the fishing zone of the Faroe Islands only. Compared to the years 1969-79 the main areas fished in later years, however, have been further north, between 63°30'N and 65°30'N (Figure 25.5), with no fishery in the main rectangle fished in the earlier period.

Sea age composition of the landings

In all spring and summer samples from the period 1969-79 the one sea-winter group predominated (62-91 per cent) with the 2 sea-winter group next in order (7-30 per cent) followed by 3 sea-winter fish and previous spawners (Struthers, 1981).

The autumn and winter samples of the same period showed the influx of the 0 sea-winter group into the area with some variation from year to year in the relative magnitude of the 0 sea-winter group. Despite the recruitment of the 0 sea-winter group the 1 sea-winter group remained predominant (Struthers, 1981).

In contrast to the early period of the fishery the sea-age composition of the salmon caught in the Faroese fishery since 1980 has been dominated by 2 sea-winter fish (Table 25.2). Assuming a birthday of 1 November 76, 87 and 87 per cent of the total seasonal catch in numbers were 2 sea-winter salmon in the seasons 82/83, 83/84 and 84/85 respectively followed by 3 sea-winter fish and to a far lesser extent 1 sea-winter fish.

Figure 25.3: The total number of salmon caught by statistical rectangle. Faroes Salmon Research 1969-79

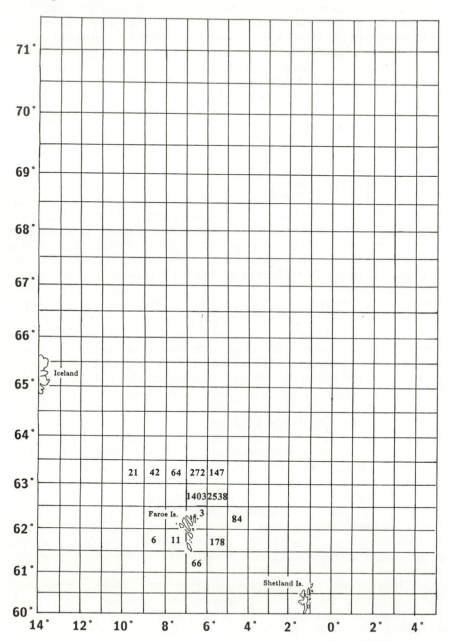

Source: Struthers (1981)

Figure 25.4: Catch (10^{-2} x number) by statistical rectangle, 1981/82 fishery season

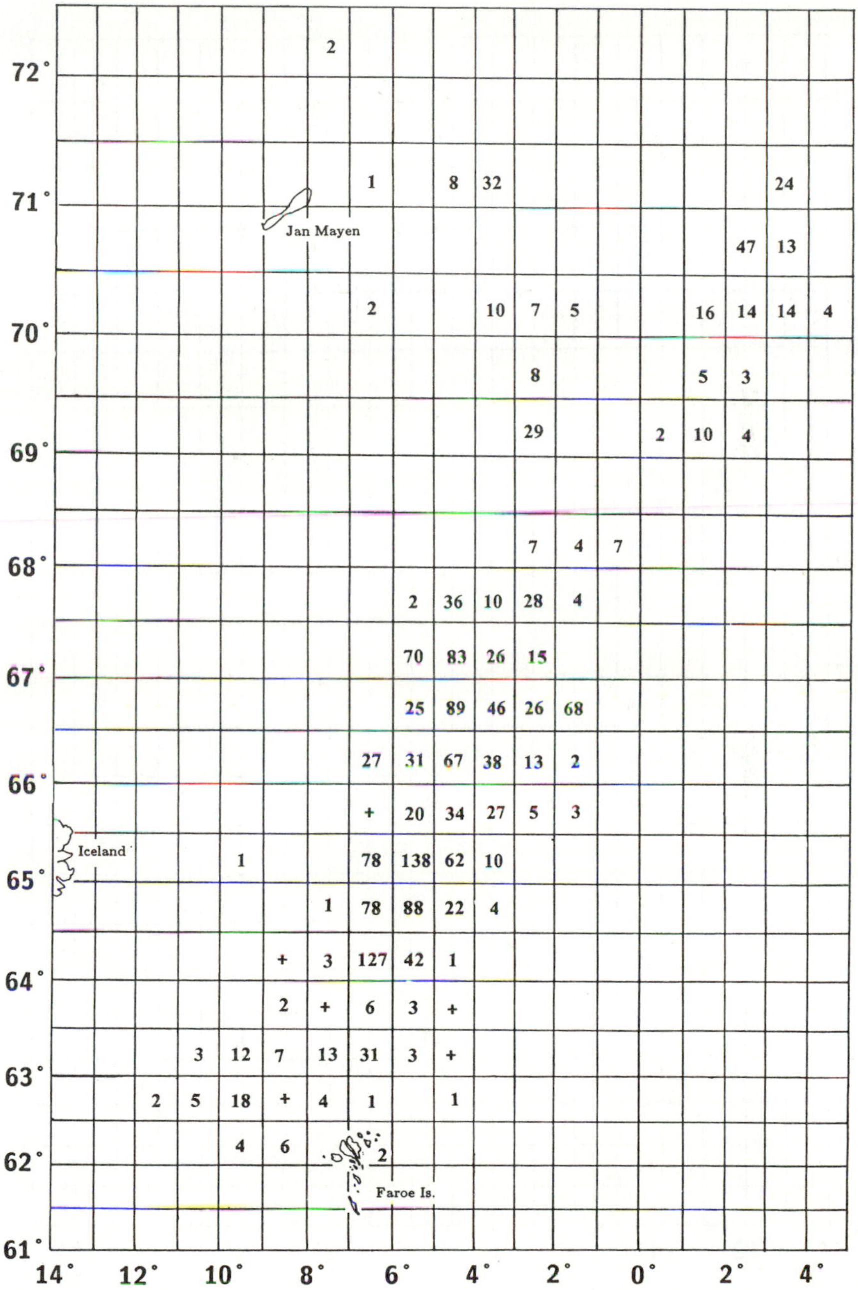

Source: Jákupsstovu (1984)

Figure 25.5: Catch (10^{-2} x number) by statistical rectangle, 1983/84 fishery season

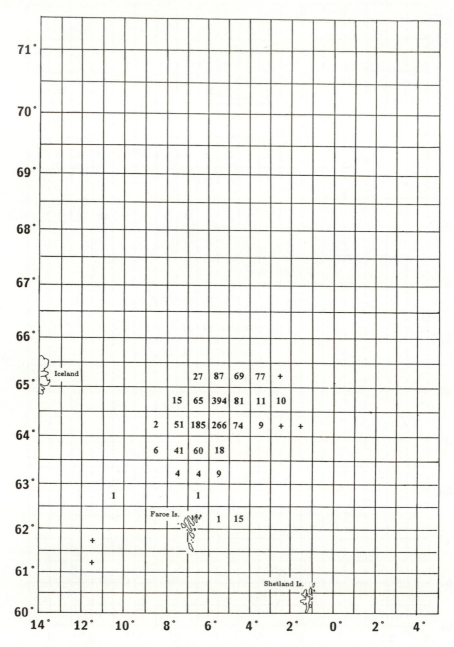

Source: Jákupsstovu (1985)

Table 25.2: Estimated catch (%) by sea age of salmon caught in the Faroese longline fishery in the seasons 82/83, 83/84 and 84/85

Season	Sea age (years)			Prev. spawn	Hatchery reared
	1	2	3		
82/83	6.9	76.2	16.4		
83/84	3.2	86.7	10.1		
84/85	0.2	87.4	7.2	1.2	3.9

Source: Jákupsstovu (1984, 1985), Anon. (1986a).

Sea-age distribution by geographical area

During the period of the expanding fishery the fishermen found multi sea-winter fish in higher concentrations further north in the Norwegian Sea compared to the area in the south first explored. Given a fixed quota in weight they preferred to fish in these areas rather than the southern ones in order to maximise the output from the quota.

During a research survey with a hired salmon longliner in February 1985 the relative concentrations of one and multi sea-winter salmon were investigated by area and in relation to the surface temperature (Jákupsstovu et al., 1985). Figure 25.6 shows the relative catch in percentage of one and multi sea-winter salmon from this investigation, together with the surface isotherms based on temperature measurements at the fishing stations. From this figure it is seen that the older fish dominated the catches in the areas where the surface temperature was less than 4°C, whereas in areas with higher surface temperatures one sea-winter fish dominated in all catches but one.

There are no systematic observations on the surface temperatures from other years covering the area normally fished by Faroese salmon fishermen during the salmon season. Investigations of the hydrography outside the salmon season (Hansen and Meincke, 1979), however, have shown a significant variation in the surface temperatures in the southern Norwegian Sea from one year to another and also within a short period. Assuming a strict relationship between the surface temperatures and the distribution of one sea-winter fish this would indicate a variation in the availability of one and multi-sea-winter fish both within a season and between seasons.

Figure 25.6: The distribution (%) of one sea-winter and multi sea-winter salmon by station. February-March 1985

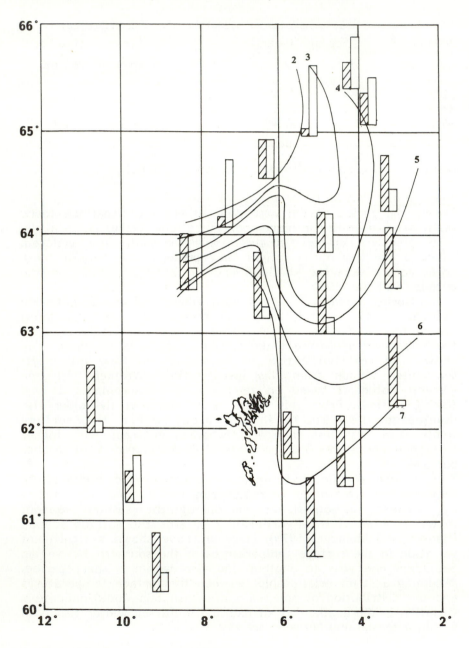

Source: Jákupsstovu et al (1985)

Catch per unit of effort

In Table 25.3 is presented catch in number per 1,000 hooks by months in the seasons 1982/83-1984/85. From this is seen the variation in availability within and between seasons. During the seasons 1982/83 and 1983/84 the catch rates were high in the early parts of the seasons and were then reduced to a lower level. In the 1984/85 season on the other hand the catch rates were at an intermediate level throughout the season with no obvious trend.

Table 25.3: Catch in number per unit effort (1,000 hooks) by month in the Faroese longline fishery for salmon in the seasons 1982/83- 1984/85

Season	Nov.	Dec.	Jan.	Feb.	Mar.	Apr.	May	Whole Season
82/83	83.9	133.7	73.2	48.5	46.0	39.1	34.1	46.9
83/84	75.1	81.0	78.6	52.5	38.9	23.1	31.5	51.3
84/85	41.7	34.6	30.7	35.0	37.4	41.5	37.0	35.8

Source: Anon. (1986b)

The catch per unit of effort by statistical rectangle (30' by latitude and 1° by longitude) generally shows a high degree of homogeneity indicating an even distribution of salmon over large areas of the Norwegian Sea. In the 1981/82 season (Figure 25.7), however, there is a trend of higher catch rates to the north compared to the southern areas. In the 1983/84 season (Figure 25.8) the same picture of an even distribution of salmon emerges with a slight indication of higher concentrations in the central part of the Faroese fishing zone north of the Faroes.

Home country

Tagging of adults.

From the 1,946 salmon tagged during the years 1969-76 off the Faroes 90 were subsequently recovered (Struthers, 1981). Of these 33 were recaptured in Scotland, 31 in Norway, 15 in Ireland, 8 in other European countries and 3 at West Greenland. Most of the fish tagged were 1 sea-winter fish (1,650). Of the recaptures, 77 were known to be recaptured the same year they were tagged and 7 the year after, while the year of recapture was unknown for the rest.

Figure 25.7: Catch in number per 1,000 hooks by statistical rectangle, 1981/82 fishery season

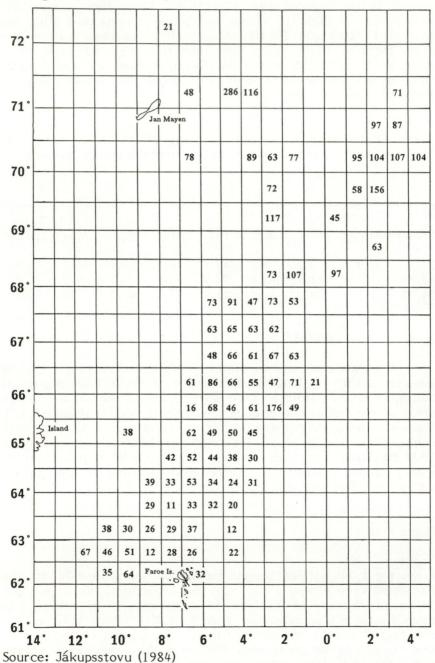

Source: Jákupsstovu (1984)

Figure 25.8: Catch in number per 1,000 hooks by statistical rectangle, 1983/84 fishery season

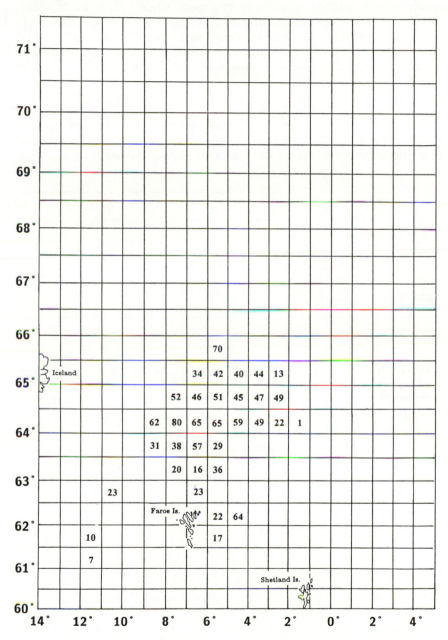

Source: Jákupsstovu (1985)

Tagging of smolts.
In Table 25.4 is shown the total number of salmon tagged as smolts with external tags and released since 1978 in Norway, Sweden, Scotland and England and the number of recoveries from the same releases in the Faroese salmon fishery. Table 25.5 shows the number of microtagged salmon released in Iceland and Ireland and the estimated numbers in the Faroese fishery in the 1984/85 season.

Table 25.4: Summary of the total numbers of tagged salmon released by country since 1978 and the number of recoveries from the same releases in the Faroese salmon fishery

Country	No. released	No. recaptured	Recaptured/ released x 10^{-3}
Norway	306,500	979	3.19
Sweden	60,200	302	5.02
Scotland	68,800	69	1.00
England	12,200	14	1.15

Source: Anon. (1986b)

Table 25.5: Summary of the total number of microtagged salmon released by Iceland and Ireland and the estimated number in the Faroese fishery in the 1984/85 season. Based on screening of landings

Country	No. released	Estimated no.	Recaptured/ released x 10^{-3}
Iceland	151,144	25	0.17
Ireland	260,816	301	1.15

Source: Anon (1986b)

From these tables it appears that the highest recovery rates are from salmon of Swedish and Norwegian origin. Compared to these countries salmon originating from the British Isles and Ireland are recovered to a much smaller extent, and only a very small fraction of the number of salmon tagged in Iceland are recovered in the Faroese fishery. Assuming the tagging and post-smolt mortality to be of the same order of magnitude in all smolt taggings it thus appears that salmon of Swedish and Norwegian origin are under the highest exploitation pressure by the Faroese fishery.

In addition to the countries mentioned above salmon tagged as smolts in Canada and USSR have been recovered in the Faroese fishery. While the salmon of Canadian origin probably can be considered as stray fish (Reddin, 1985) no conclusion can be drawn on the salmon of USSR origin as no information is available on the number of smolts tagged and released.

When recaptures of smolt-tagged salmon in the Faroese fishery are plotted by statistical rectangle in which they were caught, it is found that salmon originating from all the countries are found well mixed within the Faroese zone (Anon., 1984b) with no segregation by country. The number of tags recovered per number fished, however, appeared in the 1982/83 fishing season, to be higher towards the north and west in the zone. This would imply that the salmon stocks are not randomly mixed within the Faroese area.

Shoaling

The question is often raised, whether salmon shoal during the period of feeding. From monitoring longline haulbacks with the aim of analysing the catches of salmon for clustering on the line, it has been shown (Jákupsstovu et al., 1985) that the distribution of the salmon caught on the line is well within the expectations of a random distribution. Of all the salmon tagged as smolts in Norway in 1981 and 1982, date and position fished was known for 409 which were subsequently recaptured by Faroese vessels (Jákupsstovu, 1985). Of these only 9 were recaptured the same day as their fellows from the same release (tagged and released the same day at the same site). Of these, two pairs from the same experiment were recaptured in different statistical rectangles, and one pair in the same rectangle. Of the remaining three tagged salmon, two were recaptured in the same rectangle by the same vessel and one in a different one.

Behaviour

As mentioned previously two salmon in 1985 and two in 1986 were successfully tagged with acoustic tags and subsequently tracked for some time. As no general picture of the behaviour of salmon during feeding can be drawn from only four acoustically tagged salmon, I will only briefly describe the main things observed.

Vertical movements.

Using depth-sensitive tags, information on the vertical movements of one of the salmon tagged in 1985 and both in 1986 was obtained. After release, all salmon dived very rapidly to a depth of more than 100 m, most probably due to stress from the handling and tagging. The salmon tagged in 1985 with the depth-sensitive tag ascended slowly after the deep dive and reached the top 10 m surface layer 4 hours after release. Apart from 2-3 short dives it stayed there for the rest of the tracking period. In the top layer, however, the salmon continuously made short-term vertical movements (Jákupsstovu et al., 1985).

One of the salmon tagged with an acoustic tag in 1986 was tracked for 5.5 days. After the initial deep dive this salmon ascended to a depth of approximately 40 m (Figure 25.9), where it stayed for the next 50 hours, interrupted only by irregular visits to the surface. Following this, the salmon stayed at the surface for the next two days. The last 1.5 days it made some very rapid dives down to more than 150 m (the outer range of the depth sensitivity of the tag), and after some time made equally rapid ascents to the surface again. These very rapid vertical movements were apparently not caused by tag failure, as the salmon avoided the vessel by diving when the vessel sailed on top of it.

The second salmon tracked successfully in 1986 had a tag which did not discriminate between depths less than 12 m. This salmon after the initial dive stayed above 12 m for most of the tracking period with irregular dives of short duration, mostly down to 40 m, but also one to 80 m and one to 120 m.

Horizontal movements.

As we have no precise position of the salmon during the track, the ship's position is used as a substitute. The distance between the salmon and ship, however, was only, at most, a few hundred metres.

The horizontal movements of the two salmon tracked in 1985, when corrected for the displacement of the water masses at 15 m depth, apparently showed directed movements. The one towards west with an average speed of 0.5 knot the other towards NE with a little less speed for 12 hours after which it remained stationary (Jákupsstovu et al., 1985).

In 1986 the horizontal movements of the tagged salmon (Figures 25.10 and 25.11) could not be related to the water masses for several reasons, the most important being that the distance between the salmon and a drift buoy released at the same time very soon became too large for any meaningful correlation between the currents measured and the migration of the salmon tracked. The salmon followed for 5.5 days (Figure 25.10) had an overall direction towards SE. The other for the

Figure 25.9: The vertical movements of a salmon tagged with a depth-sensitive acoustic tag and tracked for 5.5 days in February 1986

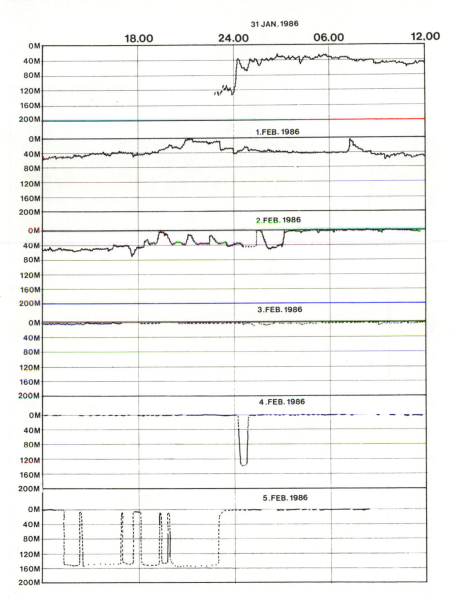

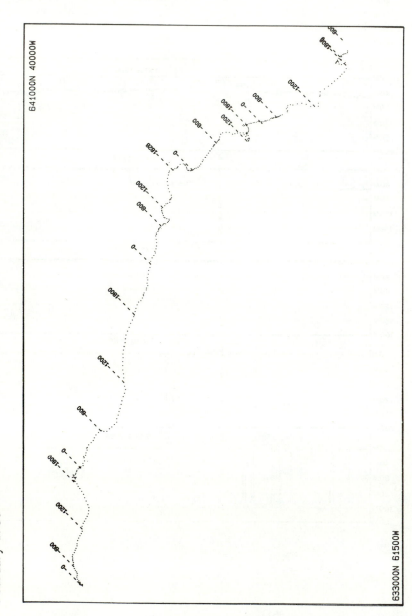

Figure 25.10: Horizontal movement of a salmon tagged with an acoustic tag and tracked for 5.5 days in February 1986

476

Figure 25.11: Horizontal movement of an acoustically tagged salmon tracked for 4 days in February 1986

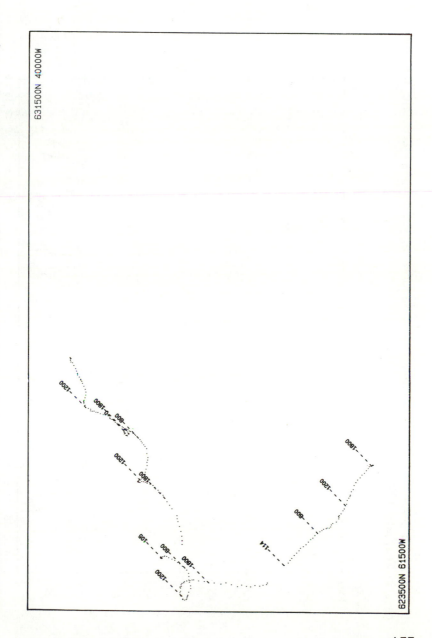

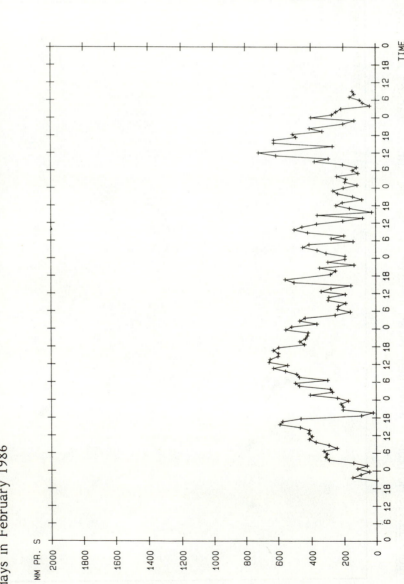

Figure 25.12: Swimming activity (mm/s) of a salmon tagged with an acoustic tag and tracked for 5.5 days in February 1986

Figure 25.13: Swimming activity (mm/s) of a salmon tagged with an acoustic tag and tracked for 4 days in February 1986

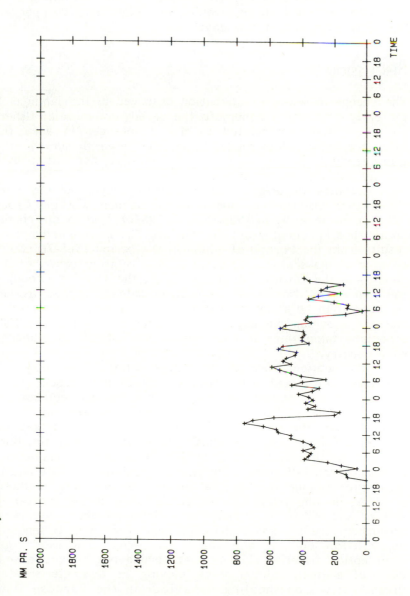

first 2 days swam towards SW and then towards SE (Figure 25.11)

Swimming activity.
Figures 25.12 and 25.13 show the swimming speed over ground of the two salmon tracked in 1986, as revealed from the vessel movement. The figures indicate periods of high activity followed by periods of low activity. The periods of low swimming activity were mostly found during evening and night and the periods of high activity mostly during daytime.

DISCUSSION

The change in sea-age distribution observed in the landings from the early exploratory fishery to the recent commercial fishery at Faroes is obviously related to the differences in area fished. From the observations made during the research survey in 1985 there seems to be a relation between sea-age distribution by area and sea surface temperatures.

Country of origin of the salmon caught in the Faroese fishery was discussed by the Atlantic Salmon Working Group of ICES at the meeting in 1986 (Anon., 1986b). Due to the change in area fished in recent years the Working Group concluded that the results from the tagging of adults in the period 1968-76 could not be used to make any conclusions on the relative contributions by country to the Faroese fishery. Further, the Working Group, from the number of Norwegian tags from salmon tagged as smolts recovered in relation to the number tagged, indicated that Norway is by far the largest contributor to the Faroese fishery, especially taking into account the number of smolts produced by each country.

As mentioned previously, tagged salmon from all countries are found well mixed in the fishing zone of the Faroes. However, in 1982/83 the number of tags recovered per number fished appeared to be higher towards the north and west in the zone, implying that the salmon stocks are not randomly mixed within the area. This apparent inconsistency is hard to explain, but one possible explanation could be found in annual variations in post-smolt mortality and differential distribution of one and multi sea-winter salmon due to sea surface temperature differences.

As homing salmon are often found in shoals in near shore waters and estuaries it has been inferred that salmon shoal throughout the sea-going phase of their life span. Based on this observation one of the main concerns expressed against interception fisheries is a fear of fishing out an entire sea-age class of a small salmon stock moving as a single shoal. The investigations on shoaling behaviour in the Faroese fishery, assuming a reasonable catchability, do not indicate that feeding salmon are found in any sizeable shoals. And, further, do not indicate that all smolts from one tag release remain in a single shoal throughout their seaphase. It may therefore be concluded that the fear mentioned above is not substantiated for the Faroese fishery by these investigations.

In a paper by Hansen (1984) a model for estimating the standing stock of a feeding fish species from catch per unit of effort by long line was provided. The underlying assumption for using this model was that the fish in its search for prey had a migrational behaviour which could be approximated to random walk. The salmon tracks of Figure 25.10 and 25.11 do not support a random behaviour, but it should be stressed, that these tracks are over ground, and although it was attempted to track the water movements this attempt failed, so the figures may very well show only the water movement and are indeed consistent with what is known about these movements (e.g. Hansen, Malmberg, Soelen and Østerhus, 1986). Thus no conclusion can be drawn on the applicability of the random walk model yet.

SUMMARY

During the period of the salmon fishery at Faroes (1969 onwards) the area fished has changed to the north. This has resulted in a change from a fishery on mainly one-sea-winter fish (1969-78) to a fishery on mainly multi-sea-winter fish.

Catch per unit of effort data have shown a variation in availability within and between seasons. These data also show a high degree of geographical homogeneity indicating an even distribution of salmon over larger areas of the Norwegian Sea.

Recaptures of salmon tagged as smolts in the major European smolt-producing countries have shown that salmon from all countries are found well mixed within the Faroes area. The highest recovery rates, however, are from salmon of Swedish and Norwegian origin. All factors alike, this indicates that Norway is by far the largest contributor to the Faroese salmon fishery, especially taking into account the number of smolt produced by country.

Detailed monitoring of salmon long-line haulbacks and analysis of recapture data on salmon tagged as smolts neither indicates that salmon during the feeding period appear in larger shoals, nor that salmon from the same smolt group (tagged and released the same day) remain in a single shoal throughout their sea-going phase.

The horizontal movements of four acoustically tagged salmon are described, the vertical movements of three of these and the swimming activity of two.

REFERENCES

Anon. (1982) Report of meeting of the working group on North Atlantic salmon. ICES, Doc. CM 1982/Assess:19

Anon. (1983) Report of informal meeting of the special study group of the North Atlantic salmon Working Group. 15 pp. Mimeo

Anon. (1984a) Report of the study group of the North Atlantic salmon Working Group. ICES, Doc. CM 1984/M:9

Anon. (1984b) Report of meeting of the Working Group on North Atlantic salmon. ICES, Doc. CM 1984/Assess:16

Anon. (1986a) Report of the meeting of the special study group on the Norwegian Sea and Faroes salmon fishery. ICES, Doc. CM 1986/M:8

Anon. (1986b) Report of the Working Group on North Atlantic salmon. ICES, Doc. CM 1986/Assess:17

Hansen, B. (1984) Assessment of a salmon stock based on long-line catch data. ICES, CM 1984/M:6

Hansen, B., Malmberg, S.A., Soelen, O.H. and Østerhus, S. (1986) Measurement of flow north of the Faroe Islands 1986. ICES, CM 1986/C:12

Hansen, B. and Meincke, J. (1979) Eddies and meanders in the Iceland Faroe ridge area. Deep-Sea Research, 26A, 1067-82

Jákupsstovu, S.H. í(1984) The Faroese longline fishery in the 1980/1981, 1981/82 and 1982/83 fishery season. Working paper for the Atlantic Salmon Working Group 1984. 6 pp

Jákupsstovu, S.H. í(1985) The Faroese longline fishery in the 1983/84 fishery season. Working paper for the Atlantic Salmon Assessment Working Group. March 1985. 5 pp

Jákupsstovu, S.H. i, Jørgensen, P.T., Mouritsen, R. and Nicolajsen, A. (1985) Biological data preliminary observations on the spatial distribution of salmon within the Faroese fishing zone in February 1985. ICES, CM 1985/M:30

Joensen, J.S. and Vedel Tåning, Å. (1970) Marine and freshwater fishes. In: S. Jensen et al. (eds), Zoology of the Faroes, volumes LXII-LXIII, 241 pp

Landt, J. (1800) Forsog til en beskrivelse over Faerøerne, Kjøbenhavn

Mills, D. and Smart, N. (1982) Report on a visit to the Faroes. The Atlantic Salmon Trust, 52 pp

Reddin, D.G. (1985) Contribution of North Atlantic salmon to the Faroes fishery. ICES, CM 1985/M:11

Struthers, G. (1981) Observations on Atlantic salmon (Salmo salar L.) stocks in the sea off the Faroe Islands 1969-79. Working document to the North Atlantic Salmon Working Group meeting, 11 pp

Svabo, J. Chr. (1782) Indberetninger fra en Reise i Foeroe 1781 og 1782. N. Djurhuus, København 1959

Chapter Twenty Six

OCEAN LIFE OF ATLANTIC SALMON (<u>SALMO</u> <u>SALAR</u> L.) IN
THE NORTHWEST ATLANTIC

D.G. Reddin

Science Branch, Department of Fisheries and Oceans, P.O. Box
5667, St John's, Newfoundland, A1C 5X1

INTRODUCTION

The exchange between Menzies (1949) and Huntsman (1938) about
Atlantic salmon (<u>Salmo</u> <u>salar</u>) life history in the sea generated
many useful ideas. Menzies hypothesised that salmon migrate far
out into the North Atlantic and would be found feeding in the
area of the Irminger Sea. Huntsman proposed the opposite
hypothesis that salmon remain near the mouth of the river from
which they originated. The answer to these opposing ideas finally
came with the beginning of deep-sea fisheries off Greenland and
in the Norwegian Sea where salmon tagged as smolts in North
America and Europe have been recaptured. Since then Templeman
(1967, 1968), May (1973), Lear (1976), Reddin (1985) and Reddin
and Shearer (1987) have demonstrated that salmon are distributed
seasonally over much of the Northwest Atlantic and thereby
make lengthy migrations far from their rivers of origin.
    The marine environment is ever-changing and conditions
can vary substantially from season to season and from year to
year (Dunbar, 1981). Therefore, it would not be surprising to
discover that this variability in the marine environment has some
influence on salmon survival and that salmon react to it with a
variety of responses. May (1973) suggested that salmon are found
in the Northwest Atlantic in relatively cool water of 3° to 8°C.
More recently Martin, Dooley and Shearer (1984), Scarnecchia
(1984a), Martin and Mitchell (1985), and Reddin and Shearer
(1987) demonstrated direct relationships between ocean climate,
principally sea-surface temperature (SST), and the abundance and
distribution of salmon at sea. Furthermore, Reddin and Shearer
(1987) demonstrated that salmon can modify their migratory path
depending on the temperature of the water that they must swim
through. They also suggested that older salmon will tolerate
colder SSTs than younger ones.

The objectives of this paper are to show:

(1)   The importance of survival during the marine phase of salmon life history to total salmon production.
(2)   The influence of environmental conditions on salmon production during the marine phase, and
(3)   To review the ocean life of salmon in the northwest Atlantic including distribution, migration routes, and feeding.

The life-history terminology used in this paper is that recommended by Allan and Ritter (1977).

METHODS

Variability in survival
Anadromous Atlantic salmon, because they have two distinct life-history phases, will spend part of their lives in two different environments. From the time of spawning until smoltification they are in freshwater and from smoltification until maturity they are in the sea. Ricker (1975) has shown that the productivity of a fish species will depend on its fecundity and rates of natural mortality and fishing mortality. Variations have been observed in total returns to many river systems (Chadwick, 1982; Cutting and Jefferson, 1985; Marshall, 1985; O'Connell, Dempson, Reddin and Ash 1985; Randall, Chadwick and Schofield 1985; Shearer, 1986) and mortality during both phases could have a significant impact on total adult production, depending on the level of mortality during each phase and its variability. The phase when most of the variability observed in total production for many river systems occurs will depend on when the variability in survival is greatest, i.e. during the freshwater or marine phases. Chadwick and Meerburg (1978) concluded from data available to them that adult salmon production was directly dependent on freshwater survival because there was little yearly fluctuation observed in sea survival. The statistic chosen to compare variability in survival between freshwater and marine phases and identify at what stage survival of salmon is most variable was the coefficient of variation as it is independent of the magnitude of the means of the two populations being compared (Sokal and Rohlf, 1969). The most complete data set available that includes estimates of egg deposition, counts of smolts produced in a system, and adults returning to the system is from Western Arm Brook (WAB), Newfoundland. The WAB data series is particularly useful for this analysis because counts of upstream migrating adults and smolts are complete (Chadwick, 1982, 1985; Chadwick, Alexander, Gray, Lutzac, Peppar and Randall, 1985). The data

series and techniques used to obtain it are described by Chadwick (1982). The only confounding factor is that adults are exploited at sea prior to return to the river. Because egg deposition is only available for year-classes 1973-1979, the same time series was used to examine survival from smolt to adult. For the reasons presented by Chadwick (1985), the anomalous 1972 year-class was omitted.

Environmental conditions and salmon abundance in the northwest Atlantic

Analyses were conducted to determine if conditions in the marine environment are related to salmon abundance and therefore the following analysis depends on the location of salmon in the northwest Atlantic and their response to water temperature. North American salmon of all sea-ages and European salmon are located in the Labrador Sea throughout the year (Reddin and Shearer, 1987). During this time they are subjected to a variety of environmental stimuli, one of which is water temperature. Shepherd, Pope and Cousens (1982) discussed three potential sources of variation in recruitment to fish stocks: direct physiological effects, disease and feeding. Sea temperature can probably have the most important physiological effect on salmon as well as other fish species because it controls the rate of enzyme activity which, in turn, controls growth. Sea temperature may also have an indirect effect by influencing the abundance of food and its quality. For salmon, it was hypothesised that the period from January to March could have the greatest influence on survival and growth and environmental variables were chosen to reflect marine conditions during this time.

Environmental variables were interpreted from the British Meteorological Office Ice Charts issued monthly from Bracknell, England from 1969 to 1984. The 4°C sea-surface temperature (SST) and 0°C air-surface temperature (AST) were used to measure the area (km^2) of ocean covered by water 4°C or warmer and air 0°C or warmer over the Labrador Sea in January and March (Table 26.1). For this purpose, the boundaries of the Labrador Sea are defined as west of 40°W longitude to the coast of North America and north of 50°N to the southern tip of Greenland including the Davis Strait. Therefore, there were four environmental variables, i.e. coverage of the Labrador Sea by 0°C air and 4°C water in January and March. The SSTs for these charts were based on 15,000 reports from ships of opportunity and ASTs from ships of opportunity and land-based weather stations. Means of SSTs and ASTs were plotted for every one-degree square and isotherms drawn at 4°C intervals for SSTs and 0°C isotherm for ASTs. The isotherms represent the mean SST and AST distributions for 10 days prior to the end of the month for

Table 26.1: The area (km^2 x 10^{-3}) of the northwest Atlantic where air was 0°C and warmer and water was 4°C and warmer in January and March, 1969-84

| | January | | March | |
Year	Air	Water	Air	Water
1969	125	98	130	92
1970	65	73	100	50
1971	62	102	90	55
1972	24	19	51	36
1973	71	83	93	57
1974	22	59	74	70
1975	0	49	100	84
1976	77	54	109	36
1977	127	64	113	52
1978	59	37	98	79
1979	150	88	184	60
1980	144	41	176	49
1981	143	45	124	54
1982	122	46	148	45
1983	82	45	98	50
1984	23	20	66	6
Means	81	58	110	55
S.D.	49	25	36	21

which the chart was issued.

Principal components analysis (PCA), as suggested by Shepherd et al. (1982), was used to summarise and provide a reduced set of variables for further analysis (Morrison, 1976). For further analysis, it is convenient to consider each principal component to be the response of a causal stimulus of the four environmental features. By using the principal component scores, it is possible to provide a smaller set of data than originally available and one that can be more easily interpreted since the component scores are uncorrelated with one another and can be individually interpreted. Such methodology is consistent with theory and Moore (1965) provides numerous biological examples of how component scores of each individual, year in this case, can become univariate raw data for analysis of variance, correlation, etc.

The hypothesis that abundance of salmon is related to environmental conditions in the northwest Atlantic during winter in the months of January and March was tested using Spearman-rank correlations. Scores from components one and two were used as together they explained 89 per cent of the variation in environment in the northwest Atlantic in January and March (Table 26.2). The higher loadings on the air temperature variables for component one indicate that this variable is measuring the relative sizes of the areas (Table 26.2). The principal component scores from components one and two (Table 26.3) were output and correlated with several sets of catch and abundance indices.

Table 26.2: The results of principal components analysis on four environmental variables from the northwest Atlantic

	Eigenvalue	Proportion	Cumulative	
Prin1	3547.35	0.7419	0.7419	
Prin2	720.65	0.1507	0.8927	
Prin3	326.15	0.0682	0.9609	
Prin4	187.10	0.0391	1.0000	
	Prin1	Prin2	Prin3	Prin4
NJ-AA	0.8035	-0.2133	-0.3758	0.4094
NJ-WA	0.1730	0.7805	-0.4745	-0.3685
NM-AA	0.5657	-0.0047	0.6204	-0.5432
NM-WA	0.0664	0.5876	0.4987	0.6337

A mixture of abundance indices including catch returns and counts of returning adults to enumeration facilities were used. Catch data if not corrected for effort may not provide a satisfactory index of abundance. However, because for most salmon statistics effort data were unavailable or unreliable catch uncorrected for effort was used. The abundance estimates of salmon came from landings by weight from the national catch statistics for Scotland, Norway, Ireland, Canada, and world catch from 1970 to 1985 (Anon., 1986). Percentage of grilse was calculated by smolt class as equal to landings of grilse in year n, divided by the sum of landings of grilse in year n, and landings of salmon in year $n+1$. Other abundance estimates used were the percentage returns by smolt class to enumeration facilities at Western Arm Brook, Newfoundland, 1971-84 (Chadwick et al., 1985); hatchery and wild returns to the Saint John River, New Brunswick, 1974-84 (Marshall, 1985); and numbers of wild fish enumerated at Millbank on the Miramichi River, New Brunswick, 1968-84 (Randall et al., 1985). When appropriate , lagged

Table 26.3: The principal component scores for components one to four. Components one and two were used for further analysis

Year	Prin1	Prin2	Prin3	Prin4
1972	-86.889	-28.488	-5.393	11.030
1984	-81.331	-44.954	-11.665	-17.010
1975	-69.263	27.047	43.136	-5.878
1974	-66.338	22.830	7.576	4.200
1978	-25.934	2.656	22.926	20.693
1971	-18.034	38.946	-26.026	-13.709
1970	-16.045	12.433	-9.897	-9.770
1973	-13.339	23.052	-17.517	-2.838
1983	-8.262	-13.096	-3.763	8.747
1976	-5.777	-12.927	-6.388	-11.839
1977	39.545	-6.031	-19.418	13.211
1982	52.062	-23.846	8.841	-6.172
1981	55.807	-23.332	-9.102	21.902
1969	56.071	44.005	-4.585	15.612
1980	84.927	-30.630	22.639	-8.008
1979	102.799	12.332	8.636	-20.171

correlations or correlations which are valid only in certain time scales were used. In total, there were 58 correlations attempted that could be biologically meaningful. Given the large number of tests of the null hypothesis (Ho) of r = 0, about 1 out of 20 tests are expected to be significant at the 0.05 level by chance alone and 2 out of 20 at 0.10 level. Therefore, I show the probability of obtaining the observed number of rejections of the null hypothesis by chance alone as determined from the binomial distribution (Peterman, Bradford and Anderson, 1986).

The principal components analysis and other statistics were executed with the aid of Statistical Analysis Systems packages (Anon., 1985b).

Oceanographic conditions in the northwest Atlantic

Inter- and intra-annual variability in oceanographic conditions in the northwest Atlantic were examined from British Meteorological Office Ice Charts issued monthly from Bracknell, England. The 4° and 8°C sea-surface isotherms were used to show the location of suitable oceanographic conditions for salmon in the northwest Atlantic during winter (January), spring (May), summer (August), and autumn (October) because salmon have been found most abundantly in water with SSTs of 3-8°C. The 4°C isotherm was selected because 3°C isotherm was not on the chart. Inter-annual variability was shown from the location of the 4° and 8°C isotherms in a cold year, i.e. 1983 or 1984, and a warm year, i.e. 1969. All of the other years typically fell between the isotherms for 1969 and 1983 or 1984.

RESULTS

Variability in survival

Mean survival from egg to smolt for the 1973-79 year-classes from WAB was 1.7 per cent with a coefficient of variation of 14 per cent (Table 26.4). Mean survival from smolt to adult for salmon returning to the counting fence on WAB from 1976 to 1983 is 5.5 per cent with a coefficient of variation of 63 per cent (Table 26.4). Therefore, variation in survival during the

Table 26.4: Survival from egg to smolt for 1973-79 year-classes (Chadwick, 1985) and from smolt to adult (Chadwick et al., 1985) for smolt classes 1976-83

Year-class	Eggs x 1,000	Smolts	Survival (%)	Year i	Smolts in yr. i	Adults to fence yr. i+1	Survival (%)
1973	428	6,074	1.42	1976	6,359	376	5.91
1974	787	13,316	1.69	1977	9,640	317	3.29
1975	667	11,855	1.78	1978	13,071	1,576	12.10
1976	827	12,743	1.54	1979	9,400	470	5.00
1977	870	18,136	2.08	1980	15,675	471	3.00
1978	737	12,735	1.73	1981	13,981	467	3.34
1979	572	8,103	1.42	1982	12,477	1,146	9.18
				1983	10,552	238	2.26
Mean survival			1.67				5.51
S.D.			0.233				3.46
C.V.			14%				63%

marine phase of salmon life-history is about four times greater than that during the freshwater phase. However, since the estimate of 5.5 per cent for marine survival included exploitation by commercial fishing gear as well as natural mortality it cannot be concluded from this analysis that natural mortality is the sole cause of variability in survival of adults returning from the sea to freshwater and total production. But, it can be concluded that variability in survival was greater in the marine environment than in freshwater, at least for the WAB, Newfoundland stock.

For WAB there is no method for separating natural mortality from fishing mortality as none of the smolts were tagged as they left the river. However, in Iceland, where there is little or no commercial fishing in the sea (Mathisen and Gudjonsson, 1978), the number of adults returning from the sea will depend only on the number of smolts produced and natural mortality. There are no published accounts of smolt and adult counts similar to WAB but for the Ellidaar (River), Iceland there is available a time series of spawner-recruit ratios (Mundy, Alexandersdóttir and Eiríksdóttir, 1978). The mean spawner-recruit ratio from 1935 to 1971 for the Ellidaar, Iceland, is 1.98 and the coefficient of variation is 62 per cent. If it is accepted that the variability in recruitment occurs in the marine phase similar to WAB, then the CV for adult production in the Ellidaar in the absence of commercial exploitation is the same as that for WAB with exploitation. Therefore, it is concluded that much of the variability in production of salmon occurs because of variation in survival rates during the marine phase of salmon life history.

Environmental conditions and salmon abundance in the northwest Atlantic

Out of the 58 correlations attempted, there were nine significant at 5 per cent and 16 at the 10 per cent level of significance. In both cases the number of significant correlations were significantly different at the 5 per cent level of significance from those expected by chance alone (for 5 per cent, $Z = 6.6$ and for 10 per cent, $Z = 5.8$). Therefore, it is concluded that salmon abundance for North American and some European stocks may be related to environment. Productivity of salmon of North American origin may be related to changes in the area covered by 4° to 8°C water in the northwest Atlantic since it has been observed that the size of the area covered by 4° to 8°C water varies annually.

Oceanographic conditions in the northwest Atlantic

SSTs of 4° to 8°C occur annually during all seasons in the surface waters of the northwest Atlantic; although the area and location of the 4° to 8°C water (between the 4° and 8°C isotherms) varies between years (Figure 26.1). In January the area of 4° to 8°C water is located mainly above 50°N and extends to the northeast up to 60°N where it turns eastwards. This area was much smaller

Figure 26.1: Seasonal oceanographic conditions in the North Atlantic for a warm year (1969) in which the area bounded by the 4° and 8°C isotherms was large and a cold year (1984) in which the area bounded by the 4° and 8°C isotherms was small

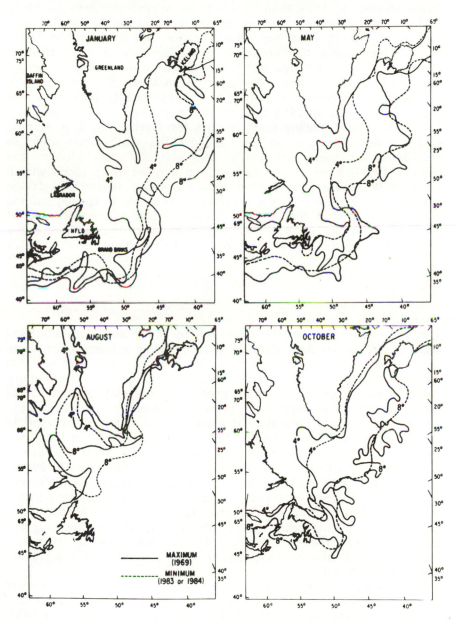

in 1984 than in 1969. In May the area of 4° to 8°C water is located from the Gulf of St Lawrence westward along the southern Grand Banks, where it turns northward and continues into the Labrador Sea and, as noted previously, this area was also much smaller in 1984 than in 1969. In August the area of 4° to 8°C water is located far north of its position in winter and spring; it extends from the mid-Labrador Sea up into the Davis Strait. Once again this area was much smaller in 1983 than in 1969. In October the area of 4° to 8°C water is spread over much of the Labrador Sea and extends from southern Newfoundland to the coast of Labrador and then to the northeast. This area was moderately larger in 1984 than it was in 1969.

The major marine feeding area for North American salmon stocks is believed to be in the Labrador Sea since water of 4° to 8°C is found there seasonally and annually and Reddin and Shearer (1987) have shown that salmon are normally found within these temperature ranges. However, there are seasonal shifts in distribution of salmon in the northwest Atlantic. In spring, salmon are concentrated in the southern Labrador Sea and in the area of the Grand Bank. In the summer, non-maturing salmon move northward into the Davis Strait off West Greenland.

DISCUSSION

Variability in survival

The analysis of survivorship from egg to smolt and from smolt to adult for Western Arm Brook, Newfoundland salmon stocks and spawner-recruit ratios for Ellidaar, Iceland indicated that survival while salmon are in the marine environment was four times more variable than survival in freshwater. This suggests that much of the fluctuation in salmon catches is related to variability in survival in the sea. Canadian Atlantic salmon catches reported to the International Commission for the Exploration of the Sea (Anon., 1986) when examined by smolt class, 1970-83 have a mean of 1,989 t (SD = 577). The coefficient of variation is 29 per cent which is half as large as the CV for survival from Western Arm Brook. This would suggest that a number of factors perhaps including climatic and hydrographic factors, marine productivity, variations in fishing effort and freshwater factors influencing the stock and recruitment process are interacting, some in phase and others out of phase, to produce short-term and long-term fluctuations in catch.

It also agrees well with recent events. Natural mortality rates of smolts returning as adults to Western Arm Brook and Saint John River were higher than usual in 1979, 1984 and 1985. Also, Greenlandic fishermen fishing the same smolt classes were unable to catch their quotas (Anon., 1985a). Reddin and Shearer

(1987) explained that these low catches at West Greenland resulted from lower than usual survival and a shift in distribution because of colder than normal environmental conditions. Anon. (1985a) also suggested that low egg depositions, at least in some Canadian rivers, was partly responsible for these low catches.

Environmental conditions and salmon abundance in the northwest Atlantic

From the time salmon smoltify and leave their natal streams for life in the sea, they are subjected to the vagaries of the marine environment and presented with a variety of environmental stimuli, one of which is temperature. Because salmon are poikilotherms, climatic conditions can be expected to have an appreciable effect on their metabolic rates (Hoar, 1965). Therefore, sea temperature can influence any activity under enzyme control, i.e. growth rates, development, reproduction, swimming speed, etc. The results of analysis of environmental conditions in the northwest Atlantic in January and March suggested that environmental conditions in the northwest Atlantic during the winter may have some influence on total salmon production. The relationship was rather weak in that only 16 out of 58 correlates tested were significant.

Others have also examined the relationship between certain aspects of salmon marine life history and climate, weather, and ocean conditions. Reddin and Shearer (1987) showed that the abundance of salmon at West Greenland is affected by environmental conditions in January and August in the northwest Atlantic and that cold water (less than 3-4°C) can act as a barrier to salmon migration. In colder years fewer salmon move into the West Greenland area from the Labrador Sea than in the warmer years.

Temperature may also be indirectly linked to sea age at maturity as has been shown for both Atlantic (Saunders, Henderson, Glebe and Loudenslager, 1983) and Pacific salmonids (Peterman et al., 1986). Martin and Mitchell (1985) demonstrated that an increase in temperature in the subarctic is associated with larger numbers of MSW salmon and fewer grilse returning to the Aberdeenshire Dee River, Scotland. This coincides with recent events in Canada where grilse:salmon ratios have decreased coincidental with a colder marine climate. Scarnecchia (1984a) found for Icelandic salmon rivers highly significant relationships between mean June-July sea temperatures and the catch of salmon in the following year. Recent events in the Icelandic area have shown that variations in climate, weather, and hydrography have been correlated with changes in primary production and salmon catches (Scarnecchia, 1984a,b).

The geographical ranges inhabited by fish species show the

493

best examples of the influence that climate can have (Shepherd, et al., 1982). The geographical ranges inhabited by most fish species and or major long-term changes in abundance have generally been attributed to the effects of climatic change. Examples of this are found for Scandinavian herring stocks (Cushing and Dickson 1976) and sardines off California (Soutar and Isaacs, 1974). Climatic causes have also been attributed to the presence/or absence of salmon at West Greenland (Dunbar and Thomson, 1979). Shepherd et al. (1982) have correlated the frequent annual variations in recruitment to various fish stocks in the northeast Atlantic area to climate. They also summarised the literature on this topic and concluded that, in spite of arguments to the contrary, climate must have a significant impact on production, because such variations have been too rapid to be caused by variations in the abundance of the parent stock.

As Shepherd et al. (1982) noted, the usefulness of any relationship between environment and catch will be somewhat curtailed because we cannot predict the weather very far into the future. On the other hand, much of the considerable variation in salmon abundance could be directly related to environment. For example, Chadwick (1985) showed that for seven rivers in Canada 46-87 per cent of the variation in recruitment could be explained by egg deposition. Recently, in Canada, fishery managers have begun a more rational approach to salmon management by adjusting harvest rates to provide adequate escapement. This of course requires accurate forecasts of how many salmon of a particular stock will be available for harvest. Therefore, stock-recruitment relationships if adjusted for the effects of environment could prove to be very useful.

Distribution

Reddin and Shearer (1987) summarised the results of research vessel surveys in the northwest Atlantic from latitude 43° to 70°N, for 1965-85. They used data from research surveys up to 1972 reported by May (1973), combined with studies in the area of the Grand Bank of Newfoundland (Reddin and Burfitt, 1984) and Irminger Sea (Møller Jensen and Lear, 1980) (Figure 26.2). Most of the fishing consisted of repetitive surface sets at single stations in conjunction with tagging experiments. Drift gill nets constructed from multi- and monofilament twine of various mesh sizes between 114 mm and 155 mm and about 3 m in depth were used. In the years that surveys were conducted stations were fished in each month from February to October. Seasons were defined as follows: spring, 22 March to 21 June; winter, 23 December to 21 March; and summer to autumn, 22 June to 22 December. Catch rates were expressed as numbers of salmon caught per nautical mile of net fished per hour.

Figure 26.2: Research vessel catches of salmon in the northwest Atlantic, 1965-85

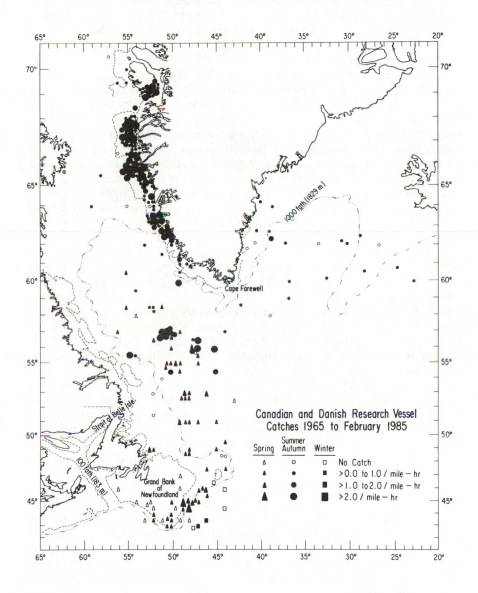

Source: Adapted from Reddin and Shearer (1987).

Salmon were found in the spring in surface waters of the northwest Atlantic from the southern edge of the Grand Bank to slightly south of Cape Farewell, Greenland (Figure 26.2). As May (1973) observed, the most westerly positions where salmon were caught closely follow the edge of the Arctic pack ice in spring. Therefore salmon are found at sea in relatively cool water from 3° to 8°C. There are two locations where salmon have been found in abundance during spring. One of them is located about 300 miles east of the Strait of Belle Isle. The other is located slightly to the east of the 200 m isobath (depth contour) along the eastern edge of the Grand Bank. In these two locations, catch rates were similar to those at West Greenland in summer, suggesting that per unit area salmon are as abundant there as at West Greenland. Reddin and Burfitt (1984) have presented some evidence based on smolt ages suggesting that the salmon stocks in the Labrador Sea may be different from those along the eastern edge of the Grand Bank. The lower smolt ages of salmon caught to the east of the Grand Bank suggests that salmon stocks there are more southerly in origin than salmon stocks in the Labrador Sea where smolt ages were higher.

In late summer and autumn, non-maturing salmon are concentrated along the West Greenland coast from the inner coastal fiords to between 45 and 60 km offshore. As well, relatively good catches have occurred in the Labrador Sea north of 55°N. Salmon have also been caught by research vessel in the Irminger Sea although catch rates were not nearly as high as at Greenland or in the Labrador Sea (Møller Jensen and Lear, 1980). In August of 1985, an experimental fishery was conducted within the coastal fiords at East Greenland from 62 to 63°N. In eight days of fishing, 400 salmon were caught (A.L. Meister, Maine Sea Run Salmon Commission, Bangor, Maine, personal communication). Some of these salmon were undoubtedly of North American origin since the capture of North American salmon tagged as smolts has been reported from this area by Møller Jensen and Lear (1980). No sets have been made in summer/autumn in the Grand Banks area.

Very few sets have been made for salmon during the winter months and these were all to the east of the Grand Bank of Newfoundland in 1985 (Figure 26.2). The low catch rates in the area of the Grand Bank suggest that salmon were located elsewhere at this time. These results suggest that since salmon were found in the Labrador Sea in the autumn and then in the following spring, that salmon of North American origin probably overwinter there.

The sea ages of salmon caught during experimental fishing in the northwest Atlantic show that multi-sea-winter (MSW) salmon range over much of the northwest Atlantic while those 1SW salmon that would mature as grilse (potential grilse) do not.

Virtually all of the salmon caught in the Labrador Sea would mature as MSW salmon. However, in sets on and to the east of the Grand Bank both MSW and potential grilse were caught. At Greenland and presumably in the Irminger Sea only salmon that would mature as 2SW and older salmon have been caught.

Regional variations in abundance

The expression of catch rate as number of salmon caught per 100 standard nets enabled comparison of the results from the research surveys in the Labrador Sea with published data from research and commercial fishing off West Greenland, in the Irminger Sea, and on the Grand Bank and vicinity (Table 26.5). The details on fishing gear and techniques are described in Møller Jensen and Lear (1980) and in Reddin (1985). Similar techniques and gear have been used during the studies in the Labrador Sea. At West Greenland from August to October, catch rates have ranged from 10.2 in 1970 to 111.0 salmon per 100 nets in 1975 with a mean of 44.3 (Table 26.5). In the Irminger Sea in July to October, catch rates have ranged from 0.7 in 1972 to 11.9 in 1975 with a mean of 5.8 salmon caught per 100 nets. In the Labrador Sea in April and May catch rates have ranged from 5.4 to 9.4 and from August to October, catch rates have ranged from 10.4 to 68.7 with a combined mean of 19.2. On the Grand Bank catch rates have ranged from 1.0 to 10.0 in May with a mean of 5.5. East of the Grand Bank, again in May, catch rates were 20.8 salmon per 100 nets although fishing occurred only in one year, 1980. Mean catch rates at Greenland were the highest, mean catch rates in the Labrador Sea and east of the Grand Bank were about 55 per cent less than that at Greenland, and catch rates on the Grand Bank and Irminger Sea were about 75 per cent lower than those in the Labrador Sea and east of the Grand Bank. Therefore, salmon were much less abundant in the Irminger Sea and on the Grand Bank than east of the Grand Bank and in the Labrador Sea than at West Greenland. Salmon were much less abundant on the Grand Bank than in the oceanic area east of the Grand Bank in 1980 but it should be noted that the 1979 and 1980 catch rates for the Grand Bank are not directly comparable because of differences in survey coverage.

Tagging results and origin of salmon

A total of 278 salmon of mixed sea ages have been tagged and released in the Labrador Sea and on the Grand Bank and vicinity from 1969 to 1980 (Figures 26.3 and 26.4). On the Grand Bank and vicinity, a total of 172 salmon were tagged and released during surveys (3 in 1973, 4 in 1979 and 165 in 1980). There have been 14 reported recaptures, two that had been tagged in 1973

Table 26.5: Average catches (numbers per 100 nets) of salmon from research and commercial fishing with surface gill nets in various regions of the North Atlantic, 1969-80

Year	Month	West Greenland[a,1]	Irminger Sea[1]	Labrador Sea	Grand Bank[2]	East of Grand Bank[2]
1969	Sep	27.4	-	16.1	-	-
1970	Apr	-	-	5.4	-	-
	Sep	10.2	-	10.4	-	-
1971	May	-	-	9.4	-	-
	Sep	21.1	-	14.1	-	-
1972	Apr	-	-	5.4	-	-
	Aug-Sep	11.3	-	-	-	-
	Aug-Oct	42.0[b]	0.7[b]	-	-	-
1973	Aug	41.0[b]	5.7	-	-	-
1974	Jul	44.0[b]	8.5	-	-	-
	Aug	72.0[b]	2.1	-	-	-
1975	Aug	111.0[b]	11.9	-	-	-
1977	Oct	-	-	23.8	-	-
1978	Aug	94.7	-	68.7	-	-
1979	May	-	-	-	1.0	-
	Aug-Sep	21.8	-	-	-	-
1980	May	-	-	-	10.0	20.8
	Aug-Sep	35.1	-	-	-	-
Mean		44.3	5.8	19.2	5.5	20.8
S.D.		32.3	4.6	20.9	6.4	-

Notes: a, Trip reports of Newfoundland Biological Station.
b, Commercial fishing.

Sources: 1, Møller Jensen and Lear (1980); 2, Reddin (1985).

and the remainder from the 1980 tagging. Five were recaptured in rivers (four by anglers in New Brunswick rivers and one at Morgan Falls fishway on the LaHave River, Nova Scotia) and nine in the commercial fishery of Newfoundland and Labrador (Figure 26.3). All except two of the recaptures occurred in the year of tagging. One exception was a fish angled in 1981 on the Kouchibouguac River, New Brunswick. This salmon was caught in the spring and was a kelt that must have entered that river the previous autumn to spawn. The other was a salmon tagged on the Virgin Rocks in 1973 and caught in the Newfoundland commercial

Figure 26.3: Numbers of salmon tagged and released on and east of the Grand Bank in June of 1973 and in May of 1979 and 1980, and subsequent recaptures in the coastal fishery and rivers. The numbers associated with recapture symbols are sea ages

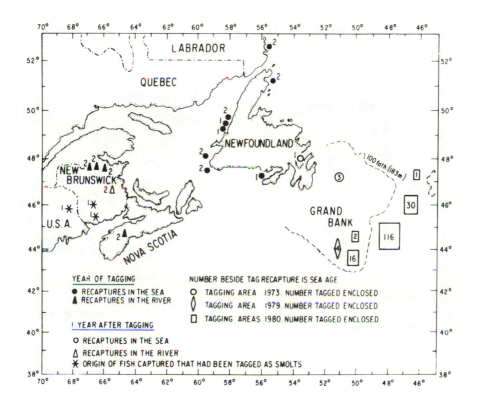

Source: Adapted from Reddin (1985)

salmon fishery in the following year. The five salmon that were recaptured in rivers were all 2SW salmon when tagged. The recaptures in the Newfoundland commercial fishery consisted of five 2SW and two 1SW salmon.

During the Grand Bank and vicinity surveys, three salmon were caught with tags already attached. These fish had been tagged as smolts, two at Mactaquac Fish Culture Station, Saint John River, New Brunswick, and one at the Green River National Fish Hatchery in Maine, USA. All were 1SW in age when recaptured.

In the Labrador Sea, a total of 109 salmon mainly 1SW and

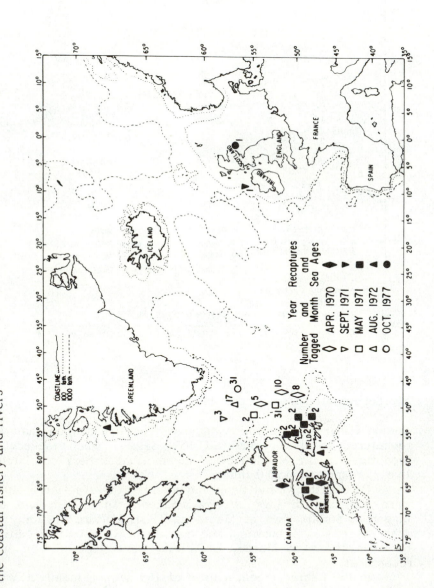

Figure 26.4: Numbers of salmon tagged and released in the Labrador Sea, and subsequent recaptures in the coastal fishery and rivers

2SW in age have been tagged and released during surveys from 1969 to 1977 (Figure 26.4). There have been 15 reported recaptures, four of which were 1SW salmon and 11 of which were 2SW salmon when tagged. Of the 52 1SW salmon tagged and released, four have been recaptured, all of these in the year following tagging. One was caught in Newfoundland, another in Ireland, another in Greenland, and the last in Scotland. Of the 55 2SW salmon tagged and released, there have been 11 recaptured, all of them in the year that they were tagged. Six were recaptured in Newfoundland, three in the Maritimes, and two in Quebec.

Tag recaptures from salmon caught in the Labrador Sea and the results of scale analysis (Reddin, Burfitt and Lear, 1979) indicates that MSW salmon from North America and Europe are found in the Labrador Sea from spring to autumn. No recaptures have been made of 1SW salmon that would mature as grilse. Templeman (1968) and May (1973) reported from the condition of the gonads that no 1SW salmon were caught in the Labrador Sea that would mature in the year of capture as grilse. In addition, it is interesting to note the differences in the recapture sites of salmon tagged on the Grand Bank and in the Labrador Sea (Figures 26.3 and 26.4). Three salmon tagged and released in the Labrador Sea were recaptured outside of continental North America; while none from the Grand Bank study were. In both cases there were recapture sites in the Maritimes and Newfoundland; but the Newfoundland recapture sites from the Labrador Sea are distributed along the northeast coast rather than west coast as from the Grand Bank study. Although the number of recaptures is small the distribution of recapture sites suggests that the salmon stocks in the Labrador Sea are rather different from those in the Grand Bank area. Reddin and Burfitt (1984) came to a similar conclusion using mean smolt ages.

Migration routes

The salmon stocks in the northwest Atlantic consist primarily of two components: one that will mature as grilse and the other that will mature as 2SW or older salmon. At the time of smolt migration, salmon of North American origin that will mature as 2SW or older salmon enter the surface waters of the Northwest Atlantic from the Connecticut River, USA in the south to the Kapisigdlit River, Greenland in the north (Figure 26.6). The immature post-smolts from rivers south of Labrador migrate northward throughout the summer, away from the warmer areas to the south up into the cooler water of the Labrador Sea, e.g. as has been shown for Maine and Quebec salmon stocks (Meister, 1984; Caron, 1983). These salmon spend much of their time feeding in water temperatures suitable for growth on the

Figure 26.5: The main surface currents in the northern part of the North Atlantic

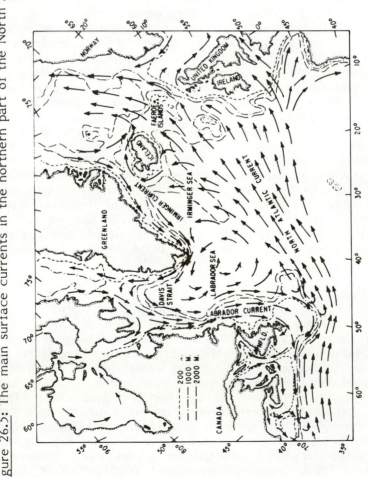

Source: Adapted from Templeman (1967) and Stasko, Sutterlin, Rommel and Elson (1973).

Figure 26.6: The migration routes for salmon smolts away from coastal areas showing possible overwintering areas and movement of MSW salmon into West Greenland

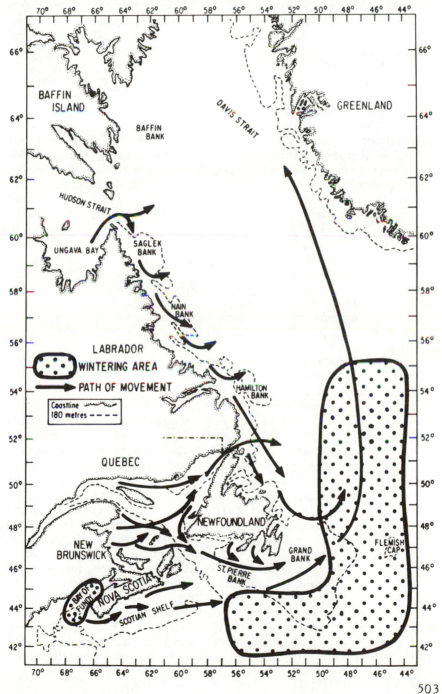

abundant food resources (Templeman, 1968) in the eddy system bounded by the Labrador, North Atlantic, Irminger and West Greenland Currents (Figure 26.5).

During their second summer in the sea some of these non-maturing 1SW salmon move northwards to an area extending from the northern Labrador Sea up into the Davis Strait and Irminger Sea. There are also non-maturing 1SW salmon caught in the Newfoundland-Labrador coastal fishery (Idler, Hwang, Crim and Reddin, 1981). In most years, if coastal water temperatures are suitable, they will move close into the fiords along the coast of West Greenland. The salmon at West Greenland have been shown by tagging studies, both of smolts in home waters and adult salmon at Greenland, to be mostly 1SW fish that will not mature until the following year. The same tagging studies have shown that these salmon originated from rivers mainly in Canada, Norway, Scotland and Ireland (Anon., 1981; Moller Jensen, 1980a, b). Reddin et al. (1979) have shown that both North American and European origin salmon occur in the Labrador Sea and Møller Jensen and Lear (1980) have reported that salmon of North American and European origin also occur in the summer/autumn period in the Irminger Sea. Although catch rates in the Irminger Sea were low compared to those at West Greenland, they speculated that because the area involved was so large, the number of salmon in the Irminger Sea could also be high. Because suitable oceanographic conditions exist annually in these areas (Figure 26.1) we conclude that salmon also inhabit these areas annually.

In the autumn, salmon move south either through the Labrador Sea or along the Labrador coast (Reddin and Dempson, 1986) to occupy an area in winter and spring about 480 km east of the Strait of Belle Isle (Figure 26.7). In the spring, salmon are also found concentrated to the east of the Grand Bank (Reddin, 1985). Reddin (1985) reported catch rates in 1980 for the Grand Bank salmon stocks that were similar to those at West Greenland in some years, suggesting the presence of a large number of salmon. Salmon were also found in lower abundance over the Grand Bank and along its southwest edge suggesting that some salmon may migrate to home rivers over the southern Grand Bank. Seasonal oceanographic conditions (Figure 26.1) suggest that salmon may not overwinter in the Grand Bank area.

Also at the time of smolt migration salmon of North American origin that will mature as grilse enter the surface waters of the northwest Atlantic. Except for the period of time when they are migrating homeward through coastal fisheries, our knowledge of the marine life history of grilse is dominated by what we don't know rather than what we know. Presumably, they must move away from their natal streams fairly rapidly since smolts are infrequently caught in inshore areas. As well, grilse

Figure 26.7: The migration routes for salmon from West Greenland and overwintering areas on return routes to rivers in North America

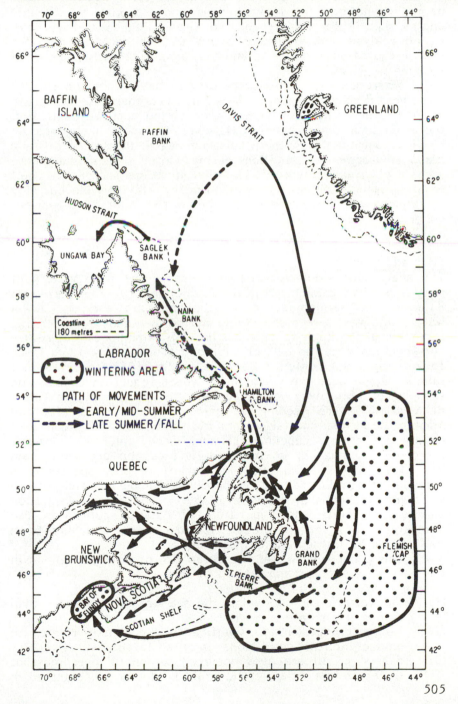

have never been captured in the Davis Strait-West Greenland area or in the Labrador Sea. In fact, the first reported capture at sea in other than coastal areas was in the spring from the area of the Grand Bank of Newfoundland and out into the oceanic depths slightly to the east of the Grand Bank (Reddin, 1985). Gonad analysis indicated that about 64 per cent of the 1SW salmon caught were maturing and thus would return to their natal streams as grilse.

Where then do grilse overwinter at sea? Since the smaller mesh gear required to capture 0+ salmon (whether potential grilse or MSW salmon) has rarely been used in the northwest Atlantic, the exact location is unknown. However, the most likely location is the area from the southern Labrador Sea to the northern Grand Bank. Environmental conditions during winter suggest that these areas are suitable (Figure 26.1). Other locations such as the Gulf of St Lawrence and Scotian Shelf area are also possible but only if salmon will move deeper in the water column to warmer water.

Food and feeding

The analysis of stomach contents indicated the wide variety of organisms that salmon prey on in the sea. In the northwest Atlantic area, salmon feed on capelin and sand lance in coastal waters off West Greenland. While in the Labrador Sea, herring, barracudina, Paralepis coregonides borealis, amphipods, euphausids and squid make up the bulk of their diet (Templeman, 1968; Lear, 1972; Lear and Christensen, 1980). Reddin (1985) reported that salmon caught over the Grand Bank were feeding on capelin and sand lance; while over oceanic depths to the east of the Grand Bank salmon were feeding on barracudina, black smelts and amphipods. In the coastal waters of Newfoundland salmon feed mainly on herring, capelin, and sand lance while in Labrador, pteropods, sand lance, juvenile cod, and capelin are their main prey (Lear, 1972). From this wide variety of prey species it is concluded that adult salmon are opportunistic feeders and prey on whatever organisms are most available in the area. Information on the opportunistic feeding nature of Atlantic salmon is also available from the northeast Atlantic area and has been summarised by Hansen and Pethon (1985).

This paper summarises the ocean life of Atlantic salmon in the northwest Atlantic and showed that:

(1)   The variation in survival in the sea was much greater than that in fresh water.

(2)   Salmon production was significantly but weakly related to environmental conditions in the northwest Atlantic.

(3)   Salmon of all sea ages occurred seasonally over most of the northwest Atlantic and could be found concentrated in

the Labrador Sea gyre throughout the year, at West Greenland in summer and autumn, and in the spring along the eastern slope of the Grand Bank.

(4)   The distribution of tag recapture sites suggested that the salmon stocks in the Labrador Sea may be different from those in the Grand Bank area.

(5)   The wide variety of prey items found in salmon stomachs and differences in prey items between areas suggested salmon were opportunistic feeders.

REFERENCES

Allan, I.R.H. and Ritter, J.A. (1977) Salmonid terminology. Conseil International Exploration de la Mer, 37, 293-9

Anon. (1981) Report of meeting of north Atlantic salmon working group. Conseil International Exploration de la Mer, CM 1981/M:10, Copenhagen, Denmark

Anon. (1985a) Report of the Working Group on North Atlantic salmon. Conseil International Exploration de la Mer, CM 1986/Assess:11, Copenhagen, Denmark

Anon. (1985b) SAS Institute Inc. SAS User's Guide: Statistics, Version 5 Edition. Cary, NC. SAS Institute, 956 pp

Anon. (1986) Report of the Working Group on North Atlantic salmon. Conseil International Exploration de la Mer, CM 1986/Assess:17, Copenhagen, Denmark

Caron, F. (1983) Migration vers L'Atlantique des post-saumoneaux (Salmo salar) du Golfe du Saint-Laurent. Naturaliste Canadian (Quebec) 110, 223-7

Chadwick, E.M.P. (1982) Stock-recruitment relationship for Atlantic salmon (Salmo salar) in Newfoundland rivers. Canadian Journal of Fisheries and Aquatic Sciences, 1496-1501

Chadwick, E.M.P. (1985) The influence of spawning stock on production and yield of Atlantic salmon, Salmo salar L., in Canadian rivers, Aquaculture and Fisheries Management, 1, 111-19

Chadwick, E.M.P., Alexander, D.R., Gray, R.W., Lutzac, T.G., Peppar, J.L. and Randall, R.G. (1985) 1983 research on anadromous fishes, Gulf Region. Canadian Technical Report on Fisheries and Aquatic Sciences, 1420

Chadwick, E.M.P. and Meerburg, D.J. (1978) Sea survival of 1SW Atlantic salmon. Conseil International Exploration de la Mer, CM 1978/M:10, 4 pp

Cushing, D.H. and Dickson, R.R. (1976) The biological response in the sea to climate changes. Advances in Marine Biology, 14, 1-122

Cutting, R.E. and Jefferson, E.M. (1985) Status of the Atlantic salmon of the LaHave River, Nova Scotia in 1984 and forecast of returns in 1985. Canadian Atlantic Fisheries Scientific Advisory Committee Research Document 85/53, Dartmouth, Nova Scotia, Canada

Dunbar, M.J. (1981) Twentieth century marine climatic change in the northwest Atlantic and subarctic regions, In Symposium on Environmental Conditions in the northwest Atlantic during 1970-79. NAFO Scientific Council Studies 5, Dartmouth, Nova Scotia, Canada, pp. 7-15

Dunbar, M.J. and Thomson, D.H. (1979) West Greenland salmon and climatic change. Meddelelser om Grønland, 202, 1-19

Hansen, L.P. and Pethon, P. (1985) The food of Atlantic salmon, (Salmo salar L.) caught by long-liner in northern Norwegian waters. Journal of Fish Biology, 26, 553-62

Hoar, W.S. (1965) The endocrine system as a chemical link between the organism and its environment. Transactions of the Royal Society of Canada, Section III (4)3, 175-200

Huntsman, A.G. (1938) Sea behaviour in salmon. Salmon and Trout Magazine, 90, 24-8

Idler, D.R., Hwang, S.J., Crim, L.W. and Reddin, D.G. (1981) Determination of sexual maturation stages of Atlantic salmon captured at sea. Canadian Journal of Fisheries and Aquatic Sciences, 38, 405-13

Lear. W.H. (1972) Food and feeding of Atlantic salmon in coastal areas and over oceanic depths. ICNAF Res. Bull., 9, 27-39

Lear, W.H. (1976) Migrating Atlantic salmon (Salmo salar) caught by otter trawl on the Newfoundland Continental Shelf. Journal of the Fisheries Research Board of Canada, 33(6), 1202-5

Lear, W.H. and Christensen, O. (1980) Selectivity and relative efficiency of salmon drift nets. Rapports et Procés-Verbaux des Reunions Conseil International pour l'Exploration de la Mer, 176, 36-42

Marshall, T.L. (1985) Status of Saint John River, New Brunswick, Atlantic salmon in 1985 and forecast of returns in 1986. Canadian Atlantic Fisheries Scientific Advisory Committee Research Document 85/104, Dartmouth, Nova Scotia, Canada

Martin, J.H.A., Dooley, H.D. and Shearer, W. (1984) Ideas on the origin and biological consequences of the 1970's salinity anomaly. Conseil International pour l'Exploration de la Mer, CM 1984/Gen:18

Martin, J.H.A. and Mitchell, K.A. (1985) Influence of sea temperature upon numbers of grilse and multi-sea-winter Atlantic salmon (Salmo salar) caught in the vicinity of the River Dee (Aberdeenshire). Canadian Journal of Fisheries and Aquatic Sciences, 42, 1513-21

Mathisen, O.A. and Gudjónsson, T. (1978) Salmon management and ocean-ranching in Iceland. Journal of Agricultural Research in Iceland, 10(2), 156-74

May, A.W. (1973) Distribution and migrations of salmon in the Northwest Atlantic. International Atlantic Salmon Foundation Special Publication Series, 4, 373-82

Meister, A.L. (1984) The marine migrations of tagged Atlantic salmon (Salmo salar L.) of U.S.A. origin. Conseil International pour l'Exploration de la Mer, CM 1984:M:27, Copenhagen, Denmark

Menzies, W.J.M. (1949) The stock of salmon. Its migrations, preservation and improvement. Edward Arnold, London, 96 pp

Møller Jensen, J. (1980a) Recaptures of salmon at West Greenland tagged as smolts outside Greenland waters. Rapports et Procés-Verbaux des Reunions Conseil International pour l'Exploration de la Mer, 176, 114-21

Møller Jensen, J. (1980b) Recaptures from international tagging experiments at West Greenland. Rapports et Procés-Verbaux des Reunions Conseil International pour l'Exploration de la Mer, 176, 122-35

Møller Jensen, J. and Lear, W.H. (1980) Atlantic salmon caught in the Irminger Sea and at East Greenland. Journal of Northwest Atlantic Fishery Science, 1, 55-64

Moore, C.S. (1965) Inter-relations of growth and cropping in apple trees studied by the method of component analysis. Journal of Horticultural Science, 40, 133-49

Morrison, D.F. (1976) Multivariate Statistical Methods. McGraw-Hill, New York, NY, XV+415 pp

Mundy, P.R., Alexandersdottir, M. and Eiríksdóttir, G. (1978) Spawner-recruit relationship in Ellidaar. Journal of Agricultural Research in Iceland, 10(2), 47-56

O'Connell, M.F., Dempson, J.B., Reddin, D.G. and Ash, E.G. (1985) Status of Atlantic salmon (Salmo salar L.) stocks of the Newfoundland Region, 1984. Canadian Atlantic Fisheries Scientific Advisory Committee Research Document 85/15

Peterman, R.M., Bradford, M.J. and Anderson, J.L. (1986) Environmental and parental influences on age at maturity in sockeye salmon (Oncorhynchus nerka) from the Fraser River, British Columbia. Canadian Journal of Fisheries and Aquatic Sciences, 43, 269-74

Randall, R.G., Chadwick, E.M.P. and Schofield, E.J. (1985) Status of Atlantic salmon in the Miramichi River, 1984. Canadian Atlantic Fisheries Scientific Advisory Committee Research Document 85/2, Dartmouth, Nova Scotia, Canada

Reddin, D.G. (1985) Atlantic salmon (Salmo salar L.) on and east of the Grand Bank. Journal of Northwest Atlantic Fisheries, 6, 157-64

Reddin, D.G. and Burfitt, R.F. (1984) A new feeding area for Atlantic salmon (Salmo salar) to the east of the Newfoundland Continental Shelf. Conseil International pour l'Exploration de la Mer, CM 1984/M:13, Copenhagen, Denmark

Reddin, D.G. Burfitt, R.F. and Lear, W.H. (1979) The stock composition of Atlantic salmon off West Greenland and in the Labrador Sea in 1978 and a comparison to other years. Canadian Atlantic Fisheries Scientific Advisory

Committee Research Document 79/3, Dartmouth, Nova Scotia, Canada

Reddin, D.G. and Dempson, J.B. (1986) Origin of Atlantic salmon (Salmo salar L.) caught at sea near Nain, Labrador. Naturaliste Canadien (Quebec), 113, 211-18

Reddin, D.G. and Shearer, W.M. (1987) Sea-surface temperature and distribution of Atlantic salmon (Salmo salar L.) in the northwest Atlantic. American Fisheries Society Symposium 1, 262-75

Ricker, W.E. (1975) Computation and interpretation of biological statistics of fish populations. Bulletin of the Fisheries Research Board of Canada, 191

Saunders, R.L., Henderson, E.B., Glebe, B.D. and Loudenslager, E.J. (1983) Evidence of a major environmental component in the determination of grilse:salmon ratio in the Atlantic salmon (Salmo salar). Aquaculture, 33; 107-18

Scarnecchia, D.L. (1984a) Climatic and oceanic variations affecting yield of Icelandic stocks of Atlantic salmon (Salmo salar). Canadian Journal of Fisheries and Aquatic Sciences, 41, 917-35

Scarnecchia, D.L. (1984b) Forecasting yields of two-sea-winter Atlantic salmon (Salmo salar) from Icelandic rivers. Canadian Journal of Fisheries and Aquatic Sciences, 41, 1234-40

Shearer, W.M. (1986) An evaluation of the data available to assess Scottish salmon stocks, In The status of the Atlantic salmon in Scotland (Symposium No. 15) Institute of Terrestrial Ecology, Abbots Ripton, pp. 91-111

Shepherd, J.G., Pope, J.G. and Cousens, R.V. (1982) Variations in fish stocks and hypotheses concerning their links with climate. Conseil International Exploration de la Mer, CM 1982/GEN: 6, Copenhagen, Denmark

Sokal, R.R. and Rohlf, F.J. (1969) Biometry. W.H. Freeman, San Francisco, 776 pp

Soutar, A. and Isaacs, J.D. (1974) Abundance of pelagic fish during the 19th and 20th centuries as recorded in anaerobic sediment of the Californias. Fishery Bulletin, 72, 257-74

Stasko, A.B., Sutterlin, A.M., Rommel, S.A. and Elson, P.F. (1973) Migration-orientation of Atlantic salmon (Salmo salar L.) International Atlantic Salmon Foundation Special Publication, series 4, pp 119-37

Templeman, W. (1967) Atlantic salmon from the Labrador Sea and off West Greenland, taken during A.T. Cameron cruise, July-August 1965. International Commission for the Northwest Atlantic Fisheries Research Bulletin, 4, 4-40

Templeman, W. (1968) Distribution and characteristics of Atlantic salmon over oceanic depths and on the bank and shelf slope areas off Newfoundland, March-May, 1966. International Commission for the Northwest Atlantic Fisheries Research Bulletin, 5, 62-85

Chapter Twenty Seven

FUTURE INVESTIGATIONS ON THE OCEAN LIFE OF SALMON

Svend Aage Horsted

Greenland Fisheries and Environment Research Institute,
Tagensvej 135, DK-2200 Copenhagen N, Denmark

INTRODUCTION

Few fish have attracted man as much as the salmon, and its anadromous nature was known in prehistory. However, that part of the salmon's life which is spent in the ocean has remained full of secrets, some of which have been unveiled, in part, in recent years. Although the marine phase has been intensively studied - or rather many attempts have been made to study it - there are still many gaps in our knowledge. This paper tries to identify the most important gaps in our knowledge of the ocean life of salmon and discusses the possibilities of filling these gaps.

MAJOR ADVANCES IN RECENT STUDIES OF THE OCEAN LIFE OF SALMON

Although the occurrence of salmon in Greenland waters was itself evidence of long-distance migration from major salmon rivers elsewhere (Nielsen, 1961) the first direct evidence of migration of salmon from a European (Scottish) river to Greenland, recorded in 1956 (Menzies and Shearer, 1957), was regarded as a sensation. In 1961, the first recapture was made of a salmon tagged in a North American river (Miramichi, Canada) (Nielsen, 1961). It was thus proved that salmon originating from the two continents would migrate over long distances in the ocean in their search for food, and that the water around Greenland provided oceanic feeding grounds for salmon. Several thousand recaptures at Greenland have followed those two.
    A small-scale commercial fishery for salmon in Greenland had taken place for many years prior to 1960, but it was the rapid technical development in the Greenland fishing around 1960 together with the modernisation of the fishing vessels and gear that was the background for the explosive development in the

salmon fishery there in 1964. The interest in the new fishery not only led to the establishment in 1965 of the ICES/ICNAF Joint Working Party on North Atlantic salmon but more particularly to salmon research being politically and, thereby, financially more favoured than before. The sea life of salmon suddenly came into focus.

The sampling programmes soon produced important results, such as salmon occurring at Greenland are the multi-seawinter component of the stocks so that the fishery would have no direct effects on grilse (Anon., 1967). Also the stock supporting the Norwegian Sea long-line fishery developing in the late 1960s was found to be composed mainly (90 per cent in 1968) of multi-sea-winter fish, although later investigations did show that one-sea-winter fish, probably of the grilse component, occur commonly in this area (Anon., 1971). However, it seems clear that some part of the ocean, such as Greenland waters, are visited by the multi-sea-winter component exclusively (excluding salmon of Greenland origin), while other parts of the ocean may be feeding grounds for a mixed salmon-grilse stock.

Although the first recaptures in Greenland of salmon tagged in Europe and North America were followed by hundreds of similar recaptures, none of these revealed any details of the migration from Greenland to home waters. One of the first recommendations for further work was therefore the tagging of salmon at Greenland.

Several methods for catching fish suitable for tagging were investigated in Greenland in the late 1960s, but only gill nets were able to supply a sufficient number of fish, some of which could be used for tagging. The tagging experiments culminated with the International Salmon Tagging Experiment in 1972 in which four large research vessels plus eight commercial vessels took part, with a total of 2,364 salmon being tagged (Horsted, 1980). Subsequent recaptures confirmed the theory that at least some of the salmon visiting Greenland return to rivers in Europe and North America, almost certainly their native rivers. The returns also confirmed that salmon at Greenland come from a very wide part of the area from which Atlantic salmon migrate to sea, although in very different relative proportions.

At the same time, methods were being developed aiming at determining the spatial origin of each individual fish sampled at Greenland. Serum transferrin polymorphism was used by Payne (1980), and electrophoretic analysis of serum proteins by Child (1980) whilst the value of parasites as biological tags was investigated by Pippy (1980). The most remarkable development was, however, the application of the discriminant function analysis of scale characters (Lear and Sandeman, 1980). These methods, especially the latter, enable the ICES Working Group on North Atlantic Salmon to estimate the relative continental contribution of salmon to the stock at Greenland year by year since 1969 (Anon., 1981). Further refinement of the method allows also for a separation of hatchery-reared fish from those of wild smolt origin (Anon., 1986).

The extraordinarily low catches at West Greenland in 1983 and 1984 in seasons after record cold winters were noted (Rosenørn, Fabricius, Buch and Horsted, 1984 and 1985) gave rise to discussion and study of the influence of climatic and oceanic conditions on the distribution and availability of salmon (Anon., 1985, 1986). It is certain that environmental factors play a very important role in the distribution of salmon in the ocean, but it has proved difficult to find correlations between distribution and density on the one hand and easily defined and measurable environmental parameters on the other. This is possibly because density is not very well correlated with catch rate. Local environmental factors may cause great short-term variation in availability and catchability.

A number of basic studies on physiology and genetics have also been undertaken in recent years, e.g. on factors determining salmonid age at maturity (Meerburg, 1986). In Canada, the Atlantic Salmon Federation and the Department of Fisheries and Oceans, jointly began the Salmon Genetics Research Programme in 1974. Similarly, development of population-dynamics models has been important (e.g. Andersen, Horsted and Møller-Jensen, 1980) (partly on the basis of Ricker, 1958); Hansen, 1984).

It may be concluded therefore that much more is known of the salmon's life at sea than was the case at the beginning of the 1960s. However, Møller Jensen (see Chapter 24) concludes that only one of the five basic questions set up by the ICES/ICNAF Joint Working Party on Atlantic Salmon in 1966 (Anon., 1967) has been answered, namely the growth of fish between Greenland and its return to home waters. This is probably too pessimistic a conclusion, but evidently a number of gaps in the knowledge of the salmon's sea life still exist.

TECHNICAL AND STATISTICAL INFORMATION

Catch data

From recent reports and recommendations of the ICES Working Group it is evident that the biggest gaps in catch statistics at present are concerned with the following problems:

(1)   Non-reported catches;
(2)   Non-observed catches;
(3)   Details on distribution of catches (locality, time, gear).

The non-reported catches fall into two major categories, namely illegal (poaching) and legal catches (e.g. local consumption). Statistics on the former can only ever be estimates, and the problem is perhaps greater for freshwater fisheries than for sea fisheries. The legal non-reported catches may be greater for the sea-going fishery (small-scale netting by non-commercial fishermen, fishermen's consumption, fish sold privately without being recorded). Again, only estimates can be produced of the magnitude of such catches.

The non-observed catches remain a problem. They contribute to what is called the non-catch fishing mortality, which has been one of the issues in recent reports of the ICES Working Group.

To achieve better estimates of this parameter it is necessary to make further observations at sea from commercial vessels. This, however, would not necessarily give information on the difficult estimation of the survival of fish after their escape from nets or hooks. For long-line fisheries the number of hooks missing when hauling might provide a measure of escapees. The number of uncertainties in all the above experiments seem, however, largely to invalidate any results obtained.

The ICES Working Group has also requested that observations relating to non-catch mortality be separated by net material, supposing that monofilament nets cause greater non-catch mortality than do multifilament nets.

The official statistics are frequently so poor that more detailed biological analyses are impossible. The major gaps at present are probably found in relation to catch by gear rather than in relation to catch by locality and time. Only better statistical systems and the understanding and co-operation of fishermen can improve such statistics.

In conclusion, therefore although the collection of catch-data statistics by itself is not considered research, such analyses of salmon stocks and populations which result in management advice are heavily dependent upon information on the true removal of fish by fishing operations. Where the official statistics are unlikely to illustrate accurately the true situation it is necessary to seek further information, for instance through sampling by observers, interviews, inspections, etc.

Effort data

While estimates of the density of fish in freshwater are obtainable by several methods, the high seas fisheries present severe problems when trying to establish measurable effort units which express not only fishing activity but also the density of fish derived from catch- per-unit-effort figures.

Some of the problems are of a technical nature. For

instance, drift nets are sometimes stretched by currents nearly to breaking point, at other times they curve and bend. The author's own impression is that the latter improves the effectiveness of the nets. Also, the area fished remains a problem. To what extent do nets and long-lines sweep a certain area and to what extent does the movement of fish bring them actively in contact with the nets?

Models to solve these problems have been formulated and discussed (e.g. Christensen and Lear, 1980; Hansen, 1984) but their practicality is not certain. As with many other models the quality of the input data is the major problem.

Even if these problems were solved the fact would still remain that all gear used for high seas salmon fisheries is surface gear. Drift nets usually extend only 5 m below the surface (Christensen and Lear, 1980), hooks somewhat more, but compared with the water depths where the gear is used only a small part of the water column is fished. Although salmon may be regarded as a surface fish there is evidence that they occur in much greater depths than those reached by the gear. Nielsen (1961) reports that for some years after 1945, salmon were commonly taken on Store Hellefiske Bank off West Greenland in bottom trawls fishing at about 30-40 m depths. Nielsen also reports salmon caught by jigs during fishing for cod at East Greenland, 1957. The author made the same observation in 1958 when the jigs were fished below 50 m depth. Although trawls as well as jigs may have caught the fish when the gear passed surface water the number of salmon caught would indicate that they were caught at the depth where the gear was operated.

Information on the three-dimensional distribution of salmon and their migrations is necessary to understand the reasons for fluctuations in catch-per-effort by the surface gear and the use of this index as an abundance index (see below).

Although the effort units used for commercial gear may not give catch-rate figures which can be taken as straightforward abundance indices it would, nevertheless, be quite informative to have the fishing activity by number of gear units and fishing time recorded. Lacking such data, the ICES Working Group spent much time discussing to what extent the very low catches at Greenland in 1983 and 1984 resulted from below-normal fishing activity. For long-term studies of fish abundance, effort figures such as fishing days, net-hours or similar simple units would be valuable and should be relatively easily obtained.

SAMPLING OF CATCHES

Sampling of both the Greenland and the Faroese salmon fishery is now carried out on a routine basis. In fact, the ICES Working

Group at its latest meeting (Anon., 1986) discussed whether the programme at Greenland should be limited, since salmon was sampled much more intensively than other fish in the area and to some extent at the expense of such other sampling. The Working Group did, however, find that as long as NASCO requests annual advice on status of stocks and on effects of varying levels of harvest at Greenland, it is necessary to maintain this high level of sampling.

Sampling of catches is therefore likely to constitute a permanent and important part of the research on salmon. At Greenland and at the Faroes the sampling programmes are probably better organised than sampling of salmon caught at sea close to home waters. Improvement in home water sampling is necessary for a break-down of catches by age groups.

The Greenland sampling programme will continue to focus on biological factors such as age, growth, sex ratios and continent of origin (scale readings), as well as distinction between hatchery-produced and wild salmon. The Faroese sampling programme has recently incorporated the distinction of escapees from fish farms.

For the Greenland programme the data base of the discriminant analysis used to identify continent of origin will be tested every two years after 1985 and will be extended.

It seems likely that scale discriminant analysis will be increasingly refined towards identifying not only the continent but also the country of origin or even separate stocks. In that case, more extensive and uniformly spread sampling of catches from the sea will be necessary as will frequent checking of the data base.

Commercial catches at Greenland are usually limited to a short part of the period of occurrence, and always to the first part of that period. Previous investigations showed some changes in stock composition with time during the season (Munro and Swain, 1980). The only possible ways of obtaining samples outside the period when the quota has been filled seem to be by research vessels (or chartered vessels to conduct research) and/or by making special allowance for one or more commercial vessels to continue fishing 'in excess of the TAC' or on a part of the TAC set aside for that purpose. The latter possibilities are, however, likely to be administratively and politically complicated, whilst the former (research vessels) is costly and/or a matter of priorities compared with other research programmes.

The variation of catch composition between areas in Greenland also complicates sampling. In the 1986 sampling programme, four of the major landing places were covered. This seems to be the minimum coverage for a thorough analysis of the total catch, although still lacking sampling of some important spatial components of the catches.

TAGGING EXPERIMENTS

Tagging experiments already conducted at Greenland and at the Faroes have resulted in a better knowledge of the migration and origin of salmon, and have also allowed for some estimates of fishing mortality rates.

Tagging of fish in Greenland waters culminated with the International Salmon Tagging Experiment in 1972 (see above). No further tagging has been conducted since, nor has it been proposed. The costs of such experiments are immense. The direct cost of tagging each of the 2,364 salmon tagged in the 1972 experiment has never been calculated, but it probably exceeded £200 (excluding the involvement of commercial vessels). Uncertainty as to tagging mortality was a major factor disturbing the analysis of the experiment (Andersen et al., 1980). It is probable that fish caught on long-lines are better for tagging experiments but that gear has been very unsuccessful in Greenland waters, at any rate in the main part of the season (August-October). It may be possible to catch fish by long-lines later in the winter, if not in the Davis Strait then at least in the Labrador Sea, where two Greenland vessels caught salmon by long-lines early in 1970.

Salmon seem to be present in the Labrador Sea not only in winter but throughout the year (Templeman, 1967, 1968). A tagging experiment conducted here, preferably every quarter, over a year, might give very interesting results.

Likewise, tagging experiments off East Greenland and in the Irminger Sea would be interesting. Where, for instance, is the eastern limit for sea-going salmon of North American origin, and where do Icelandic fish have their major feeding areas?

Tagging experiments in freshwater, mainly of smolts, are of course of great importance for studies of ocean migrations, and it seems most likely that these relatively inexpensive experiments will continue. As far as micro-tags are concerned, continuation of the newly introduced scanning programmes at Greenland seems to be necessary, together with similar scanning at the Faroes. It would be an advantage if such micro-tagging experiments were concentrated on rivers where estimation of the freshwater stock is possible.

PRACTICAL EXPERIMENTS TO SUPPLY DATA FOR DENSITY MODELS

In a paper presented to ICES, Hansen (1984) developed two models to measure stock size for a given area by means of long-lines. One of the models, the drift model, assumes that the long-line sweeps an area where the density of fish remains constant during the experiment. The other model, the diffusion model, assumes that feeding salmon move randomly in their search for food. The latter model is more complicated than the former. The theory and assumptions of the models are beyond the scope of this discussion. As for many other assessment models, the essential thing is to supply sufficient good data to test them and get them into operation.

Some experiments to supply data for Hansen's models were performed in Faroese waters in 1985 (Jákupsstovu, Jørgensen Mouritsen and Nicolajsen, 1985). Besides sampling of catches, notes of ship's movements, drifting of long-lines, current measurements etc., interesting observations were made on the movements of two salmon tagged with acoustic tags. Such limited observations cannot, of course, lead to general statements about the behaviour of salmon (and the authors did not make such statements), but there seemed to be indications of a non-random movement of the two fish. One of the tags registered the depth, and the results indicated that fish spend the first 1-2 hours after release at depths of about 100 m. Thereafter, for the last part of the total 19 hours' observation, the fish mainly occurred at depths of 3-6 m.

This technique seems promising, not only for supplying information for Hansen's models but also for allowing detailed study of the vertical and horizontal movement of salmon and their responses to gear, to the distribution of prey and to local environmental factors. The techniques can perhaps be developed further as a remote sensing technique using buoys and satellites.

Other proposals for studies of vertical distribution include fishing experiments in areas where commonly used salmon gear indicate good concentrations of salmon. Fishing with two-boat pelagic trawls might give information on the distribution of salmon at various depths. Also the use of large, deep purse seines may yield interesting results on salmon density but such experiments are very costly if carried out for that purpose only. Direct acoustic observations may be a way to observe salmon at different depths but seem to require further basic studies to allow identification of signals as being salmon.

Part of the future research at sea must be devoted to such studies which will confirm whether the numerous assumptions built into every assessment model hold good.

SOME DREAMS

Extremes of catches at Greenland
The recent (1986) salmon season in Greenland was very short. A catch of approximately 650 tons was taken in about 10 days. What would the result have been if fishing had been completely

free? It is quite likely that continued freedom to fish would have resulted in a record catch at Greenland. It is possible, at least in some years, that the stock of 1-sea-winter fish at Greenland is well above the estimate of some 2 million fish found to be a probable total for 1972 (Horsted, 1980) when taking into account results of the 1972 International Tagging Experiment.

Simultaneous fishing throughout the area of distribution

Salmon occur at East Greenland as well as West Greenland. In fact, if fishing activity in the two areas is compared, one might find that the amount of salmon present at East Greenland is of the same magnitude as that at West Greenland _in the same period_. At the same time of the year salmon are known to occur in the Labrador Sea (Templeman, 1967, 1968) and in the Irminger Sea (Møller Jensen and Lear, 1980). Some home waters are reported to have a run of multi-sea-winter fish at the same time as the Greenland fishing season.

An interesting experiment might be based on simultaneous fishing, preferably by drift nets, over a very wide part of the area in which salmon are likely to occur, ranging from the water off Maine and eastern Canada to the coastlines of the European continent from Norway to and including the Bay of Biscay, through which some salmon of French and Spanish origin must pass. It would also be interesting to tag any salmon caught in waters between Cape Farewell and the British Isles and the waters around Iceland, since the border line for distribution of European origin grilse must occur somewhere in this area. Again, however, the enormous cost of launching such an experiment, allied perhaps with some political reluctance, seems to make this no more than a dream.

SUMMARY

Although the anadromous nature of the salmon has long been known, research on the sea life of salmon was limited until the fishery at West Greenland developed in the mid-1960s.

Recaptures at Greenland of fish tagged in Europe and North America proved Greenland waters to be a major oceanic feeding ground. Sampling showed that fish occurring there were destined to become multi-sea-winter fish, not grilse. The International Tagging Experiment at West Greenland in 1972 demonstrated that some of the salmon present at Greenland return to home waters, most likely to their native rivers.

Application of discriminant function analysis of scale structure has allowed distinction at Greenland of fish of European origin from those of North American origin.

The fisheries in the Norwegian Sea and later, in Faroese waters harvest fish of almost exclusively European origin, including both grilse and multi-sea-winter fish.

Though a much wider knowledge of the salmon's sea life has been gained, there are still a number of gaps. These are identified and some proposals to fill the gaps are put forward.

Catch and especially effort statistics for the distant-water fisheries as well as for home water fisheries need to be improved. Until such statistics are sufficiently good it may be necessary to acquire information by sampling. For catch-rate figures to be used as density indices, studies of catchability are necessary. Studies of the three-dimensional movements of salmon have shown promise.

Tagging experiments at sea are costly, and if carried out at Greenland and in Faroese waters are unlikely to contribute much to what is already known. However, tagging of fish in the Labrador Sea, off East Greenland and around Iceland may give further information on the migration and distribution of salmon in the North Atlantic.

A number of practical experiments to elucidate the reaction of salmon to fishing gear, to determine their prey and examine the environmental conditions they prefer seem necessary in order to supply good data for assessment models and to find out whether the assumptions for such models are valid. Acoustic tags may be a tool for such experiments, but also experimental fishing by pair-trawls and purse seines should be considered.

Long-term analysis of distribution and migration of salmon in relation to environmental conditions, primarily water temperatures, should be encouraged.

For a full understanding of the sea-life of the salmon, a knowledge of the physiology and genetics of the fish is relevant.

REFERENCES

Andersen, K.P., Horsted, Sv. Aa. and Møller Jensen, J. (1980) Analysis of recaptures from the West Greenland Tagging Experiment. Rapport et Procés-Verbaux de la Reunion, Conseil International pour l'Exploration de la Mer, 176, 136-41

Anon. (1967) Report of the ICES/ICNAF Joint Working Party on North Atlantic Salmon, 1966. International Council for the Exploration of the Sea, Co-operative Research Report Series A, no. 8

Anon. (1971) Third Report of the ICES/ICNAF Joint Working Party on North Atlantic Salmon, Dec. 1970. International Council for the Exploration of the Sea. Co-operative Research Report Series A. no. 24

Anon. (1981) Report of the ICES Working Group on North Atlantic Salmon, 1979 and 1980. International Council for the Exploration of the Sea. Co-operative Research Report Series A, no. 104

Anon. (1985) Report of the ICES Working Group on North Atlantic Salmon. International Council for the Exploration of the Sea, CM 1985/Assess: 11 (mimeo)

Anon. (1986) Report of the ICES Working Group on North Atlantic Salmon. International Council for the Exploration of the Sea, CM 1986/Assess, 17 (mimeo)

Child, A.R. (1980) Identification of stocks of Atlantic salmon (Salmo salar L.) by electrophoretic analysis of serum proteins. Rapport et Procés-Verbaux de la Reunion, Conseil International pour l'Exploration de la Mer, 176, 65-7

Christensen, O. and Lear, W.H. (1980) Distribution and abundance of Atlantic salmon at West Greenland. Rapport et Procés-Verbaux de la Reunion, Conseil International pour l'Exploration de la Mer, 176, 22-35

Hansen, B. (1984) Assessment of a salmon stock based on long line catch data. International Council for the Exploration of the Sea, CM 1984/M:6 (mimeo)

Horsted, Sv. Aa. (1980) Simulation of home-water catches of salmon surviving the West Greenland fisheries as a method for estimating the efforts of the West Greenland fishery on home water stocks and fisheries. Rapport et Procés-Verbaux de la Reunion, Conseil International pour l'Exploration de la Mer, 176, 142-6

Jákupsstovu, S.H.I., Jørgensen, P.T., Mouritsen, R. and Nicolajsen, A. (1985) Biological data and preliminary observations on the spatial distribution of salmon within the Faroese fishing zone in February 1985. International Council for the Exploration of the Sea, CM 1985/M: 30 (mimeo)

Jensen, J. Møller and Lear, W.J. (1980) Atlantic salmon caught in the Irminger Sea and at East Greenland. Journal of Northwest Atlantic Fish Science, 1, 55-64

Lear, W.H. and Sandeman, E.J. (1980) Use of scale characters and discriminant functions for identifying continental origin of Atlantic salmon. Rapport et Procés-Verbaux de la Reunion, Conseil International pour l'Exploration de la Mer, 176, 68-75

Meerburg, D.J. (ed.) (1986) Salmonid age at maturity. Canadian Special Publications on Fisheries and Aquatic Science, 89 (14 papers), 118 pp

Menzies, W.J.M. and Shearer, W.M. (1957) Long-distance migration of the salmon. Nature, 179, 790

Munro, W.R. and Swain, A. (1980) Age, weight and length distribution and sex ratio of salmon caught off West Greenland. Rapport et Procés-Verbaux de la Reunion, Conseil International pour l'Exploration de la Mer, 176, 43-54

Nielsen, J. (1961) Contribution to the biology of the salmonidae in Greenland. I-IV. Meddelelser om Grønland, 159, no. 8

Payne, R.H. (1980) The use of serum transferrin polymorphism to determine the stock composition of Atlantic salmon in the West Greenland fishery. Rapport et Procés-Verbaux de la Reunion, Conseil International pour l'Exploration de la Mer, 176, 60-4

Pippy, J.H.C. (1980) The value of parasites as biological tags in Atlantic salmon at West Greenland. Rapport et Procés-Verbaux de la Reunion Conseil International pour l'Exploration de la Mer, 176, 76-81

Rosenørn, S., Fabricius, J.S., Buch, E. and Horsted, Sv. Aa. (1984) Isvintre ved Vestgrønland. Forskning/Tusaut i Grønland, 2, 2-19, see below

Rosenørn, S., Fabricius, J.S., Buch, E. and Horsted, Sv. Aa. (1985) Record hard winters at West Greenland. North Atlantic Fisheries Organisation, Secretariat Document 85/61, Ser. No. N1011 (mimeo) (English translation of the above mentioned publication)

Templeman, W. (1967) Atlantic salmon from the Labrador Sea and off West Greenland taken during A.T. Cameron Cruise, July-August 1965. Research Bulletin of the International Commission for Northwest Atlantic Fisheries, 4, 5-40

Templeman, W. (1968) Distribution and characteristics of Atlantic salmon over oceanic depths and on the banks and shelf slope areas off Newfoundland. Research Bulletin of the International Commission for Northwest Atlantic Fisheries, 5, 62-5

Chapter Twenty Eight

THE IMPACT OF ILLEGAL FISHING ON SALMON STOCKS IN THE FOYLE AREA

W. Gerald Crawford

Foyle Fisheries Commission, 8 Victoria Road, Londonderry, Northern Ireland, BT47 2AB

INTRODUCTION

This is not a scientific paper, but rather a collection of some facts and observations collated and interpreted in the light of experience. All opinions held and statements made, are those of the author and not necessarily those of the Foyle Fisheries Commission.

THE FOYLE AREA

The Foyle Area is the name given to the catchments of all rivers entering the Atlantic Ocean between Malin Head, the most northerly point in Ireland and Downhill near Coleraine. These rivers, of which the River Foyle and its tributaries are by far the highest component, flow in a generally northerly direction and drain an area of some 1,545 square miles. This comprises part of County Donegal in the Irish Republic and also parts of Counties Londonderry and Tyrone in Northern Ireland.

Responsibility for the conservation, protection and improvement of the Fisheries in the Foyle Area is statutorily vested in the Foyle Fisheries Commission. The Commission consists of four members, two appointed by the Minister for Tourism, Fisheries and Forestry in the Republic of Ireland and two by the Department of Agriculture for Northern Ireland. It has the power to make Statutory Regulations, subject to the approval of the Minister and the Department. It employs a full time staff of 25 plus some temporary staff to enable it to carry out these statutory duties.

The vast majority of the fish stocks in the system are salmonid species - with Atlantic salmon the predominant single species although there are significant stocks of sea-trout (known locally as white trout) and brown trout although the latter are

524

usually small.

'The Foyle has the reputation of being one of, if not the most productive salmon system in Europe' - so stated Dr Paul Elson in his report to the Commission (Elson and Tuomi, 1975). He also computed the maximum equilibrium yield at 143,000 adult salmon/year, a production equivalent to 317 lbs/acre of stream bed, representing the surviving progeny of 8,000 spawning fish, evenly spread over the nursery areas of the system.

It will be apparent therefore that in terms of production there is considerable scope for exploitation of salmon stocks both by legal and illegal means.

CURRENT RESTRICTION ON LEGAL SALMON FISHING

Each year the Commission licenses some 115 drift nets and 150 draft nets to fish in the area. Licences were also issued to 5,567 anglers in 1985.

In order to try to attain the necessary escapement of 8,000 spawners per year, the Commission has had to introduce considerable restrictions on licensed fishermen over the years. These include a very short commercial netting season of about 26 days spread over six weeks from late June to early August. Other restrictions are imposed on the hours of fishing, length of net, depth of net, size of boat, an extended weekly close time, suspension of netting in special circumstances, reduced length of angling season, limits on the breaking strain of line size and number of hooks (lures) etc.

It has been argued that because of all these regulations, legal fishermen are being forced to fish unlawfully in order to make any profit. These restrictions are necessary and must be accepted by all as in the interest of the long-term continuation of the fishery as a viable commercial and sporting system. It is still possible to make a reasonable profit while fishing within the constraints imposed by the regulations.

Illegal fishing can take many forms and before considering the impacts of these, the known types and methods in use in the Foyle Area should be set out.

SOME ILLEGAL FISHING METHODS

First there is illegal fishing by licensed fishermen either by using illegal equipment (e.g. monofilament nets) or by fishing at an illegal time - before or after the season, during the weekly close time (which is 72 hours in the Foyle Area) or, in the case of drift nets, outside the daily 10 hour fishing period. These methods have the advantage that the fisherman has a licence under which illegally caught fish may be sold during the season. The licence also provides him with a reason to have a boat and net and to be present on the fishery. He can claim that mechanical breakdown had kept him out beyond the permitted fishing hours. The disadvantage from his point of view is that he can lose his licence and if he is apprehended and successfully prosecuted it is likely that even on a first offence he will not be granted a licence for the following season.

Secondly, there is the unlicensed or illegal fisherman fishing either by legal or illegal methods. The illegal fisherman fishing by illegal methods is by far the most common form of illegal fishing in the Foyle Area and it is practised by both netsmen and anglers.

Who are the people concerned? With very high rates of male unemployment (up to 28-30 per cent in some areas) in addition to the usual crop of 'entrepreneurs' and 'enthusiastic amateurs' there is a large pool of prospective practitioners, all with varying reasons to justify their actions.

What equipment do they use? Usually it fulfils the requirements of being cheap, easily replaceable and efficient.

For illegal netting, which is usually carried out on the lower and tidal portions of the major rivers, a cheap boat is used (possibly home-made and 3-4 metres long), at a total cost of about £100 including oars. A small outboard motor may also be used but this is not common. On some narrower stretches of river an inflatable rubber dinghy or even a large inner tube from a tractor wheel may be used instead.

Nets are usually made from monofilament or multistrand material with stones for weights and short lengths of pipe insulation material for floats. They vary from about 15 metres to 2,000 or more metres in length according to the location in which they are to be used.

The nets are usually set at night and may be left (usually unattended) anything from half an hour to 24 hours before being lifted. Obviously, the longer they are left the better the chance of detection.

The favourite method of illegal fishing by anglers is to snatch or foulhook fish, frequently downstream from a weir or dam, using a treble hook and heavy line with a cheap rod and reel.

The advantage of these type of operations is that equipment costs are comparatively low, and if detected the operator can flee, leaving the equipment to be seized at little financial loss. He does not hold a licence and so has nothing to lose if apprehended. The only punishment is therefore a fine or a gaol sentence, if successfully prosecuted. The disadvantages are that all fish caught have usually to be sold on the 'black market', and they thus command a much lower price than legally caught fish and also there is probably an increased chance of detection, as in Ireland the onus of proof of legal capture in law lies with the person in charge of the salmon.

In Ireland, legislation also requires any person, other than a fisherman, selling salmon or trout, to be licensed. Under the terms of the licence a record of all purchases and sales must be kept and this reduces the ability of illegal fishermen to dispose of their catch to the trade.

A form of illegal fishing that is practised from time to time is that of 'poisoning', where a poisonous substance, usually magnesium cyanide (known locally as 'rabbit gas'), is introduced into a river. Not only does this product kill adult salmon by preventing the function of the gills, but it also kills juvenile fish, thus wiping out three or more year classes from a stretch of river. Unfortunately this cyanide product is sold in 3 kilo tins, where one tin has the capacity to destroy all fish in a considerable area of river.

Another method of illegal fishing is the use of lamp and gaff for adult fish on the spawning beds during the long dark nights of November and December. Why people persist with this method seems to be a complete mystery, as by this time in their life cycle the flesh of the salmon is of very poor quality and virtually inedible. Fish of this quality are not usually sold, but are eaten by the poacher, his family and friends.

ANIMAL PREDATION

There is another form of unlicensed fishing that should be mentioned, and although it is not illegal it has increased dramatically in the Foyle Area during the past few years. This is the predation of salmon stocks by animals - other than the humans. Since the recent introduction of conservation legislation in both the Republic of Ireland and Northern Ireland, increasing the level of protection afforded to many animal species, there has been a dramatic increase in the seal population off the north coast of Ireland, especially during the summer months. This has led to substantiated reports from drift net fishermen of an increased incidence of salmon heads being found in their nets, indicating that a seal has eaten the remainder of the fish. One would imagine too that if there is a substantial increase in netted fish being eaten by seals, there must be some increase in mortalities from the same source amongst all returning salmon.

Another fisherman that is causing increased concern is the cormorant. Whilst no substantial breeding colony exists in the Foyle Area, there are significant numbers of non-breeding birds in and around Lough Foyle and at least one large breeding colony within striking distance.

Cormorants have always been recognised as predators of salmon smolts and for a number of years after its inception the Foyle Fisheries Commission offered a reward for dead cormorants. This practice was discontinued about 10 years ago.

Recent investigations (as yet unpublished) on the River Bush in Northern Ireland indicate that predation by cormorants on salmon smolts during riverine migration is very substantial, and while the River Foyle is not as close to a major breeding colony as is the River Bush, this is a factor that may cause increased concern in the coming years.

CURRENT LEVELS OF SALMON STOCKS

Dr Paul Elson's estimate, already referred to, set a target of 8,000 spawning fish (males and females) in the Foyle system. He also assumed that each hen fish cut two redds, so we are looking therefore for 8,000 redds evenly spread throughout the system each year. While it is accepted that counting redds is an inefficient exercise, dependent as it is on ideal weather conditions, it is an indicator of stock levels. Since the publication of the Elson Report in 1975 the magic figure of 8,000 redds has never been obtained although many of the individual rivers in the system have exceeded their quota in one or more years. This, combined with other evidence such as the returns from the two rivers in the system on which a count of ascending salmon is made and the generally low level of catches, leads one to believe that the system is largely understocked, although in recent years a recovery in the earlier downward trend has been noted.

IMPACT OF ILLEGAL FISHING ON STOCKS

It is an impossible task to quantify the effects of illegal fishing and to separate its impact from the impact of other factors which reduce the stock of salmon, such as water pollution and water abstraction as well as sand or gravel abstractions; none of these fall within the scope of this paper.

Table 28.1 illustrates the amount of illegal fishing equipment seized by the Commission's staff over the past 10 years and it is apparent that the general trend in seizures over this period is upwards. This may not be due to an increased level of poaching, but to more efficient detection, combined with a shift in the major area of illegal fishing from the lower/middle reaches of the River Foyle to the upper reaches where both the length of net required is shorter and the 'investment' is smaller. It is suspected that the level of illegal fishing has shown some actual decrease over the period. This does mean, however, that

Table 28.1: Equipment seized in the Foyle Area 1975–1985

Equipment	1975	1976	1977	1978	1979	1980	1981	1982	1983	1984	1985
Illegal Nets	245	261	366	439	499	660	710	485	689	809	683
Boats	21	15	41	42	28	31	34	61	68	69	44
Rods	12	11	26	53	26	26	42	15	30	20	47
Vehicles	2	2	4	2	1	NIL	NIL	NIL	6	NIL	2

even these figures for seizures do not help us to quantify the impact of illegal fishing. There is no doubt that there is a considerable amount of illegal fishing in the Foyle Area, the volume of equipment seized and numbers of persons successfully prosecuted (Tables 28.2, 28.3 and 28.4) tell us that much. There is also little doubt that they are to some degree successful and catch salmon (Tables 28.5 and 28.6).

Table 28.2: Number of successful prosecutions for fishery offences by Foyle Fisheries Commission and the police forces in the Foyle Area each year 1977-1984

Year	Police	Commission
1977	95	190
1978	53	190
1979	41	80
1980	52	137
1981	12	102
1982	124	127
1983	154	177
1984	30	158

Table 28.3: Number of netting offences successfully prosecuted in the Foyle Area 1976-1984

	1976	1977	1978	1979	1980	1981	1982	1983	1984
Using a net in a prohibited area	2	4		4	7	4	13	21	12
Possession/use of unlawful net	17	56	53	32	43	38	89	74	38
Fishing a net during close or suspension periods	7	21	14	13	26	14	33	43	22
Using a boat as an aid to an offence	3	18	38	16	24	8	26	30	7
Total	19	99	105	65	100	64	161	168	79

The Impact of Illegal Fishing on Salmon Stocks in the Foyle Area

Table 28.4: Number of angling offences successfully prosecuted in the Foyle Area 1976-1984

	1976	1977	1978	1979	1980	1981	1982	1983	1984
Angling without a licence	3	13	15	10	6	11	13	6	15
Angling during close times		8	3	1	1	3	3		1
Angling in a prohibited area	3	2	6		5	2	3	3	5
Failing to produce a licence			9	4	2		1		1
Possession of illegal instrument (gaff etc.)	4	4	9	4	17	1	6	13	11
Snatching or foul hooking	2	2	5		1	2	3	25	5
Trespass on a Several Fishery			6	2	7	7		17	18
Using line in excess of permitted breaking strain							2	4	3
Total	12	29	53	21	39	26	31	68	50

Table 28.5: Number of salmon seized in the Foyle Area each year 1976-1984

1976	97
1977	162
1978	56
1979	36
1980	99
1981	208
1982	234
1983	311
1984	146

Table 28.6: Number of illegally netted salmon seized compared with the number of nets seized annually

Year	No. of salmon	No. of nets seized
1976	34	261
1977	114	366
1978	51	439
1979	46	499
1980	81	660
1981	148	710
1982	119	485
1983	159	689
1984	95	809

Since the system is presently understocked, it therefore follows that part of the reason for the low stock is that each year, x number of salmon are caught by illegal means that would otherwise survive to spawn.

The nett loss to the spawning escapement is not of course x as we have to deduct from this total:

(1) the number of fish that would be caught by licensed fishermen;
(2) the number of fish that die before spawning;
(3) the number of infertile fish.

The expression for the increased number of spawners if illegal fishing were to be curtailed is therefore x -(1)-(2)-(3) = ?

This is not very conclusive as we cannot put a value on any of the components. At least we know that if illegal fishing were to cease there would be an increased spawning escapement. That is the major impact that illegal fishing has on salmon stocks, although not the only one. If there were no illegal fishing the increase in population would have a multiplier effect on succeeding generations. If it were possible to increase the potential spawning stock each year by even 10 per cent this would, in the course of a few generations, have a very marked effect on the stock level.

There is another effect of illegal fishing which, whilst it may be less important in respect of the depletion of adult fish numbers, has a significant effect on the biomass of fish in the system. This is the effect of the use of poison on a stretch of river. Most poisons are not selective and kill all fish with which they come in contact, and some have a severe effect on the invertebrates on which fish rely for food. A large quantity of poison may be added at a point where the river is some five metres wide, with deep holding pools interspersed with fast flowing nursery areas and some gravel spawning grounds. The poison is usually added to the river at the head of a pool known to contain adult fish but unfortunately the pool acts as a giant mixing bowl, gradually releasing the poison downstream and thereby increasing the time that fish and invertebrate life are exposed. It may be 1.5 to 2 kilometres downstream before this deadly mixture is sufficiently diluted not to cause death. The result is that a large area - up to a hectare of stream bed - will be denuded of fish and possibly invertebrate life. This is a substantial loss, involving as it does at least three year classes of salmon, plus resident brown trout. The salmon are, of course, irreplaceable except by restocking and this is not always possible or indeed advisable.

Recent work has indicated that a well-stocked salmon nursery in a productive river has a potential value of £85,000 per hectare (Kennedy, 1987).

Considering the impact of illegal fishing on the work of the Foyle Fisheries Commission, which has a statutory duty to conserve, protect and improve the fisheries of the Foyle Area, undoubtedly the prevention of illegal fishing and the apprehension of those involved in it comes within that remit. In fact it is a primary function, in that it is of little use improving the runs of salmon in other ways if these, at a later stage, are going to be removed from the system by unlawful means.

If, however, less time and effort had to be spent on law enforcement, more resources would be available for other work involving stock enhancement, such as the improvement of spawning beds, the control of water pollution, or for basic research work in the area of salmon movement. Most of this type of work has had to be reduced over the past 10 years and now forms only a very minor part of the Commission's activities.

In conclusion it is possible to paraphrase this entire paper by stating that the impact of illegal fishing in the Foyle Area is a complex web extending far beyond the taking of a few fish from a river. If not kept under strict control at reasonable cost, illegal fishing could directly and indirectly lead to a significant depletion in salmon stocks throughout the system.

SUMMARY

The Foyle Area is delineated, and the constitution of the Foyle Fisheries Commission and its statutory responsibilities are set out. The main fish species and the high level of productivity are discussed, as are the targets set by Dr Paul Elson in his report to the Commission in 1975 and the Commission's approach to achieving these targets. Various types of illegal fishing are described, including both nets and rods, and involving licensed and unlicensed fishermen. The use of poisonous substances to kill fish and 'lamping' on spawning beds are also noted. Concern is also expressed over the increase in animal predators, especially cormorants and seals.

The impact of illegal fishing is discussed, not only in respect of the removal of adult salmon by net and rod but also by the use of poison. Attention is drawn to the effect of illegal fishing on the wider work of the Commission.

REFERENCES

Elson, P.F. and Tuomi, A.L.W. (1975) The Foyle Fisheries - New Basis for Rational Management. p. 1
Kennedy, G.J.A. (1987) Silage effluent pollution costs and prevention. Agriculture in Northern Ireland, 60, no. 12

Chapter Twenty Nine

THE INDIAN ATLANTIC SALMON FISHERY ON THE
RESTIGOUCHE RIVER: ILLEGAL FISHING OR ABORIGINAL
RIGHT?

Stephen D. Hazell
Counsel, Canadian Wildlife Federation, 1673 Carling Avenue,
Ottawa K2A 3Z1, Canada

INTRODUCTION

In the summer of 1986, the salmon fisheries on both the Atlantic
and Pacific coasts of Canada were plagued by turmoil and
confrontation. In British Columbia, recreational and commercial
fishers vehemently opposed bylaws enacted by four Indian Bands
that purported to establish a commercial fishery and an
independent management regime for salmon stocks in waters of
the Skeena River adjacent to Indian reserves. For their part, the
Bands threatened violence to fisheries officials attempting to
enforce fisheries laws against Indian salmon fishers (Anon. 1986a).
Confrontations on the Fraser and Skeena Rivers of British
Columbia were frequent throughout July and August, culminating
in a violent incident on 30 August in which 18 Indians were
charged with offences, including assault and obstruction of a
peace officer (Parfitt, 1986).

The establishment of a licensed salmon food fishery for the
Conne River Indian Band in Newfoundland was also protested by
non-Indian fishers to no avail, on the grounds that prior to the
nineteenth century the ancestors of the Band had lived, fished
and hunted in Nova Scotia, but not in Newfoundland (Anon,
1986b).

On the Restigouche River, which straddles the boundary
between the provinces of Québec and New Brunswick,
recreational fishers and the New Brunswick government charged
that the Restigouche Indian Band, whose reserve is situated on
the Québec side of the river, was flagrantly violating a salmon
fishing agreement between the Band and the province of Québec,
which set quotas on the Band's harvest. Band members stopped
fishing only after the Atlantic Salmon Federation launched
injunction proceedings against the Restigouche Band for its
failure to abide by the terms of the agreement and fisheries laws
(Anon, 1986c).

On both coasts, the debate between aboriginal and

non-aboriginal fishers over rights of exploitation and the status and best use of salmon stocks often descended to vitriol and invective. In such an emotional and conflicting climate, the great danger is that certain Atlantic and Pacific salmon stocks will be sacrificed in the political fight.

The current impasse among salmon fishers and fisheries officials is rooted in the steadily growing harvesting pressures on salmon species, the complicated regime for regulating anadromous fisheries in Canada, and the claims of aboriginal peoples for preferential salmon fishing rights and/or ownership of the resource. This paper will examine these factors in the context of one fishery in one drainage basin - the Atlantic salmon fishery in the Restigouche River.

Legal aspects of the alleged overfishing by the Restigouche Band are explored, and the aboriginal and treaty fishing rights claimed by the Band are discussed. Finally, several policy considerations are noted, and elements of an Indian Atlantic salmon fishery policy that could lead to a resolution of the dispute are outlined.

THE RESTIGOUCHE SALMON FISHERY

The drainage basin of the Restigouche River is shared by the provinces of Québec and New Brunswick, and the river serves as the interprovincial boundary for some 80 km along its lower stretches (Figure 29.1). Upstream from the confluence of a major Québec tributary, the Patapédia, the Restigouche flows exclusively through New Brunswick, aside from the headwaters of one tributary, the Kedgwick.

Next to the Miramichi, the Restigouche River system historically has been the most productive Atlantic salmon river in Canada. Between 1951 and 1970, total commercial and recreational landings of multi-sea-winter salmon and one sea-winter salmon (grilse) ranged from 18,000 to 46,000 fish per year, with an average of 32,000 fish per year. Since 1970, total commercial and recreational fishery landings have averaged roughly 10,000 per year, with 1985 landings at 5,228 fish. However, the commercial fishery was closed between 1971 and 1980, restricted from 1981 to 1984, and closed again in 1985 and 1986.

In 1983, Atlantic salmon egg deposition in the Restigouche river system was estimated to be as low as 10 per cent of the level required adequately to seed the river, compared to 25 per cent during the previous decade.

The serious deficit in spawning stocks of fish in the Restigouche is but one example of the decline in Atlantic salmon populations throughout eastern Canada. For example, the New

Figure 29.1: The drainage basins of the Restigouche and Grand Cascapédia Rivers

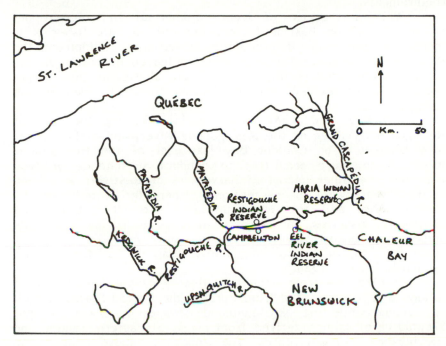

Brunswick commercial fishery as a whole caught 260,000 kg of Atlantic salmon in 1970, 66,235 kg in 1980 and only 4,000 kg in 1982, whereas the New Brunswick recreational fishery caught 56,937 kg in 1970, 32,014 kg in 1980 and 19,444 kg in 1983. The 1983 report of the Atlantic Salmon Task Group bleakly reported that:

> The Atlantic salmon resource appears to be at a point where drastic measures are deemed necessary in order to reverse the alarming decline in abundance ... Expectations for salmon abundance in 1984 give little hope for improvement in the short term or longer term unless actions are taken to allow more fish to spawn. Even if major changes were to occur in the coming season, we can expect to see very dismal returns for the next five (5) years (Anon, 1983).

The 1984 Atlantic Salmon Management Plan imposed severe restraints on commercial and recreational fishers to address the problem of underescapement. The 1985 report of the Canadian Atlantic Fisheries Scientific Advisory Committee recommended that exploitation levels of large (multi-sea-winter) salmon in the major salmon rivers of New Brunswick and Nova Scotia not be increased in 1986 due to the absence of a surplus above spawning requirements.

While the access of local commercial and recreational fishers to salmon has decreased, demand for all fisheries is increasing. But even if production of Atlantic salmon in the Restigouche and other Maritime rivers achieved its potential, the demand from commercial, recreational, aboriginal and illegal fisheries would exceed that production (Beamish, Healey and Griggs, 1986).

In 1985 and 1986, two Indian Bands, located at Restigouche, Québec on the Restigouche River, and at Eel River, New Brunswick, on Chaleur Bay just east of the Restigouche River mouth, were permitted to conduct subsistence or food fisheries. The other important harvesters of Restigouche salmon in these years were the recreational fishery (grilse only) on the Restigouche, and the Newfoundland and West Greenland commercial fisheries. The commercial fisheries in New Brunswick and on the Québec shore of Chaleur Bay were not permitted under the Atlantic Salmon Mangement Plan to fish for Atlantic salmon in 1985 or 1986.

In 1985, the West Greenland fishery caught 864,000 kg of salmon that had migrated from Europe, the United States as well as Canada, while the Newfoundland commercial fishery caught 870,000 kg of salmon derived from rivers of eastern Canada and the northeast United States. The impact of these fisheries on Restigouche stocks has not been quantified but must be assumed to be considerable.

The Indian fisheries at Restigouche and Eel River are the only local Restigouche fisheries to have increased their harvests in recent years. The quota for the Restigouche Band was 4,545 kg in 1977, rising to 11,363 kg in 1980, and falling to 6,995 kg (1,186 fish) in 1986 (Anon, 1981). But in 1980, the actual harvest by the Band was estimated at 58,815 kg while in 1986 various estimates place the harvest at a minimum of 29,500 kg (Bielak, personal communication).

According to official estimates, the catch of the unregulated fishery of the Eel River Band was 214 Restigouche salmon in 1984 and 241 in 1985. In summary, the limited evidence suggests that at least since 1984 the Indian fishery has dominated the local harvest of Restigouche stocks, with catches at higher levels than the restricted recreational fishery.

Indian Bands on the Saint John and Miramichi rivers of New Brunswick defy documentation of catch levels and are also believed to exceed permissible fishing effort and/or quotas (see Chapter 7).

MANAGEMENT OF ATLANTIC SALMON FISHERIES

In Canada, the authority to legislate over matters pertaining to fisheries is divided between provincial and federal governments. Under S.91 of the Constitution Act, 1867, the federal government has exclusive legislative authority over both sea coast and inland fisheries, whereas provinces have exclusive jurisdiction over property and civil rights under S.92. The federal constitutional responsibility has been judically restricted to enacting laws governing the management and conservation of fisheries (e.g. limiting fishing seasons, restricting types of gear). The federal Department of Fisheries and Oceans (DFO) carries out this responsibility under the Fisheries Act. Federal control over fisheries normally ends once fish have been legally caught.

Provincial governments, as the owners of fisheries resources in their respective provinces, may enact legislation concerning such matters as licences, leases, and commercial transactions, subject to federal fisheries conservation laws. Provincial jurisdiction over fisheries is restricted to non-tidal waters where property rights in fisheries exist.

The federal government has delegated the administration of the Fisheries Act and regulations to some provinces. Through orders-in-council and informal arrangements, the province of Québec has been delegated administrative powers over anadromous fisheries. No such delegations have been implemented for the province of New Brunswick, so that both the federal and provincial governments exercise some administrative powers over Atlantic salmon in this province.

As indicated, the Restigouche River acts as the boundary between New Brunswick and Québec from the confluence of the Patapédia River to Chaleur Bay. Over 50 per cent of Atlantic salmon spawning in the Restigouche drainage basin do so in New Brunswick waters, the remainder in Québec tributaries of the Restigouche, such as the Patapédia and Matapédia. If the 'Québec' Atlantic salmon kept to their side of the river on their spawning run up the Restigouche, perhaps the difficulties of having two distinct management regimes on the river would not be as acute. But as we know, Atlantic salmon, like all wild creatures, are not mindful of political jurisdictions.

To review briefly the existing regime for Atlantic salmon in these provinces, a starting point is the five-year Atlantic Salmon Management Plan developed by the Department of Fisheries and Oceans. Announced on 6 April 1984, the Plan applies to New Brunswick and the three other provinces of Atlantic Canada, but not Québec. The Plan prohibits anglers from taking salmon greater than 63 cm (multi-sea-winter salmon), and commercial fishing with the exception of the Newfoundland commercial fishery. Anglers are permitted to take up to two salmon less than 63 cm in length (grilse or one sea-winter salmon) per day and ten grilse per season, except in Newfoundland where there is a seasonal limit of 15 grilse. Anglers are permitted to fish for salmon greater than 63 cm, but are required to release the fish. No restrictions are placed on Indian salmon fisheries under the 1984 Plan.

The Québec Ministere du Loisir, de la Chasse et de la Pêche (MLCP) announced similar restrictions on 12 April 1984, including a grilse-only fishery for the Restigouche and Patapédia rivers (but not on the Matapédia river) and a ban on the Québec commercial fishery in the Gaspé and upper St Lawrence north shore and delta regions. However, in 1985 anglers on all Restigouche tributaries in Québec were permitted to take one salmon per day and seven per season.

Regulation of Indian salmon fisheries has been accomplished by regulation under the Fisheries Act accompanied by agreements between the Indian Band and the government administering the regulation (in the Restigouche drainage basin, either the Québec or federal government).

On 13 April 1970, the government of Québec, the federal Minister of Indian and Northern Affairs and the Québec Indian Association agreed that Indians in Québec are entitled to special status with respect to hunting and fishing (Moisan, 1979). Indians holding valid permits or hunting or fishing under an appropriate agreement have the right to hunt or fish for the subsistence of themselves and their families on their hunting or trapping grounds. The right to fish included the right to fish with nets.

Thus, in 1971, the Restigouche Indian Band commenced an Atlantic salmon gill net fishery on the river, after many decades of little fishing activity. In 1975, the Department of Fisheries and Oceans officially sanctioned Indian Bands adjacent to several Maritime salmon rivers to fish for salmon for subsistence only. In the same year, Québec issued similar subsistence fishery permits to the Restigouche and Maria (Grand Cascapédia) Indian Bands.

Current policy of the Department of Fisheries and Oceans permits the issuance, under Fisheries Act regulations, to Indian Bands of licences for subsistence fishing only. However, Indians are entitled to commercial fishing licences on the same basis as non-Indians. The Indian subsistence fishery has priority over recreational and local commercial fisheries to salmon stocks that are surplus to spawning escapement requirements. The policy of the province of Québec is similar.

The Restigouche Indian Band fishery is currently regulated under section 18 of the Québec Fishery Regulations, which provides that a holder of a salmon fishing licence issued to an Indian Band for subsistence purposes must affix a yellow seal issued by MLCP to any salmon he catches and keeps (S. 18), and that no person shall have in his possession an anadromous salmon to which a seal has not been affixed (S.18).

In 1985, the MLCP and the Restigouche Indian Band signed an agreement for the 1986, 1987 and 1988 fishing seasons ('the Restigouche Agreement') setting out the terms under which a salmon fishing licence would be issued to the Band under the Regulations. The Restigouche Agreement was developed without input from the federal or New Brunswick governments.

The Agreement allows a quota of 6,995 kg or 1,184 salmon (based on an average weight of 5.9 kg salmon), permits nets of a length not exceeding 150 metres, and allows fishing from 1800 h to 0800 h five nights per week commencing 2 June. All salmon captured are to be tagged and weighed at one of three landing sites by one of the twelve auxiliary agents employed by the Restigouche Band Council to assist MLCP in enforcing the Agreement.

The Band is also entitled to hire four helpers and six watchmen to assist MLCP with the implementation of the research programme for the three-year term of the Agreement and four boat operators to assist with a salmon tracking programme in 1986 only. The Band also is to hire a liaison agent to coordinate the implementation of the Agreement. The Band receives $303,100 in grants for the salaries of these employees and other costs in 1986 and similar amounts (with adjustments for cost-of-living increases) in 1987 and 1988. In addition, MLCP is obliged to hire four Band members as conservation agents for the three years of the Agreement.

EVIDENCE OF FISHERIES INFRACTIONS BY THE RESTIGOUCHE BAND

During June of this year, conservation officers of the New Brunswick Department of Natural Resources maintained observation posts across the Restigouche River from the Indian reserve. Observations of Restigouche Indian fishing activity were made using high-powered telescopes, but only a small percentage of the Indian fishing activity was recorded in the logbook. The reasons for this were that most of the fishing took place at night and at any one time 40 to 50 fishers might be working up to 100 nets.

New Brunswick conservation officers recorded numerous incidences of salmon not being tagged after capture contrary to S.18 of the Regulations and Article 3.3 of the Agreement. For example, on 14 June 79 fish were observed being landed, but only five of these were tagged by auxiliary agents as required under the Agreement. After capture, fish were rarely brought to one of the three designated landing sites nor were they registered with, or weighed by, auxiliary agents as required under Annex III of the Agreement.

541

The New Brunswick officials observed several incidences of fishing prior to the 2 June opening of the fishery, and numerous incidences of fishers starting to fish prior to the 1800 h start time or continuing to fish after the 0800 h finish time. Restigouche Band nets have also been set in New Brunswick waters, although it must be stated that the Québec-New Brunswick boundary in the Restigouche-Campbellton area is in dispute.

Other observers, including Dr A.T. Bielak of the Atlantic Salmon Federation and Ms Susan Shalala of the Campbellton (New Brunswick) Tribune made observations similar to those of the New Brunswick officials, particularly the failure to register, tag and weigh fish that were landed. During a three-hour period on the morning of 14 June, Ms Shalala observed a total of 75 salmon being landed but not being tagged, and not once did she see a salmon being properly tagged at a designated landing site.

The most serious infraction however, is that the total catch of Atlantic salmon by the Restigouche Indian Band in 1986 has been estimated by various sources to be at least 5,000 fish - far in excess of the quota of 1,184 provided under Article 3.3. However, the MLCP estimated the Band's catch at 1,149 fish, which estimate is presumably based on records gathered by the Band liaison agent and/or the four Band conservation agents hired by MLCP under the Agreement.

Another infraction was the breach of S.14 of the Fisheries Act, which provides that nets must not be set within 250 yards (230 metres) of another net. Restigouche Indians would often string two to five nets of up to 150 metres in length end to end across a section of the river in violation of this provision. This 'stringing' may block the river and drastically reduce escapement for spawning.

Finally, Band members violated subsection 24.(1) of the Act, which provides that not less than two-thirds of the width of the main channel of tidal rivers at low tide shall be left open, free from nets or other fishing apparatus.

The evidence summarised points to repeated violations of the Fisheries Act, the Québec Fishery Regulations and the Restigouche Agreement, although the size of the Band's 1986 catch is in dispute. No charges have been laid by the MLCP against Band members for any infractions committed during the 1986 salmon season.

ABORIGINAL, TREATY AND STATUTORY FISHING RIGHTS OF THE RESTIGOUCHE INDIANS

In the eyes of non-aboriginal recreational and commercial fishers, these breaches by Restigouche Indians of fisheries laws and the Agreement are unconscionable. But one Band representative has dismissed the allegations of Indian breaches of the Agreement as 'nothing but a pack of lies'.

Perhaps a more convincing rationale for the current Indian fishing activity on the Restigouche is that the Indians strongly hold that their aboriginal and treaty fishing rights take or ought to take precedence over federal and provincial fisheries laws. A 1982 brief of the Union of New Brunswick Indians entitled 'Indian Fishery for the the 1980s' states as follows:

> Indian people are aware that fishing is a treaty and a traditional right. They are aware that regulations and laws were developed to take away those rights and that the courts and law enforcement are there to ensure no Indian violates the White man's laws. ...

> Indian people are seen, to being punished (sic) for something they have not control over, construction of dams, pollution of the waterways, over-fishing by commercial fishermen, acid rain, forest development, to the point of being shot or imprisoned and harrassed, this is ironic as Indian people have been generally accepted to be the true conservationists.

Similar sentiments were conveyed by the Restigouche Band to the Atlantic Salmon Advisory Board earlier this year:

> The passage of various regulations by federal and provincial governments governing the harvesting and management of wildlife resources early this century adversely affected the Restigouche Micmac in exercising their aboriginal right to hunt and fish as usual. ...

> The legislation passed particularly by provincial governments were and are contradictory to aboriginal rights confirmed in various treaties.

The views expressed above reflect a deep-seated alienation from the current regime of fisheries management and a sense that existing laws are enforced to benefit non-aboriginal fisheries to the detriment of aboriginal fisheries in violation of aboriginal and treaty rights.

Discussion of the specific rights of the Restigouche Indian Band to fish for Atlantic salmon must begin with an overview of the rights of the Indian, Inuit and Metis peoples to hunt and fish in Canada. At the outset, it must be recognised that such rights at common law and under treaties signed with the Crown have received clear judicial recognition and acceptance by the federal government and most, if not all, provincial governments.

543

Further, the existing aboriginal and treaty rights of the aboriginal peoples of Canada are recognised and affirmed in the Constitution of Canada, under S. 35 of the Constitution Act, 1982. Constitutional conferences involving provincial and federal governments and aboriginal organisations to negotiate the constitutional entrenchment of rights of aboriginal peoples, particularly the right to self-government, were held in 1983, 1984 and 1985 with a final conference to be held in 1987.

This section explores the Restigouche Indian claims to aboriginal and treaty fishing rights, not to establish conclusively their status and scope at law, but rather to determine the plausibility of their rights claim as a moral if not legal basis for disregarding Canadian fisheries laws.

Aboriginal rights

At common law, the rights of aboriginal peoples to possess or use lands that they traditionally occupied have been described as usufructury and personal in nature, as burdens upon the Crown's underlying title to those lands and as rights that must only be surrendered to the Crown. Although the content of aboriginal rights has not been clearly defined, there is no doubt that such rights include the right to fish and hunt on traditional hunting grounds.

The rights of aboriginal peoples to fish and hunt on lands that have not been surrendered by them to the Crown either explicitly under treaty, or implicitly by operation of law has been recognised by Canadian courts in such cases as Calder v. Attorney-General of British Columbia and Baker Lake v. Minister of Indian and Northern Affairs.

The Royal Proclamation of 1763, which continues to operate in parts of Québec possibly including the north bank of the Restigouche River where the Indian reserve is situated, recognises that aboriginal peoples have legal possession of their traditional lands at common law and that the proper process of acquisition is by surrender of title to, and purchase by, the Crown.

Apparently, no pre- or post-Confederation treaties extinguishing the rights of the Restigouche Band to fish in Québec or New Brunswick have been signed. The Restigouche Reserve was created by order-in-council of the province of Canada in 1853, which followed an 1851 statute authorising the setting apart of lands for the use of certain Indian tribes in Lower Canada (Anon., 1851). That this legislative mode of reserve creation was employed by the province of Canada rather than extinguishment of aboriginal title under treaty indicates that the Band has at least some grounds for claiming a continuing aboriginal right to fish in the Restigouche River.

The Indian Atlantic Salmon Fishery on the Restigouche River

A key objective of the Assembly of First Nations (the national organisation representing most Canadian Indians) is to obtain the recognition of the federal and provincial governments that self-government is an inherent aboriginal right of Indian peoples. This concept of self-government as an aboriginal right has not yet achieved judicial approval in Canada. However, the federal government indicated its willingness to further the self-government of Indian Nations in 1984, and more recently declared its approval for constitutional entrenchment of the self-government principle.

Indian Nations exercising rights of self-government could have authority to legislate over education and family relations, revenue-raising, economic and commercial development, justice and law enforcement and most importantly land and resource use. Presumably, powers over resource use might allow Bands such as the Restigouche to control access to traditional fishing waters in the Restigouche, and to manage this fishery as they see fit. The recognition of the principle of Indian self-government as an aboriginal right by federal and provincial governments could then lead to the establishment of numerous independent Indian fisheries management authorities on or near the Restigouche, Miramichi, St John or other Atlantic salmon rivers that have traditional Indian fisheries.

Treaty rights

The Indian Nations of Canada have signed numerous treaties that have extinguished, limited or expanded their rights to fish and hunt on or occupy their traditional lands. Two treaties relevant to the Restigouche area are the friendship treaties of 1752 and 1779.

These treaties were intended to secure the neutrality or assistance of the Indian Nations and to ensure that they are not disturbed in their traditional activities. The Treaty of 1752 between the Governor of Nova Scotia or Acadia and the Micmac tribes of the region provides that:

> 4. It is agreed that the said Tribe of Indians shall not be hindered from, but have free liberty of Hunting and Fishing as usual and that if they shall think a Truckhouse needful at the River Chibenaccadie or any other place of their resort they shall have the same built and proper Merchandize, lodged therein, to be Exchanged for what the Indians shall have to dispose of and that in the meantime said Indians shall have free liberty to bring for Sale to Halifax or any other Settlement within this Province, Skins, feathers, fowl, fish or any other thing they shall have to sell, where

> they shall have liberty to dispose thereof to the best
> Advantage. (Dougherty, 1983).

Although it is not certain that this treaty applies to the Indians of the Restigouche area, the treaty clearly recognises the right to sell fish - a right currently denied to them under the fisheries laws of Québec and New Brunswick - as well as to catch them.

The Treaty of 1779 entered into by the Micmac Indians of Nova Scotia (and New Brunswick) from Cape Tormentine to Chaleur Bay, including the Restigouche Tribe, provides:

> That the said Indians and their Constituents shall remain in
> the Districts before mentioned Quiet and Free from any
> molestation of any of His Majesty's Troops or other his
> good Subjects in their Hunting and Fishing. (Dougherty,
> 1983).

Neither of these treaties involved surrenders of land to the English Crown nor the establishment of reserves for the various Indian tribes as was the case for more recent treaties.

To my knowledge, the Restigouche Micmac Indians have never agreed to their aboriginal rights being extinguished under the Royal Proclamation or their rights under the two treaties mentioned, such as the right to fish for salmon in the Restigouche River. In a 1985 decision of the Supreme Court of Canada, Simon v. The Queen, Chief Justice Dickson, speaking for the Court, wrote:

> Given the serious and far-reaching consequences of a
> finding that a treaty right has been extinguished, it seems
> appropriate to demand strict proof of the fact of
> extinguishment in each case where the issue arises.

Chief Justice Dickson went on to hold that the Treaty of 1752 was validly created, never terminated, and cannot be limited by provincial laws. This decision confirms earlier cases that determined that fishing and hunting rights provided in treaties cannot be limited or abrogated by provincial legislation such as fish and game laws. Federal legislation such as the Fisheries Act, on the other hand, does apply so as to limit the aboriginal or treaty fishing rights of Indians. However, there is now judicial support for the proposition that Indian Band bylaws properly enacted under the Indian Act take precedence to the provisions of the Fisheries Act on Indian reserves (R. v. Barnaby).

Further, the fishing rights under the Treaty of 1752 may be existing treaty rights within the meaning of S.35(1) of the Constitution Act, 1982. If such fishing rights have been so elevated to constitutional status, they may now enjoy priority over the federal Fisheries Act.

The issue of the priority of treaty rights over legislation is of considerable interest, particularly because the Treaty of 1752 recognises the right of the Micmacs to sell fish they have caught. Section 70 of the Québec Wildlife Conservation and Development Act, 1983, prohibits the purchase or sale of fish species, such as Atlantic salmon, that are listed under regulations to the Act. From the earlier discussion, a conclusion that this section does not apply to Indians covered by the Treaty of 1752 is probably warranted.

To summarise, there is considerable judicial support for claims by the Restigouche Indians that their aboriginal and treaty rights to fish for salmon have never been surrendered and enjoy priority over fisheries laws. For this reason - and of course there may be others - many Indians feel little obligation to abide by the various provincial and federal fisheries laws.

Statutory rights

Under the federal Indian Act, 1870, Indian Bands are empowered to enact bylaws to regulate certain local affairs on reserves including the conservation and utilisation of wildlife on reserves. Section 81 of the Indian Act provides, in part, that:

> S.81. The council of a band may make bylaws not inconsistent with the Act or with any regulations made by the Governor in Council or the Minister for any or all of the following purposes, namely:
>
> (o) the preservation, protection and management of fur-bearing animals, fish and other game on the reserve.

Indian Bands in British Columbia and New Brunswick have enacted bylaws under S.81 of the Act to control and manage the harvesting of salmon migrating up rivers adjacent to the reserves of these Bands on the assumption that these river waters are part of their reserves. Under subsection 82.(2) of the Indian Act, the Minister of Indian and Northern Affairs has authority to disallow such bylaws, but they enter into force if not disallowed within 40 days of receipt. If allowed to enter into force, the decision in R. v. Barnaby indicates that such bylaws take precedence over the Fisheries Act and regulations.

The four bylaws recently passed by the British Columbia bands were not disallowed by the Minister of Indian and Northern Affairs, but are currently subject to an injunction granted to the Attorney-General of British Columbia. The Bands are appealing that injunction. The crucial legal question is whether or not such

Band bylaws apply to waters of rivers, such as the Restigouche River, that are adjacent to Indian reserves, such as the Restigouche Reserve. If the reserve boundary terminates at the high-water mark of the river, then presumably Band bylaws cannot govern Atlantic salmon fishing on the river because the river bed is not 'on the reserve'.

No general answer to this issue is possible, because the legal description of the boundaries of a reserve are particular to that reserve, and may or may not include the beds of rivercourses that border the reserve. However, at common law the beds of navigable waters below the high water mark were generally not granted by the Crown. In the case of the Restigouche River, an 1857 survey indicates that the reserve does not include the bed of the Restigouche.

However, the salmon fishing bylaws of at least one Band, the Kingsclear Band on the St John River in New Brunswick, have not been disallowed by the Minister of Indian and Northern Affairs, and are currently being enforced by the Band on the river. Some of these bylaws contradict the Fisheries Act and regulations under the Act, and the issue as to whether or not the reserve includes the St John River is currently being prosecuted in a New Brunswick provincial court.

The purpose in describing in some detail the aboriginal, treaty and statutory rights of Canadian aboriginal peoples such as the Restigouche Micmac Indians has not been to circumscribe definitively the scope of these rights or their impacts on fisheries management, but rather to demonstrate the complexities of the legal issues relating to Indian fishing rights.

However, conclusions can be drawn. The Restigouche Band has defensible if not powerful claims under the Royal Proclamation and the Treaties of 1752 and 1779 to priority fishing rights. The Treaty of 1752 provides a right to sell or barter salmon, which right appears to have priority over provincial, if not federal fisheries legislation, and therefore threatens legislated prohibitions of commercial sales of salmon by Indians. Indian Bands, such as the Restigouche, may also have the right (as an aboriginal or statutory right under Section S.81(o) of the Indian Act) to manage Atlantic salmon stocks that spawn in rivers bordering on or adjacent to their reserves.

AN INDIAN ATLANTIC SALMON FISHERY POLICY

As argued above, the issue of Indian exploitation of Atlantic salmon in the Restigouche River has foundered over Indian assertions that their aboriginal and treaty rights take precedence over 'the white man's fisheries laws and demands by non-Indian fishers that the Restigouche Band fishery must be regulated under law as are recreational and commercial fisheries.

The thrust of the previous section is that the scope of Indian rights to fish for salmon is not at all clear and furthermore is in legal and political flux. The constitutional and legal issues are diverse and complex, as our sampling of some of the issues affecting one Indian Band, the Restigouche, has shown. The resolution of these issues at law could take decades. The Atlantic salmon of the Restigouche River may not endure that long.

Further, it is unlikely that even another 'salmon war' such as the 1981 Québec Provincial Police raids on the Restigouche Band fishery would significantly reduce the level of Indian fishing, or induce greater compliance with the Fisheries Act and regulations. Imposed solutions, through prosecutions or additional regulatory action, are unlikely to alter the belief of Restigouche and other Indians that these laws are enacted to benefit non-Indian fishers to the detriment of Indian fishers, in violation of the their aboriginal and treaty rights. Increasing the number of MLCP conservation officers enforcing fisheries laws will not work nor will laying more charges against Indian 'poachers'.

In the long term, a new policy for the Indian Atlantic salmon fishery, developed by the federal Department of Fisheries and Oceans is required if the Restigouche salmon is to be restored to its former abundance and the conflicts resolved. The policy must establish regulations for the conduct of Indian community fisheries, reiterate priorities for allocations among Indian and non-Indian fisheries, and identify processes for the verification and enforcement of harvest levels.

The process of policy development must be of negotiation with the Bands, primarily, but with other stakeholders such as the federal Department of Indian and Northern Affairs, Québec and the Atlantic provinces, and recreational and commercial fisheries as well. This is a formidable task, given the level of mistrust among the different groups. However, the time may be right for such a policy insofar as the Union of New Brunswick Indians has recently spoken out publicly in favour of an Indian fishery policy.

It is crucial that the policy has general applicability to Indian Atlantic salmon fisheries throughout eastern Canada, and possibly should include Indian Pacific salmon and other aboriginal fisheries. Allocations to individual Indian Bands would then be made, preferably by agreement, as part of overall salmon management plans, such as the five-year Atlantic Salmon Management Plan.

Clarification of DFO's mandate to negotiate on behalf of the federal government is desirable before the process gets underway. The Department should have full authority to manage and administer all Atlantic salmon fisheries. The most serious challenge to this authority are Indian Band bylaws validly enacted under S.81(o) of the Indian Act, which as indicated currently enjoy priority over the Fisheries Act. DFO must rely on the Minister of Indian and Northern Affairs to disallow Band bylaws that establish independent fisheries management regimes on reserves, a task he is often reluctant to undertake. To overcome this problem, the federal cabinet should enact a regulation under sections 81 and 73 of the Indian Act, requiring that Indian Band bylaws concerning fisheries be made subject to the Fisheries Act.

In the Restigouche, the federal government should consider revoking the current power of the Québec government to administer the Atlantic salmon fisheries under the Québec Fishery Regulations of the Fisheries Act, at least insofar as the Restigouche River is concerned. The content of the Québec regulations as well as the vigour of their enforcement differs from the New Brunswick regulations applicable to the Restigouche river, even though they apply to the same population of Restigouche salmon. The powerlessness of conservation officers in New Brunswick to act against repeated infractions of fisheries laws they have observed on the Québec side has created frustration and heightened the tension.

A second reason for revoking the Québec government's Fisheries Act powers is to neutralise a dispute about the position of the Québec - New Brunswick interprovincial boundary in the vicinity of the Restigouche Reserve. The dispute between the governments of Québec and New Brunswick is important because a fishing ground, used by the Band, lies in the disputed territory. If DFO administered both the Québec and New Brunswick regulations, the dispute would be academic. To reiterate, a single Indian Atlantic salmon fisheries policy developed and implemented by a single agency - the Department of Fisheries and Oceans - is essential to the conservation of the Atlantic salmon.

The Restigouche Band fishery illustrates the key concerns that must be addressed in preparing an Indian Atlantic salmon fishery policy acceptable to all salmon fishery stakeholders. Despite the frustrations about the Restigouche Agreement, it does represent progress. By agreeing to quotas and a system of tagging for all salmon caught under the Agreement, the Restigouche Band at least implicitly accepts the principle that their take must be regulated. Another very positive feature of the Restigouche Agreement is that Indians are involved at a variety of levels (enforcement, research, liaison with MLCP) in the management of Atlantic salmon stocks.

If a workable Indian fishery policy is to be developed, common ground among governments, and Indian and non-Indian fishers must be found and conflicts between the fundamental interests of the stakeholders avoided. To conclude this paper, the basic elements of such a policy are identified.

Elements of an Indian Atlantic salmon fishery policy

An Indian Atlantic salmon fishery policy must incorporate several elements crucial to the public interest and the interests of Indian people. These are to:

(1)   conserve and enhance Atlantic salmon populations in the Restigouche and other rivers of eastern Canada;

(2)   respect the aboriginal and treaty rights of Indian people to fish for Atlantic salmon;

(3)   increase the economic benefits derived from the fishery that Indian people of Atlantic Canada enjoy; and,

(4)   encourage respect for and compliance with the fisheries laws of Canada.

Conservation and enhancement of Atlantic salmon

The conservation of stocks is the paramount objective. Given the precarious state of Restigouche and other Atlantic salmon stocks, an essential element of a policy must be to regulate the harvest of all Indian fishers under allocated quotas set by the relevant senior government agency charged with the administration of the policy in a particular region (i.e. DFO, or MLCP) in consultation with individual Bands. To ensure adequate spawning escapement, the first allocation of the administering agency must be to the resource (the conservation allocation) which takes priority over all other allocations. If and when surpluses to spawning escapement needs are available, then allocations are made first to the Indian community fishery, and then other fisheries.

The Assembly of First Nations has claimed an aboriginal right of Indian peoples to manage their fisheries, while other Indian organisations such as the Restigouche Band have advocated comanagement of fisheries resources. An independent right of Indian Nations to manage anadromous fisheries is not desirable from a conservation standpoint. If Indian Bands acquire the political authority and physical capability to set quotas and harvest salmon independently of DFO and without regard to the interests of other users, a 'tragedy of the commons' would likely result.

The murky concept of comanagement offers greater promise. As elucidated in a joint policy proposal developed by the federal Ministers of Fisheries and Oceans and Indian and Northern Affairs for the British Columbia Indian salmon fishery, comanagement refers to a process wherein Indian Bands are involved in fisheries management processes, such as the setting of allocations and the development of community fishery plans for specific rivers. Within the context of the overall plans, Bands are entitled, through bylaws, to develop fishing plans for their reserve and determine how the allocation for their reserve is to be used (e.g. subsistence, recreational), and suballocated among individual Band members.

The principle of comanagement could provide a workable element in an Indian fisheries policy, so long as DFO retains responsibility for the overall management of Atlantic salmon fisheries, and conservation of stocks is the paramount consideration.

Aboriginal and treaty rights

An Indian Atlantic salmon fishery policy must reflect, if not recognise, the existing aboriginal and treaty rights of Indians such as the Restigouche Band. The current DFO policy and the Québec and New Brunswick regulations under the Fisheries Act provide for the issuance of permits to Indian subsistence fisheries, presumably on the basis that Indians are entitled, whether at common law or under treaty, to harvest fish for their personal consumption and that of their families, but not to treat the harvested fish as commercial commodities. At least with respect to the Treaty of 1752, this presumption appears to be incorrect. The treaty clearly provides a right to sell and barter fish, and this right should be reflected in the policy, albeit subject to regulation.

The policy must legitimise the Indian Atlantic salmon fishery, and the first step is to discard the bureaucratic fiction of licensing Indian subsistence fisheries in favour of Indian community fishery licences. Under community fishery licences, the Band could decide how to use its salmon allocation to best economic advantage, whether through a subsistence, recreational or commercial fishery.

Economic benefits of the salmon fishery

The crucial objective of an Indian Atlantic salmon fishery policy must be to maximise the economic benefits to Indians, given a certain allocation of Atlantic salmon. Although the current Restigouche Indian fishery provides significant economic benefits to the Band, there is little doubt that these economic benefits could be increased if the Band's allocation was utilised by a Band-operated recreational rather than a licensed commercial fishery. Comparisons of economic benefits of non-Indian recreational and commercial salmon fisheries are definitive in concluding that the former delivers economic benefits on a per fish landed basis at least several times that of the latter.

An Indian recreational fishery could provide permanent jobs as guides, outfitters and lodge employees to members of the Restigouche reserve, which has an unemployment rate of about 95 per cent. Unfortunately, Atlantic salmon apparently do not respond well to the artificial flies of anglers in the brackish, tidal waters of the Restigouche near the reserve. The development of a Band-operated recreational fishing industry might, then, depend on negotiated leases or purchases of riparian lands from non-Indian recreational fishing interests further upstream.

The licensing of Indian community fisheries permits a shift in focus by Bands to economic development away from food-gathering. The Restigouche and other Indians of 'settled' eastern Canada do not depend on Atlantic salmon for subsistence nearly as much as aboriginal people in remote regions depend on local fish and wildlife. The key issue for these Indian Bands is economic development, as they themselves are perfectly aware. The subsistence fishery has been an indirect, controversial but ultimately unsatisfactory means of achieving that economic development. If the concept of Indian community fisheries is adopted as an element of an Indian Atlantic salmon fisheries policy, the door is open to encouraging Indian Bands to focus on recreational fishing as a means to achieving the economic development they seek. Organisations such as the Atlantic Salmon Federation could play an important role in providing financial, technical and training support for development of Band-operated recreational fisheries, substitution of trap for gill nets, and feasibility studies of salmon aquaculture on reserves.

Respect for fisheries laws

A final objective of an Indian Atlantic salmon fishery policy is to gain respect for federal and provincial laws. The present fiction of a licensed Indian food fishery, tacitly accepted by MLCP as a commercial fishery, encourages contempt for fisheries laws among Indians and non-Indians. An Indian fishery policy must as a minimal requirement have the virtue of honesty and not continue the present charade. But all the values of the policy and fisheries laws must be recognised and believed to be fair and reasonable by the Restigouche Micmacs and other Indians who fish for salmon.

This element of the policy must provide for verification of catch levels by an agency not subject to political influence by Indians or non-Indians and for even-handed enforcement of breaches of quotas and fisheries laws. Draconian and arbitrary enforcement of laws perceived to be unjust as violating aboriginal rights cannot succeed. However, some fisheries law enforcement, in concert with a fair Indian fishery policy, solidifies respect for the law and is a continuing demonstration of public concern for fisheries conservation.

Other concerns

An Indian Atlantic salmon fishery policy would provide a framework for the rehabilitation of Atlantic salmon stocks in the Restigouche and other rivers. The policy would allow fisheries managers to focus on the key issue of allocation, and certainly no one can expect that even with the best possible policy, disagreements among Indian, recreational and commercial fishers over allocations will end.

The development of an Indian community salmon fishery that allows regulated commercial sales of Atlantic salmon would likely meet with resistance from non-Indian commercial fishers who currently hold licences but are not permitted to fish. Non-Indian commercial and recreational fishers must clearly understand that the reasons for the priority allocation to Indian fishers is based on aboriginal and treaty rights not on the allocation principle that the highest and best economic use of the resource should prevail. Given that recreational fisheries are demonstrably a better economic use of wild Atlantic salmon than commercial fisheries, mandatory buyouts of the licences of local commercial fishers may be desirable.

A final point. Gainsayers of an Indian Atlantic salmon fishery policy might forcefully argue that the Restigouche Band presently has nothing to gain from such a policy. It is certainly unlikely that allocations to the Band in the near future would exceed its 1986 salmon catch. This catch probably provided economic benefits in the hundreds of thousands of dollars in addition to over $300,000 per year in payments to the Band under the Restigouche Agreement. But the better arguments are that the Band does have much to gain from an Indian fishery policy. Many Restigouche Band members probably cherish their reputation as good conservationists and thus must be somewhat disconcerted by the adverse media publicity.

In addition, there are Band members such as the auxiliary and conservation agents, helpers, and watchmen who profit directly from the Restigouche Agreement, and thus have an economic interest in ensuring that quotas under such agreements are respected. If violations of the Agreement continue, MLCP simply may not renew it.

A crucial point is that if the Restigouche Band fishery is not controlled over the long term, stocks may be irretrievably devastated and opportunities for a thriving fishery-based economy on the reserve lost.

But as an extra sweetener, perhaps the policy should provide affected Indian Bands with a fixed percentage of permitted salmon harvest, which percentage would not change from year to year. If the Restigouche Band was allotted, say, a 20 per cent share of the total allocation in any given year, the Band's allowable catch would rise if Restigouche salmon populations increased and fall if they decreased. Such a mechanism would provide a good economic incentive for the Band to practise conservation.

CONCLUSION

All of the considerable conservation efforts of governments and non-government organisations and the sacrifices of recreational and commercial fishers could be wasted if the Indian Atlantic salmon fishery on the Restigouche and other eastern Canadian rivers is not brought under control. To achieve this goal, Indian rights to fish for salmon must be recognised, their economic development requirements accommodated, the complicated salmon fisheries management regime overseen by the Department of Fisheries and Oceans streamlined, respect for fisheries laws improved, and the legitimate concerns of other fisheries taken into account.

The task is enormous and the disposition of the interested fisheries increasingly unfavourable to conflict resolution. Much goodwill has already been lost. In a final analysis, whether or not one describes the Restigouche fishery as illegal fishing or the exercise of legitimate Indian fishing rights is a matter of perspective, and the question assists not at all in protecting Atlantic salmon. The liberal western democracies, including Canada, must rely on voluntary acceptance of, and compliance with, laws by the vast majority of citizens, if tyranny is to be avoided. Fisheries laws are no different than others. In urging Indian people to respect Canadian fisheries laws that have often ignored or trampled on aboriginal and treaty rights, the only acceptable path is negotiation and reasoned argument.

REFERENCES

Anon. (1851) An Act to authorise the setting apart of lands for the use of certain Indian tribes in Lower Canada. Statutes of the Province of Canada c.106. 14 and 15 Victoria

Anon. (1981) International Atlantic Salmon Foundation, Coalition for the control of Indian Fisheries. The Indian Food Fishery for Atlantic Salmon, p.3

Anon. (1983) Atlantic Salmon Task Group. A Report to the Assistant Deputy Minister Atlantic Fisheries Service from the Atlantic Salmon Task Group of Canada. Summary Memorandum of Dr B. Muir, Task Group Chairman

Anon. (1986a) B.C. Indians warn of fishing violence. The Globe and Mail, 30 June, p.A5

Anon. (1986b) Canadian Press, 5 June; The Evening Telegram, 13 June; Siddon says Conne River decision will enhance salmon conservation, p.1

Anon. (1986c) ASF seeking injunction to halt Indian salmon fishing. Atlantic Salmon Federation News Release, 20 June

Beamish, F.W.H., Healey, P.J. and Griggs, D. (1986) Atlantic Salmon Task Force Review. August 1978. Biological conservation Subcommittee report prepared for the Atlantic Salmon Review Task Force. Appendix B, p.28

Dougherty, W.E. (1983) Treaty of 1752 - Treaty or Articles of Peace and Friendship Renewed in Maritime Indian Treaties in Historical Perspective. Treaties and Historical Research Centre, Department of Indian and Northern Affairs, Canada. Appendix 1, pp. 84-5

Moisan, G. (1979) The Amerindians and Hunting and Fishing Activities - Government Policy Proposal (translation). Secrétariat des activités governmentales en mileu amérindien et inuit (SAGMAI), Governement du Québec, pp. 29-31

Parfitt, B. (1986) 18 Natives face charges in Fraser confrontation. The Vancouver Sun, p. A10, 29 August

Chapter Thirty

ILLEGAL NET FISHING FOR SALMON IN NORWAY

Sven Mehli,
Directorate for Protection and Management of Nature, Norway

INTRODUCTION

Recent research work has shown heavy exploitation of Norwegian salmon stocks. This is due partly to the situation in Norwegian home waters where fishing with drift nets, bag nets and bend nets in the sea are the most important gear. Apart from the fact that this exploitation is heavy, most of the catch is also taken in mixed stock fisheries, which we consider a biologically incorrect way of exploitation which does not allow for proper management of the population. This problem is especially marked with the drift net fishery. We have also noticed increasing pressure on the salmon stocks in the rivers, caused not only by fishing but also by eutrophication, water power development and, last but not least, the fluke, Gyrodactylus salaris. This fluke causes the death of all salmon parr in an infected river. The fluke is now present in 28 rivers, some of which were once among the best salmon rivers in Norway.

In Norway, we have a fast-growing cage culture industry for salmon, producing about 40,000 tonnes in 1986, more than 20 times the production of wild salmon. The price of this farmed salmon has fallen about 30 per cent in the past year. Taking into account this large quantity of farmed salmon and other factors that have an influence on the wild stocks, one might expect less interest in fishing for wild salmon. So far, however, there has been no evidence of such decline in interest in salmon fishing either by commercial fishermen or by general-interest groups using various methods of fishing. This raises the question of illegal fishing, where in Norway we have great practical problems concerning the control of salmon fishing because of the large number of salmon rivers, the great distances involved and a sparse human population. Not only nets are used for illegal fishing, but in certain areas of the sea we have experienced the use of long-line fishing, which is not allowed in the Norwegian salmon fishery.

THE FISHERY

'Net fishing in Norway' means any salmon fishing with commercial equipment in Norwegian waters. The main types of equipment here are drift nets, long nets and bend nets. Long nets and bend nets are classified as 'fixed nets' and the right to use them belongs to the landowner. This special fishing right may be leased out. In principle, any number of bag nets and bend nets may be used, but there must be a minimum distance of 100 m between each net. As well as fixed nets, set nets and seine nets were used to fish for salmon, but these were banned in 1979.

The use of drift nets requires a special licence, the number of which for the period 1986-88 is 580. Individuals are granted licences which also apply to a particular boat and they can lose their licence for repeated illegal fishing. The number of nets that may be used under each licence varies according to the size of the boat, i.e. the number of fishermen. A one-man boat is allowed 20 nets, boats with two men can use 35 nets and a boat with three or more men, 50 nets. Drift nets can only be used between the baseline and 12 nautical miles offshore. The use of drift nets is prohibited in the areas east of North Cape and from Utsira to the Swedish border.

The bag net fishing season starts on 15 May and bend nets and drift nets start on 1 June. The salmon sea fishing season lasts until 5 August. The different dates of opening bag net and bend net fishing creates problems in supervision. Our Salmon Act gives no precise definition of bend nets and bag nets. As the former are cheaper and easier to handle, almost every day we see modified bend nets which look like bag nets. The fishermen with modified bend nets then start fishing on 15 May instead of 1 June. The first days of the fishing season are considered very valuable for catching the larger salmon.

There is a weekly close season from 1800 h on Friday to 1800 h on Monday.

As will be seen later, the use of salmon nets gives different control problems. The drift netters may cross the baselines, use a boat without being registered, fish without a licence, and the users of fixed nets may not observe the weekly close period. Also, there are different types of nets which are designed to catch marine fish which can also be used to catch salmon. The most important of these are those for catching mackerel, Scomber scombrus L. There is commercial mackerel fishing along the Norwegian coast up to 62° North, but that type of fishing is unlikely to cause any problems for the salmon. In the fjords, however, we have an increasing amount of leisure fishing for mackerel where both drift nets and set nets are used. Mackerel nets are allowed a smaller mesh size than salmon nets and can more or less 'incidentally' catch certain quantities of salmon.

THE CONTROL OF SALMON FISHING

The Directorate for Nature Management is responsible for the control of all activities concerning salmon. The total budget for this work during recent years has been 5.5 MU NKR or about £500,000 per annum. The work is organised into four main categories.

(1)   The Central Aircraft Patrol, used mainly for controlling the drift net fishing. They employ twin-engined aircraft and often work in cooperation with patrol boats along the coast from Utsira in the Stavanger area to the North Cape.

(2)   The Local Aircraft Patrol. The coast is divided into seven areas. Each of these areas is allocated one single-engined sea-plane. In northern Norway, a helicopter has also been used during bad weather, with great success. The local chief of police organises the work and the main task is to control the weekly close time of long nets and bend nets.

(3)   Patrol Boats. The Directorate has eight high-powered launches between 21 ft and 37 ft long. They are used mainly to control drift net fishing, often working with the aircraft. The boats patrol for 500-800 hours each during the season.

(4)   In the different regions, about 350 bailiffs are engaged on a part-time basis. They cover the fjord areas and the rivers, and concentrate on the closing of fixed nets, illegal net fishing, especially in river mouths, and the use and placing of mackerel nets.

The work is organised and carried out in cooperation with the Directorate and local police chiefs in 43 districts. In addition, the Directorate works closely with the Norwegian Coast Guard. They mainly oversee drift net fishing.

NETS WITH INCIDENTAL SALMON CATCH

Since 1981 we have noticed an increase in the use of monofilament set nets and drift nets for fishing mackerel. This is particularly evident in the fjords where there are river outfalls. The mesh size of mackerel nets is between 37 mm and 45 mm knot to knot. The minimum mesh size for salmon nets in Norway is 58 mm knot to knot. Everyone can use mackerel nets for recreational fishing without limitations on number of nets and time of use.

To obtain information about the catching efficiency of mackerel nets for salmon and grilse, we carried out some preliminary experiments in certain areas on the west coast of Norway in 1984 and 1985. The results are given in Table 30.1.

Table 30.1: The catch of different groups of fish by set nets and drift nets in different localities. Mesh size from 35-58 mm. The figures are given in % of total catch (kg)

Type of net	Locality	Mesh size (mm)	Mackerel	Other marine fish	Salmon	Sea trout
Set net	Fjord	35-45	27	30	40	3
Drift net	Baseline	40	69	30	1	0
Drift net	Fjord	40	34	0	44	22
Drift net	Fjord	37-44	37	24	37	3
Drift net	Baseline	58	0	0	100	0

The use of mackerel nets is most pronounced in the areas where the experimental grilse population is dominant. From te experimental fishing given in Table 30.1, we observed that the average size of grilse from the different stations varied between 1.2 and 2.1 kg. Salmon larger than 2 kg were rarely caught in the mackerel nets.

In order to see how effective mackerel nets are in catching salmon we can compare the catch with the registered catch for typical salmon nets. From the statistics given by the Central Bureau of Statistics we find that the salmon catch on single bend nets in the same areas are 1.2-1.3 kg of salmon each day. The use of mackerel set nets shows a salmon catch in the range 0.2-2.4 kg per net per day, with an average of 1.6 kg per net per day. For 1984 the result was 0.6 kg per net per day. On this basis the mackerel set nets may catch at least as many salmon as a single bend net.

In these experiments, the location of the mackerel nets was not limited to places where the salmon catch had previously been good. A local net-fisherman can use his knowledge of migration times and routes of salmon and methods of setting nets to increase his illegal salmon catch by mackerel nets well above the amounts noted above.

In our opinion, it is the recreational use of mackerel nets in the fjords that causes the problems. This is because commercial mackerel fishing is on the whole not very profitable in the fjord areas. In view of this we could introduce special measures for the recreational fishing of mackerel in order to decrease their salmon catch. Lowering the nets from the surface could be one solution. In the investigation carried out in 1984 and

1985, the depths at which different types of fish were caught were noted. The average catch depth for mackerel was 2.0 m $\pm$ 0.7 m ($\underline{n}$ = 162), with the greatest catch depth being 3.5 m. For sea trout the average catch depth was 1.3 m $\pm$ 0.7 m ($\underline{n}$ = 33).

A special working group has been set up to discuss the problem of salmon caught in mackerel nets. This group has proposed that the mackerel nets used for recreational purposes should be lowered 3 m from the surface during the summer period, not including commercial mackerel nets. The Norwegian Government has gone further with this and has now, as part of the new measures for the salmon fishery, decided to secure the necessary legislative background for the lowering of mackerel nets.

The observed net mark frequency on grilse populations has been high; up to 80-90 per cent of the fish have net marks. We have reason to believe that the use of mackerel nets is a factor in this respect.

RESULTS

The result of the supervision activities may be difficult to measure accurately. Firstly, effective supervision may deter people from illegal fishing. If the chance of being discovered is high, they may be wary of using illegal methods. This may be the situation in some districts, but is not necessarily the case.

In the summer of 1986 we sent a questionnaire to 43 local police chiefs, asking for the number of reports to the police and the number and size of the fines concerning salmon fishing for the years 1981-1985. The results, given in Table 30.2, show a

Table 30.2: The number of reports to the police and the subsequent number of fines concerning salmon fishing in Norway in the years 1981-1985

	1981	1982	1983	1984	1985
Reports	338	427	419	468	511
Fines	155	117	132	121	122
Fines, % of reports	46	27	32	26	24

slight increase in the number of reports to the police chiefs for the years 1981-1985 but the number of fines, did not increase. In fact, there was a gradual reduction in the percentage of reports resulting in fines. There are various reasons for this. One is that the police do not have the personnel to give priority to such reports. Another is that the reports are increasingly of illegal set nets. These nets are not marked with the owner's name and address.

A number of replies to the questionnaire show that illegal fishing for salmon with nets has increased in recent years. This trend seems to have continued in 1986. To illustrate this, we can examine reports from one of the most important salmon areas in Norway, where for the period 1981-1985 the number of reported cases were 16, 29, 62, 54 and 93, respectively.

The level of fines has increased in recent years from 500 NKR to 5,000 NKR (£150 to £500). In addition to fines, catches and nets are regularly confiscated, and in some instances, boats.

CONCLUSIONS

It may be that there has been a slight increase in illegal fishing for salmon in recent years. The extent of this, however, is uncertain. The main reasons for the increase are ill-defined and inadequate regulations. The Norwegian Government has now decided to bring in new measures relating to salmon fishing. The most important are a total ban on the drift net fishery, a licensing system for fixed nets and a ban on the use of monofilament nets. The new measures, combined with a new Salmon Act, include regulations that will give better possibilities for supervision in the future. It is also important that control is strengthened by more resources and that organisation is tightened.

In spite of doing the best we can, we may still see some irregularities. We have to accept this and bear in mind that the main task of the supervision service must be prevention. Through this, the service can create great respect for the measures which have been taken.

Closing Address

STEWARDS OF THE SALMON

Richard A. Buck, Chairman, Restoration of Atlantic Salmon in America, Inc.

The wild Atlantic salmon is a shared resource. Shared by the producing nations with the primary responsibility of managing the stocks, and by the nations which only harvest the fish, accepting this as reasonable recompense for permitting the salmon to forage in the waters of their continental shelfs.

Ten years ago, the Atlantic salmon-producing nations had just secured a phasing out of the destructive high seas salmon fishery off West Greenland. Then came the extension of fisheries jurisdiction for all species generally to two hundred miles. And now the salmon begin to come under further control as we set regulatory measures under the new salmon treaty. Jobs well done. And the future can look bright indeed, but there are a number of 'Ifs'.

Man has the ability to produce salmon artificially in any desired numbers, but he has not shown an ability to hold his selfish instincts in check - with his nets in the ocean. Thus, it is not in the scientific area, but in the socio-economic and geopolitical arena that the struggle to save the wild salmon must be engaged. How to divide up any harvestable surplus over and above the spawning requirements to perpetuate the species is the question that today plagues the governments of both the producing and user nations. These nations now realise that a continuation of the present overexploitation can lead only to exhaustion of the species, as being of any practical use to man. We are on a slippery slope, indeed, with the world catch in home waters off some 40 per cent in twenty years.

Whether or not these fish will be restored to their former abundance - and we should require nothing less - depends upon, first, a complete understanding of the nature of the interdependence of world stocks, and second, our willingness to reduce to a minimum ocean interceptions of salmon of foreign origin. Perhaps this message begins to get through, because, salmon-wise, a new wind sweeps over the North Atlantic today -

a clearing wind from the West, carrying with it the seeds of a new spirit of conservation. We want to see it become the prevailing wind, sweeping away the dense fog which envelops the Northeastern waters of this great ocean which becomes ever more small, more interdependent. Today most nations are enthusiastically engaged in conserving stocks and restoring rivers - exciting ventures. We may have turned the corner towards important restoration. Where drastic measures have been introduced, salmon runs are increasing. Where not, there is no progress - and, occasionally, chaos.

Let's take a kaleidoscopic view of what the salmon nations bordering the North Atlantic Ocean have done recently to conserve in-river stocks and reduce ocean interceptions. These are the basic objectives of the new North Atlantic Salmon Conservation Organisation (NASCO), the operating mechanism of the new multi-national treaty to which we are all signatory.

In the United States, regulations permit the taking of salmon by angling only; salmon taken in US waters may not be sold or offered for sale; angling is permitted only in rivers where restoration programmes approach maturity. In Canada, the in-river angling catch limits have been drastically reduced; drift netting has been banned; inshore commercial netting banned entirely in certain areas, and seasons cut back in others; a government netting buy-back programme has been in force; for Newfoundland, under a NASCO agreement, Canada has agreed to terminate the autumn commercial fishery as of 15 October, designed to reduce interceptions of US stocks.

For the interceptory fishery at Greenland, Denmark, since the treaty became operative three years ago, has accepted two reductions in quota under NASCO agreements. The Faroe Islands, under pressures from within NASCO, has come to two separate cut-back agreements within the EEC. Iceland continues its exemplary example; drift netting is not permitted, neither is commercial in-river netting, with the exception of several small traditional stations. In Norway, the government, influenced by steep declines in stocks and also undoubtedly by progress in NASCO, has announced that, from January 1989, drift netting for salmon will no longer be permitted; inshore netting seasons are to be shortened from January 1988, and the use of monofilament nets prohibited. And in France, efforts are underway in a number of conservation and restoration projects in rivers capable of supporting runs of salmon. Spain, like the United States at the tail end of a migration and intercepting chain, has dedicated its salmon fishery entirely to angling.

Russia has now become signatory to the treaty, inasmuch as its salmon often frequent the North Atlantic Ocean, migrating around the North Cape to feed in the Norwegian Sea. Like the United States, Russia has never permitted fishing for salmon in the ocean, and permits the taking of salmon in rivers only by rod and line.

Now we come to the United Kingdom and the Republic of Ireland, for both of whom the European Economic Community negotiates under NASCO. Sad to say, in both these nations which produce around one-half of the world supply, there has not been major forward movement. Drift netting is condoned and illegal operations continue unabated. In the UK, the only governmental action consists of a new Salmon Bill before Parliament, which as recently structured provides no important reductions in catch, commercial or for angling. It does tighten up current law. But the indiscriminate drift netting off Northumberland is not banned, nor are drastic measures taken to reduce substantially the inshore and in-river catch. The long downward trend in catch of salmon stocks demands a more urgent response than these timid proposals tabled in Parliament.

Is it possible that the fisheries bureaucracies in the EEC-UK-Irish complex do not fully comprehend that the other NASCO members have already begun the long and arduous task of reversing the trend towards overexploitation? Do they realise their obligations to other signatories to the salmon treaty? Does the recalcitrance of this triumvirate in adopting modern principles of sound management have any serious effect on further progress under NASCO? The answer is definitely 'Yes'. Denmark, for instance, has indicated that it is not disposed to make further concessions until there is 'fair-sharing of the burden of conservation'. And this, in fact, could kick up quite a storm.

There is no question as to the direction in which Denmark points the finger. So what might be the answer of these original States of Origin such as the UK, for instance, to the question 'What are you going to do about all this?' It might very well be that the Fisheries Ministers would point out that NASCO is an area of the EEC's responsibility, and that it would be inappropriate to predict what the EEC approach might be. Are we to believe that these salmon-producing nations can so easily walk away from problems and responsibilities, hiding forever behind the cloak of community representation?

And what might be the EEC's answer to the same question? It might be to the effect that all is as it should be. There is legislation in the producing States of Origin with respect to conservation, restoration and enhancement; money is being spent on management measures, and there is continuous reappraisal of their effectiveness. In other words, the claim is made that salmon conservation has been for centuries a successfully realised objective of the Member States concerned.

If indeed all this can be considered a likely scenario, and if that great bard William Shakespeare were put down in the salmon world of today, he might even be tempted to exclaim once again

- 'A plague on both your houses'.

All this 'Let-George-Do-It' business, appears clearly to be in contravention of the intent of Article 9(c) of the NASCO Treaty. It states that 'in exercising its functions, a Commission shall take into account the efforts of States of Origin to implement and enforce measures for the conservation, restoration, enhancement and rational management of salmon stocks in their rivers and areas of fisheries jurisdiction.'

True conservation is not fossilised, not static. The challenge has been accepted by most signatories. What is needed now is for these others to cooperate in designing and embarking on a long-term plan for the future enhancement of salmon stocks throughout the whole of their area of competence, including strict interim regulatory measures on an emergency basis wherever necessary.

All this is not to say that Denmark's territories (Greenland and the Faroe Islands), which harvest but do not produce wild salmon, are entitled to annual catch levels which currently equal about one-quarter of the world catch in the home waters of the producing nations. How much is too much?

Now as to NASCO itself - the over-arching bridge of mutual support and management regulations for the benefit of all. Those of us who played a part in conceiving, designing and negotiating this new treaty did not do so with the thought of perpetuating an administrative status quo. There was the vision to perceive that arresting and reversing the decline in the world stocks requires the adoption of new directions and the acceptance of new disciplines. We look to the Council and Commissions of NASCO to develop creative new ideas. Thus any lack of progress under the treaty would be due to people, not the process. Actually NASCO has already established itself on a sound basis. Never before, between all the Atlantic salmon-oriented nations, have we had dialogue - that useful lubrication that comes from personal association and exchanges among negotiators. Conservation measures have been adopted. Dramatic progress will always appear to be too slow, yet important breakthroughs will be achieved from time to time. It was ever thus with international institutional arrangements, and particularly so with fisheries treaties. Examine the more important of these fisheries treaties, and it will be realised that NASCO ranks high with respect to forward progress in early years.

Looking to the future, what can be the role of research in achieving our objectives? The crying need here today is to be able to determine just what is the normal harvestable surplus of salmon stocks, and, when we have that answer, a means must be found to divide up that surplus on a rational, reliable and equitable basis. A Working Group should be set up in NASCO to study and recommend regulatory measures which give promise of accomplishing this objective - 'fair-sharing of the burden of conservation'. The various contributions by states of origin and host states would be identified and quantified. An end to pulling tonnage catch figures out of a hat and attempting to negotiate an acceptable compromise. There is considerable support for such an effort. We also need scientific studies on such priority matters for instance as: smolt mortality; the detailing of migration routes to show where stocks separate and to identify overwintering areas; the evaluation of net mesh sizes as a management tool; and the effects of varying levels of harvest at the feeding grounds of salmon.

We also must be constantly on guard to ensure that science is not used for the purpose of furthering existing political positions among nations. Science and biology are not a substitute for rational management - indeed they are the servants of rational management. Too often a lack of complete evidence is used as an excuse for delaying rational management decisions. Thus it often becomes necessary to take prudent reasonable action based upon the best scientific information available at the time.

So, here we are, bound together by a dependence on each other, yet split apart by the overlapping of jurisdictions and conflicting objectives of management and harvest. There simply will have to be less for all and more sacrifice by each, until we have the fish back in abundance. The challenge is, as usual in any common undertaking, for each to do his part. Some are accepting; some hold back.

In conclusion, one final point should be made - philosophical and ideological. The wild Atlantic salmon is the most valuable fish that swims - aesthetically, recreationally, nutritionally, economically, and its continued overexploitation demands that we address a basic question. Even though a producing nation may have 'primary responsibility' for the salmon throughout its range in the ocean, such is not an exclusionary right. So then who does own the salmon? The answer lies in another question. Who does not own these salmon? These fish have been around for centuries. They have served the imagination of men and women during all that time. Yet it is only in the last hundred years that they have been 'discovered', that is, increasingly exploited commercially and recreationally. It is clear that, whatever can be said about who owns them, the wild Atlantic salmon do not belong to the present generations alone. We are the stewards of the salmon, holding the resource in trust. As trustees do we then have the right within a few decades to deplete it in such a way that future generations can never turn to it to be enriched, ennobled and sustained?

The opinions expressed herein are personal ones, and do not necessarily reflect the official position of the three United States Commissioners to NASCO, of which the author is one.

SUMMARY AND RECOMMENDATIONS

Derek Mills,
Department of Forestry and Natural Resources, University of
Edinburgh, Scotland

Time and again throughout this Symposium we have heard from various speakers of the concern they have over the reliability of catch data. The future of any fishery is dependent on the knowledge of its stocks, much of which has to be built up from data on catch figures, recruitment estimates and fishing effort levels. The problems of catch data collection were particularly well highlighted by Graeme Harris who referred to change in methods for data collection and how it was found that in some instances the true catch was 300-400 per cent greater than the declared catch. I am sure a number of us can recall instances which revealed inaccurate records being submitted to the appropriate authorities. However, I cannot agree with his remarks that the annual catch figures he depicted for Wales are entirely meaningless. Even if they are underestimates, they will still be of some value in indicating years of scarcity and abundance. The series of catches over long periods described by Geoff Power and Alex Bielak confirm the value of even incomplete records in revealing significant trends. One should not be surprised to find marked fluctuations in fish populations over time, after all the cyclical abundance of animal populations has been recorded by animal ecologists like Charles Elton, while similar phenomena have been recorded for marine fish populations in the excellent treatise Climate and Fisheries by Cushing. In past years the cyclical abundance of Atlantic salmon has been recorded also by Berg, Huntsman and Dunbar who related it to climatic factors. It was therefore of no surprise when Dave Reddin showed that salmon production in the northwest Atlantic was related to environmental conditions during the winter.

Ideally, of course, we should demand the most complete data we can obtain for stock assessment and Willie Shearer referred to the need for information on effort and, indeed, knowledge of availability of fish during the fishing season when assessing effort. It was most revealing to see that a significant proportion of the run of fish into the North Esk came <u>after</u> the netting season, so that a knowledge of the stock based on the commercial catch would be most misleading. The North Esk is not unique and a number of other salmon rivers in Scotland and elsewhere, also experience this phenomenon. When one knows how detailed data on fishing effort are for marine fisheries and how carefully such data are collected, it makes our own attempts at recording effort seem rather poor. We also need better estimates of mortality at certain stages in the salmon's life cycle and no one seems to have included these in their assessment of spawning escapement, exploitation rates, yes, but natural mortality from disease and predation, no, although Martin Ventura revealed the extent of the recent loss of adults from disease (probably UDN) in Spanish rivers.

The only realistic method of assessing individual stocks of anadromous fish is in the river with traps and counters, as Gareth Edwards pointed out so forcibly. Neither method is foolproof, but we have heard from Willie Shearer, Patrick Prouzet and Thor Gudjonsson that traps and counters can produce invaluable data on the numbers and seasonal distribution of fish and exploitation rates. Gareth Edwards' remarks about anglers welcoming the installation of traps on their rivers reminds me of my own early experience of traps and anglers some 30 years ago which makes me say that the anglers may welcome traps provided they are not on <u>their</u> rivers. So many restrictions were placed on the way we operated our traps or counting fences and handled the fish that the operation was almost valueless. We could not handle the fish in case it deterred them subsequently from taking the anglers' lures and our traps were always accused of delaying their ascent. On one river, the Axe, in southern England a magnificent trap, operated by scientists of the Ministry of Agriculture, Fisheries and Food, had to be removed because of the objections of anglers upstream of the traps. Those scientists that can operate counting installations with impunity should consider themselves lucky. Unfortunately, there seems to be little financial input to the further development and wider use of counters and much valuable money is being diverted elsewhere in the salmon world, and the important aspects of stock assessment and stock-recruitment relationships tend to be ignored by administrators and actions taken, such as the closure of net fisheries, without due regard to these. As Richard Buck just said 'Too often a lack of complete evidence is used as an excuse for delaying rational management decisions. Thus it often becomes necessary to take prudent reasonable action based upon the best scientific information available at the time.'

NASCO, through the ICES Working Group on North Atlantic Salmon, is now collating catch data. This year a new catch category has appeared in the statistics for the first time, namely 'unreported catch', which is given as 3,000 tons, a vague and perhaps underestimated 'guesstimate' I would suggest. This 'unreported catch' refers to the size of the illegal fisheries off where? - Scotland and Ireland, or does it also include the undeclared catch of the northeast England drift-net fishery, the illegally caught fish in Scottish, Welsh and Canadian rivers? To whatever it refers we should support NASCO in its endeavour to achieve comparability in the collection of catch statistics.

How can one overcome these weaknesses in the collection of catch statistics? There are at the moment in eastern Canada, including Newfoundland, and Spain and shortly in France, as we have heard, salmon-tagging schemes which make it an offence to possess any salmon which has not been tagged. As each tag has a serial number it means that every fish caught, whether by net or rod, is accounted for, although as Stephen Hazell has told us not all the indigenous human population of his country abides by this scheme. But still it is possibly as foolproof as it can be and, after all, as long as there are banks there will be bank robbers - all one can do is to make it as difficult as possible to rob the bank. One can, of course, reduce the size of the unreported catch by banning the method by which this catch is chiefly made and which produces the greatest source of error - namely enmeshing nets such as fixed hang nets and drift nets - excepting the fishery off Greenland. An MEP on the EEC sub-committee on fisheries who is the rapporteur for the Draft Report on the Protection and Management of Salmon Stocks surprisingly recommends the continued use of monofilament drift nets but with stricter control, giving as her reason that it enables salmon fishing to be carried out effectively and safely. I am sure there are few present who would agree with this, although the word control does provide us with some ray of hope. Not only does this form of fishing produce a significant loss of fish through 'drop-out' from the nets after death, but results in loss of fish that escape from the nets through damage and subsequent disease and/or predation and, as Lars Hansen pointed out, results in size selection, and maybe even sex selection, both of which could cause a genetical change in the stocks. The chances of an increase in salmon ranching would, I think, be considerable were drift netting and the use of mono- and multi-filament to be banned, and we should follow the example of Norway which, as Sven Mehli told us, is taking firm action in this direction. Such action would tend to direct us, as HRH, the Prince of Wales, so forcibly told us in his address at the Second International Symposium, 'towards a sensible alternative to the hunting of salmon at sea'.

The main problem to the effective control of illegal fishing is surveillance. Scotland has managed to contain her illegal drift net fishery by surveillance from aircraft and helicopters in addition to surface craft. However, land-based inspection is not so easy and sometimes downright hazardous, as it is in even more conventional commercial fisheries in eastern Canada, if my past experience of the Canadian lobster fishery in the Maritime Provinces is anything to go by and, indeed, from what Stephen Hazell has told us about the Indian fishery on the Restigouche. Our before-mentioned rapporteur Ms Joyce Quinn of the European Parliament believes that Community funds should be made available to improve and co-ordinate national salmon fisheries inspectorates, here I am sure we would agree with her.

At this stage, I think it is appropriate to propose the first three Recommendations:

Recommendation 1:

That NASCO investigates the value of a salmon-tagging scheme such as is in operation in eastern Canada and Spain, with the view that it recommends its adoption by all member countries for both a more reliable collection of catch data and a more effective control of illegal fishing.

Recommendation 2:

That all methods of enmeshing salmon, such as drift nets and fixed hang nets (excepting the operations off Greenland), should be phased out and the fishermen relying on these methods be given other opportunities to participate in the salmon fisheries or related industries including salmon farming and ranching.

Recommendation 3:

That we should support the recommendation of the Rapporteur to the EEC Sub-committee on Fisheries that Community funds should be made available to improve and co-ordinate national salmon fisheries inspectorates.

Some discussion has been centred on the future of commercial salmon fishing - some would like to see it phased out completely, others would like to see it reduced. We must realise that the salmon is a resource that must be harvested wisely and therefore harvesting rates should be adjusted to cater for periods of scarcity and abundance to provide adequate escapement for spawning. This can be achieved in one way, as Dave Reddin said, by a more flexible regulation of commercial fishing based on more accurate forecasts from stock-recruitment relationships. Acquisition of these takes time and we are not yet in the enviable position of the International Pacific Salmon Fisheries

Commission who can promote bye-laws at the drop of a hat to stop the fishing of specific runs of fish as they come within the influence of the Fraser River. At the moment, various measures are being taken in different countries to reduce commercial fishing from mandatory closures in the Maritime Provinces, reductions in the numbers of licences compulsorily or voluntarily through purchase, reduction in the length of netting seasons as in Newfoundland - these were listed fully by Larry Marshall and discussed by Jack Fenety. One problem in achieving unanimity on a decision on the future of commercial fisheries for salmon involves the question of ownership and this varies widely from country to country as we have heard from Robert Williamson and Graeme Harris among others, and nowhere would it be more difficult to resolve this problem than Scotland where a large proportion of the coastal fisheries belong to the Crown. It says much for the recently established Atlantic Salmon Conservation Trust (Scotland) that it is purchasing those commercial fishings over which the Crown has relinquished its hold, not to curtail commercial fishing for ever and a day, but until such time as the Trust considers that the salmon stocks have once again reached a commercially exploitable level.

To phase out all netting in some countries would result in unemployment and few governments are likely to want that, but there is a feeling that a reduction in commercial harvesting could be beneficial to stocks. If, however, there was alternative employment which fishermen could be offered, the chance of a reduction in commercial harvesting being acceptable to the industry might be greater. However, the owners of commercial fisheries have to be considered and substantial financial compensation might be necessary. It has been said frequently in the last two or three years that with the rapid escalation of salmon farming the proportion of wild salmon on the market is markedly less. That is an obvious statement and quite irrelevant to the argument that, as a result, commercial fishing for salmon should be curtailed. If the commercial fisherman wishes to take the risk of obtaining less for his harvest that is his business. So compensation would be necessary if commercial fishing was reduced compulsorily. After all, if a farmer decides it is no longer profitable to grow one crop or rear certain animals, he turns to another measure or as a last resort, sells his farm; but if his land is taken for road construction he expects compensation. The difficulties are enormous and, as Stephen Hazell rightly said about this sort of situation, the only acceptable path is negotiation and reasoned argument. However, in the case of certain commercial fisheries it should be possible to direct the fishermen into related industries such as salmon farming and salmon ranching. Ken Whelan of the Irish Central Fisheries Board suggested, in the Board's film on salmon management, that this might go a little way to reducing the Irish drift-net fishery.

The type of ownership of salmon fishing rights in Iceland, where the farmers or landowners own the fishery, is one we do not all possess. As the farmers can obtain more for their fishing rights by letting it for angling than for commercial harvesting it was easy for the Icelandic government to ban coastal commercial fishing and allow netting in lakes and glacial rivers only. This has led to a salmon angler's paradise. However, paradise does not come cheaply as anyone who has fished in that country is aware. Salmon angling generates both considerable revenue and employment and there is no doubt that a resource exploited for sport fishing only is a situation that many would prefer, and if this occurred probably some commercial fishermen would be employed as wardens to combat the resulting increase in poaching. Those of you from Canada will remember how this escalated during the commercial fishing ban in New Brunswick in the 1970s - it was no picnic for fishery officers who went about their duties in fear of their lives and had to be trained in the use of firearms.

What are the advantages and disadvantages of a 'sport-fishing-only' exploitation of salmon stocks - underexploitation? Perhaps, although, according to Thor Gudjónsson, the rate of exploitation by anglers on Icelandic rivers can be very high indeed, and in Scotland, as Willie Shearer pointed out, at certain times of the year, especially in the spring, the rods on some rivers (i.e. Aberdeenshire Dee, Spey and Tweed) catch more fish than the nets. Increased poaching in the rivers and illegal fishing at sea would most certainly occur. So, too, would the demand for the improved fishing, resulting in increases in the price of licences and rents. Increased catches and the absence of funds arising from commercial fishing rates would lead to increased rate assessments for sport fisheries (this happened on the Tweed this year, where an agreed late start to the netting season led to higher rate assessments for the rod fisheries). Lastly, as HRH the Prince of Wales warned us at the Second Symposium, angling could become too commercialised. There is already some evidence for this on some of our Scottish rivers, and we have heard from some speakers of the misgivings they have about 'time-sharing'. To overcome this I feel there is some need, following in Canadian footsteps, to have some form of bag limit to prohibit anglers from selling their catch and to encourage the 'catch-and-release' philosophy mentioned by Jack Fenety which is commonly adopted by North American and some European anglers. It has been pointed out that there could be problems of released fish dying from damage inflicted inadvertently by anglers and this should be investigated. This leads me to proposing a fourth Recommendation:

<u>Recommendation 4:</u>

> Following the action taken by Canada, resolve that each
> nation, through its salmon conservation bodies and
> sport-fishing organisations, encourage the introduction of a
> reasonable daily and season rod-catch quota, prohibit the
> sale of their catch and consider the adoption of a
> 'catch-and-release' philosophy.

So far I have said little about research and therefore I
would like to refer to Svend Horsted's thought-provoking paper on
this subject. Most government salmon research is discussed at the
North Atlantic Salmon Working Group at ICES from whom
NASCO seeks information. However, I would like to think that
Svend's dream of an extensive survey of salmon-feeding areas in
the North Atlantic might come true. Research vessels are of
course costly animals to keep at sea and it costs £6,700 to keep
one ocean-going research vessel at sea for one day, but I would
have thought that there was always the possibility of chartering
commercial vessels.

Salmon fishing and ranching has been mentioned in several
contexts over the last three days. Its future impact on wild
salmon exploitation must not be ignored. We are warned of the
genetic repercussions that interbreeding between 'farmed' and
'wild' salmon may have on their offspring and the gene pool, but
what will be the cumulative effects on salmon stocks of the
increasing numbers of escapees from farms which is put at 4-5
per cent of the farm stock in Norway? When one considers that
the annual production of farmed salmon in Norway is predicted to
rise in the not-too-distant future to between 100,000 and 125,000
tonnes it can be seen that 4-5 per cent of escapees means
between 4,000 and 6,200 tonnes of 'farmed salmon' swimming
around the ocean, plus what escapes from Scottish salmon farms.
It is no wonder that about 8 per cent of the salmon taken in the
Faroese fishery are of hatchery origin. These escapees, after
feeding in the ocean in competition with 'wild' salmon, have to
return to some river or other to spawn and, if there are no nets,
what happens? Next year at ICES it is planned to discuss this
topic of the inter-relationships of wild, farmed and ranched
salmon and this could be most valuable.

Perhaps rather surprisingly, the effects of man on the
salmon's freshwater environment have received little attention at
this Symposium. Sven Mehli mentioned the problems of
acidification in Norway. This problem affects us all, some eastern
Canadian salmon rivers have been severely damaged and there is
also evidence of a decline in salmonid stocks in southwest
Scotland and mid-Wales as a result of acidification. French and
Spanish rivers seem to have been particularly badly affected by

hydro-electric schemes and pollution and both these countries will no doubt be pressing for more effective legislation in this direction.

Fifteen nations are represented here today. We all have our problems, some are common to all of us while others are only the concern of one or two. However, we are all united in one aim and that is the conservation of salmon. As Allen Peterson, head of the US delegation to NASCO and its Vice-President said at the opening Council meeting of NASCO this year - if we are to conserve the salmon we should respect each other's problems and endeavour to work together. With this aim in mind a small committee has, during the course of this Symposium, framed a Resolution for your consideration. I now call on Monsieur Higgins, the President of the Association Internationale de Defénse du Saumon Atlantique, to propose this Resolution.

THE RESOLUTION

It was resolved that:

In view of the greater income and employment potential of salmon angling and its appreciably smaller harvest of limited salmon populations, each national government of salmon-producing countries is urged to declare a salmon policy which will institute as a conservation measure, within its area of jurisdiction, management programmes to reduce commercial harvesting of salmon with a view to increasing salmon stocks and improving recreational salmon fisheries.

The Resolution was seconded by Mr David Clarke, Chairman of the Atlantic Salmon Trust.

After comments from the assembly the Resolution was passed unanimously. However, some of those participating at the Symposium who were government or inter-government employees were unable to associate themselves with this Resolution.

There was considerable discussion of the recommendations which were welcomed widely.

The Symposium then closed after a Vote of Thanks from Mr T. Wills of the Wessex Water Authority.

INDEX

Aberdeen Harbour Board 244
abundance forecast 336-7
acidification 143, 575
acoustic observations 519
acoustic tags 463, 473-80,
    519, 521
adult tagging programmes
    446, 461, 469, 500-1
adult transfers 354-5, 407-8
aerial counts 329
Aeromonas hydrophila 222
afforestation 112
age at return (see sea-age
    groups)
aircraft patrols, 559, 571
air-surface temperatures
    485-6
Altamira Grotto 210
Altnahinch 349-50
American Shad (Alosa
    sapidissima) 415-17, 419
Amerindiens 402
amphicycline 222
amphipods 506
Anadromous Fish Conserva-
    tion Act (1965) 415
Angerolles 392
angling lures and baits 85,
    172-3, 196, 198, 206
angling offences 531
Arctic charr (Salvelinus
    alpinus) 149, 162, 376
arterial drainage 233

artificial redds 349
Assembly of First Nations
    545, 551
Association Internationale
    de Défense du Saumon
    Atlantique (AIDSA) 32, 576
Association pour la Protection
    du Saumon 32
Atlantic eel (Anguilla
    anguilla) 149, 162
Atlantic Salmon Advisory
    Board 22, 543
Atlantic Salmon Conservation
    Trust (Scotland) 98, 573
Atlantic Salmon Federation
    23-4, 514, 535, 553
Atlantic Salmon Management
    Plan, 1984 538-40
Atlantic Salmon Task Group
    327
Atlantic Salmon Trust 577

bag limits 76, 134, 139-40,
    564, 574-5
bag nets 93, 95-6, 146, 154,
    174, 231, 256, 557-8
Barracudina (Paralepis
    coregonides borealis) 506
Basle 38, 45
Bay of Biscay 520
Bay of Fundy 380
Bec d'Allier 389-90
Belleville 389-90

Bellows Falls Dam 417-18
bend nets 146, 154-5, 557-8
Berkshire National Fish
      Hatchery 420
Berwick Salmon Fisheries
      Company 241
Bethel 420
Black bass (Micropterus) 203
black salmon (see kelts)
Black smelt 506
Bleach 182
Blois 389
Boswell, Joseph Knight 403
Brioude 391-2
British Columbia 535
Brook trout (Salvelinus
      fontinalia) 276, 355
Brown trout (Salmo trutta)
      149, 162, 203, 374, 376
buy-back of licences 23, 126,
      138-9, 554, 564

cage-rearing (see salmon
      farming)
Canada Fisheries Act, 1920 24
Canadian Atlantic Fisheries
      Scientific Advisory
      Committee 138
Cape Farewell 438, 441, 496,
      500
Capelin (Mallotus villosus)
      506
carrying capacity 326, 374
Cartier, Jacques 400
catch-and-release 23, 63, 434,
      574-5
catch-per-unit effort 327,
      322-3, 469-71, 497-8,
      515-16
catch quotas of salmon 9-10,
      15, 19-20, 128, 258, 441-2,
      444-5, 451, 458, 538, 541
catch reminders 81
Central Fisheries Board 573
Centre d'Études du
      Machinisme Agricole, du
      Génie Rural, des Eaux et
      des Forêts 31-2, 38, 62

Centre National d'Exploita-
      tion des Oceans 32
Cerzat 391
Chilac 391
Chinou 389
chloramphenicol 222
chlorbutol 149
Chum salmon (Oncorhynchus
      keta) 375-6, 382
close seasons 74-6, 94-5, 138,
      146, 231, 256-7, 373, 558
closure of fisheries 139-40
coastal drift nets exploita-
      tion rates 295, 364
cod 440, 445, 450, 506
cod jigs 516
Coho salmon (Oncorhynchus
      kisutch) 203
Comité Central des Pêches
      Maritimes 32, 63
Comité Interprofessionel des
      Poissons d'Estuaires 32,
      36, 63
commercial fishery closures
      130, 139-40
Common Fisheries Policy 12
Common Law 434-5, 544, 548,
      552
Connecticut Department of
      Environmental Protection
      418
Connecticut River Atlantic
      Salmon Commission 416-
      17, 423-4
Conseil Supérieur de la
      Pêche 31-2, 36, 62-3
Constitution Act, 1867 539
Constitution Act, 1982 544,
      546
control of fishing effort 95,
      97
control on sale of salmon 88,
      232
coracle net fishing 74
corgones 210
cormorants 527-8
costs of stocking and restor-
      ation (see revenue)

cotos 183
counters (see fish counters)
counting fences 310, 312, 331
Crown Estate 92
Cunner (Tautogolabrus
 adspersus) 276

daily distribution of rod
 catches 194-6
daily water discharges and
 sport catch of salmon 313
damage to fish by nets 151,
 154, 157
Dampiere-en-Burly 389
Debane, Pierre 22-3
densities of fry and parr 279-
 83, 347-8, 392
Department of Fisheries and
 Oceans 539, 549, 555
Department of Indian and
 Northern Affairs 549
Départment de la Nievre et
 du Cher 397
Dery Gorge 403
Directorate for Nature
 Management 559
discriminant function
 analysis 513, 517, 520
disease (see UDN)
Disko Bay 442
Disko Island 438, 441
Domesday Book 236
Donnacona 400, 402, 409, 411
draft nets (see net and coble
 fishing)
drift ice 442
drift nets 74-5, 84-95, 93-4,
 124, 130, 146, 154, 230,
 232, 288, 516, 525, 557-8,
 562, 565, 571-4
Dunalassstair Dam 377, 379

Economic and Social Research
 Institute 231
education 410-11
egg deposition 293, 297, 301,
 310, 315, 321
Electricité de France 390

Electricity Supply Board 230
enforcement procedures 232
euphausids 506
European Economic Comm-
 unity 12-20, 61, 66-7, 444,
 565, 571-2
exploitation rates 147-8,
 151-2, 158, 169-77, 174,
 264, 269-70, 272, 295,
 364, 452, 455
explosives 182

farm slurry 233
Faroese fishery 9-10, 15, 18-
 20, 458-81
fecundity 286, 293, 326-7
Federal Energy Regulatory
 Commission 417-18
Federal Power Act 417
fin-clipping 36, 406, 462
fines 562
fish counters 78, 169-72,
 175, 256, 266-9, 330-2
fish farm escapees 575
fish passes 56-9, 61, 170-2,
 181, 330, 390-1, 408-9,
 417-18
Fisheries Act 539-40, 546-7,
 549-50, 552
fishing close times 74-6, 94-5,
 138-46, 231, 256-7, 373,
 558
fishing effort 138-9
fishing equipment budget 51-2
fishing mortality rates 311
Floy tags 263, 284
Fly nets (see stake nets)
food of salmon 506
foul-hooking 526, 531
Foyle Fisheries Commission
 524-5, 528, 530, 533
Fraser, John 22-3
free zones (see zonas libras)
freeze-branding 35-6
Freshwater Fisheries Laboratory, Pitlochry 461
Friendship Treaty, 1752 545,
 552

Fry survival 291, 293, 347-8
Fylla Bank 450

Galway 284-6, 288
gaffs 531
gene pool 406, 420
genetic differentiation 374-6,
    382, 419, 575
genetic diversity 381
genotypes 375-6, 382
gentamycin 222
gill nets (see drift nets)
Grand Bank 494, 496-7, 499,
    504, 506
gravel extraction 390-1,
    397, 528
grilse error 260
Grotte de la Riera 210
Groupe Permanent de
    Concertation 62
Gulf of Finland 381
Gyrodactylus salaris 143, 557

haaf netting 75
hatcheries 53, 345, 406-7,
    419, 420, 422
haulbacks 463
Herring (Clupea harengus) 506
holding facilities 420
Holyoke Dam 417, 419, 421
Holyoke Water Power
    Company 418
Hunter Committee 98
hydro-electric projects 186,
    188, 418

ice cover 450
ICES (International Council
    for the Exploration of the
    Sea) 6, 8, 66, 138, 440,
    461, 513-14, 518, 575
ICES Working Group 513-17,
    570, 575
ICNAF (International
    Commission for Northwest
    Atlantic Fisheries) 9,
    21-2

IFREMER (Institut Francais
    de Recherche pour
    l'Exploration de la Mer)
    32, 36, 62
Île aux Coudres 406
Île d'Orleans 406
illegal fishing prosecutions
    530
illegal salmon fishing 14, 74,
    77-8, 88, 232, 524-34, 557-
    62, 574
index rivers 88
Indian Act, 1870 547-8, 550
indian bands 535-6, 538, 540,
    542-3, 545-7, 549, 551,
    553-4
indian fishery 123-4, 535-54
indian reserves 535-6
Institute for the Conservation
    of Wildlife 183
Institut National de la
    Recherche Agronomique
    31-2, 62
Institut Scientifique et
    Technique des Pêches
    Maritimes 32
International Pacific Salmon
    Fisheries Commission
    572-3
International Salmon Tagging
    Experiment 513, 518, 520
invernizos 189
Irminger Sea 483, 485, 494,
    497, 509, 518, 520
Issoire 391

'Jack' salmon 319
Jacques Cartier National Park
    409
Jacques Cartier River Fishing
    Club 402
Joseph Johnston & Sons 264

Kamehatka 375
kelts 121, 135, 137, 498
kelt reconditioning 410
Kinnaker Mill fish trap 262,
    266

Labrador Sea 485, 492, 496-7, 499-501, 504, 506, 518, 520
Lake Azabachye 376
lake-rearing of smolts 355-6
lamping 527
landing sheets 462
Langeac 391
Langogne 392
Laurentian Park 405
lave netting 74
Law of the Sea Conference 2, 13
Leesville Dam 417-18
licences by-catch 130, 232
    commercial 73-5, 81, 83-4, 117, 124, 126-8, 138-9, 231-2, 431-2, 445, 458, 462, 525-7, 540, 558
    rod 75-6, 83-4, 122, 138, 146, 423, 525
lift nets 146
Loch Rannoch 377-83
log books 462, 541
Logie fish counter 266-9
long-line fishery 458-60, 513, 515-16, 518, 557
long nets 558
losses to home water stocks 17-18, 452, 455, 459
Lough Corrib 277, 279, 293
Lough Erne 374
Lough Feeagh 333
Lough Foyle 527
Lough Mask 277
Lough Melvin 377
Lough Neagh 377

Mackerel (Scomber scombrus) 558
mackerel nets 558-61
Mactaquac Fish Culture Station 499
magnesium cyanide (cymag) 527
Marché d'Intérêt National 48
marinas 398

mark and recapture 264, 277, 332, 334, 452
mature female parr 201
mesh size of drift nets 444
Micmac Indians 545-6, 548, 553
micro-tags 277, 284-5, 287-9, 356, 364, 462, 518
migration (smolts) 285, 290, 294, 308, 392-3
migration (adults) 269, 393-4, 499-504, 501, 503-6
migration routes 501-6
Ministry of Agriculture, Fisheries and Food 433
Ministère du Loisir, de la Chasse et de la Pêche 402, 540-2, 549-50
Montrose 261
Morphie Dyke 261, 269

NAFO statistical divisions 439, 441-2, 444
NASCO 1-10, 13, 15, 18-20, 21, 25, 26-7, 66, 444, 461, 517, 564-7, 570-2, 575-6
National Association of Bailiffs 182
National Service for Freshwater Fisheries and Game (1960-72) 182
native fishery (see Indian fishery)
nature conservation volunteers 412
Naussac reservoir 392
net and coble fishing 93, 96-7, 111, 230, 257, 259, 261, 269, 286, 525
net damage (see damage to fish by nets)
Net Limitation Order 74, 84, 431-2
netting offences 530, 535, 542-3
Nettle, Richard 403
New England Power Company 418

Newfoundland commercial
fishery 22-5, 137-8, 564
non-catch mortality 515
non-observed catches 515
non-reported catches 260-1,
515
Norse Udal Law 92
North Cape 558-9
Northumberland drift-net
fishery 565
Norwegian Coastguard 559
Norwegian Sea 463, 467, 469,
483
Nuuk, Greenland 438, 450

observers 461, 515
ocean ranching 164, 571-3,
575
optimal egg densities 321
ova deposition (see egg
deposition)
ovarian development in
smolts 319, 321

palisade nets 394
panjet 263
parasites, (see Gyrodactylus)
parr (see densities)
patrol boats 232, 559, 572
peat silt 233
pelagic trawls 519
Perigord caves 236
poaching 77, 179, 182, 186,
232
poisoning 527
pollution 182, 225, 233, 392,
405, 419
Pont-du-Chateau 391
Pont-Rouge 403, 408-9, 411
Pontès 391
postas 179, 181
post-smolts 380
primary school participation
410-11
principal component analysis
486-8
programme administration
400, 402, 416-17

prohibition of sale of
illegally-caught fish 61,
88
pteropods 506
purchase of netting rights
97-8, 573
putts and putchers 74

Qaqortoq 438, 441
quotas (see catch quotas)

radio tracking (see tracking)
Rainbow Dam 417-18
Rainbow Trout (Salmo gairdneri) 149, 203
recreational fisheries 121-2,
130-5
redd counts 78, 170, 329, 395,
528
redd destruction 349
Regional Water Authorities
72, 88
repeat (previous) spawners
314-17
resistivity fish counter 266-9,
331-2
Restigouche Angling Club 246
return rates (see smolt return
rates)
revenue 13-14, 46, 48-52, 64,
67, 412, 421, 518, 541,
574-5
Reykjavik 169
rivers
    Canada
    Betsiamites 118
    Big Salmon 303, 310-11,
    314-15
    Cachee 409
    Cascapédia 118, 307-8,
    537
    Castors 313
    Chibenaccadie 545
    Conne 118, 132
    Coté 319
    Eagle 118, 132
    East 354
    Eel 538
    Epaule 409

Escoumins 118
Exploits 118, 132, 350, 363
Fraser 535, 573
Gander 132, 136
Genevieve 313
George 118, 133
Godabout 248
Grand Codroy 303, 314
Hayeur 319
Humber 132, 136
Jacques Cartier 400-13
Jupiter 118, 133
Kedgwick 536
Koksoak 118
Kouchibouguac 498
La Have 118, 135, 498
Little Codroy 302-3, 306, 311-12, 314, 319
Matane 118, 133
Matapédia 539-40
Middle 118
Mingan 118
Miramichi 124, 130, 135, 138, 158, 307-8, 334, 352, '376, 486, 538, 545
Moisie 118, 133
Natashquan 118
Olomane 118
Patapédia 536, 539
Pinware 313
Pollet 310, 352
Rattling Brook 354, 363
Restigouche 124, 130, 135-6, 244-48, 307-08, 535-55
Richibucto 118
River of Ponds 313
Robinson 303
Robitaille 319
St. Augustin 118
Saint John 124, 130, 136, 138, 307-08, 380, 486, 538, 545
St. Genevieve 118, 132, 313
St. Marguerite 245, 248
St. Mary's 118, 135

Sand Hill 303
Sautaraski 409
Shooner 318
Skeena 535
Torrent 355
Western Arm Brook 138, 302-3, 306, 308, 310-11, 316, 319, 355, 484, 486, 489-90, 492

England and Wales
Avon 69
Dee 69
Eden 69
Frome 332
Rheidol 79, 362
Severn 69, 74
Thames 362
Towy 354
Usk 69, 84, 353
Wye 69, 79, 84
Ystwyth 79

Finland
Teno 236

France
Adour 44
Allier 34, 329, 363, 389-99
Arques 34
Aulne 363, 396
Bresle 34, 53, 66
Calonne 34
Cher 390
Dordogne 34, 55, 362
Ellé 34
Elorn 34, 362
Garonne 34, 36, 38, 55
Gartempe 55, 390
Gave d'Oloron 329, 396
Léguer 34
Loire 34, 36, 44, 363, 389-99
Maine 390
Nivelle 34, 354, 363
Orne 34
Rhine 38, 45
Scorff 34, 362

Selune/Oir 34
Vienne 390

Greenland
Kapisillit (Kapisigdlit)
438, 501

Iceland
Blandá 167, 172, 175-6
Ellidaar 167, 169-70, 174,
176, 490, 492
Hólmsá 175
Hvítá 162, 166, 170-2, 177
Laxa 164
Nordurá 167, 170-2, 175-7
Olfusá-Hvítá 162, 166-7,
173, 176-7
Svarta 176
Thjorsa 176
Úlfarsá 167, 170, 176

Ireland
Abbert 282-3
Bealinabrack 279, 281
Black 282-3
Boyne 233, 276
Burrishoole 293
Bush 230, 293, 353, 528
Comamona 279-80, 283
Corrib 276-7, 284, 290-1,
293, 298
Failmore 279, 281
Foyle 352, 354, 524-33
Grange 279, 282-3
Moy 233
Owenriff 279-80, 283
Sinking 279, 282-3

Norway
Eira 147
Imsa 149, 151
Laerdal 147
Orsta 154
Vefsna 155

Scotland
Clyde 91

Dee (Aberdeenshire) 98,
244, 493, 574
North Esk 261-2, 264, 266,
269-72
Spey 98, 574
Tay 97
Tummel 256, 332
Tweed 94, 241, 574

Spain
Asón 183, 188, 190, 192,
194, 196, 201, 203, 206-07
Bedon 211
Besaya 188
Bidasoa 186
Canero (or Esva) 188,
211-12
Cares 183, 211
Deva 183, 190-1, 193-4,
196, 206
Duero 183
Eo 183, 211
Miera 188
Mino 188
Nansa 188, 190-1, 193-4,
196, 206
Narcea 183, 211
Naton 225
Navia 188, 203, 206, 211
Pas 183, 188, 190-2, 194,
196
Porcia 211
Sella 183, 188, 211-12,
225
Ulla 188

Sweden
Lainio 376
Torne 376

USA
Connecticut 362, 377,
415-25, 501
Farmington 417, 420
Penobscot 420
Salmon 417-18

USSR
Kalininka 375
Naiba 375, 383
Neva 381
Royal Proclamation Act, 1763
544, 546, 548

St. Etienne-du-Vigan 391
St. Laurent-des-Eaux 389
St. Marguerite Salmon Club
250
Salmon Commissions 258
salmon cruives 93
salmon dealer's licence 232
salmon farming 114, 233, 310,
380, 553, 557, 572-3, 575
Salmon Fishery Act, 1865 73
Salmon Fishery Boards 92-3
salmon fishery districts 92-3,
257-58
Salmon and Freshwater Fisheries Acts 1923 79
1951 257, 271
1972 80
1975 73, 431-2
Salmon Research Trust of
Ireland 86, 228, 230-2
salmon stamp 46, 50, 52
salmon stock enhancement 98,
345-88
salmon tracking 55, 473-80
sanctuaries 394, 408-09
sand lance 506
satellites 519
Scotian Shelf 506
Scottish east coast drift net
fishery 93-4
sea-age distribution 467-8,
480
sea-age groups in salmon
catches 380, 393-4, 396,
447, 449, 463, 467
sea surface temperatures
450-1, 467, 480, 483-6,
488, 490
Sea trout (Salmo trutta) 66,
69, 71-80, 83, 86, 561
seals 527

seasonal uses of baits and
lures 197-8
Second International Atlantic
Salmon Symposium 1, 27
seine nets, 74, 85 (see also
net and coble fishing)
seizure of illegally-caught
salmon 532
seizure of illegal fishing
equipment 529-30
selective fishing 158
serum transferrin poly-
morphism 513
shad restoration 417
shoaling 473, 480-1
Siddon, Tom 23
silage 233
Sisimiut, Greenland 438,
440-1
size limits 76, 540
smolt ages 285, 291, 392,
405, 448
smolt migration (rate of)
392
smolt production 280-4,
347-8, 351, 393
smolt release ponds 353, 419
smolt return rates 284, 287-9,
295, 302, 306, 311, 345,
361-2, 407, 421-3, 489
smolt sizes 317-20
smolt traps 284, 286, 288,
330, 377-9
smolting 377, 379
snatching 526, 531
Sockeye salmon (Oncorhynchus
nerka) 376
Sonor counters 332
Spanish Civil War 182
spawner-recruit ratio 490
spawners required 326-7, 344,
409-10
spawning potential 409-10
squid 506
stake nets 93, 95-6, 146,
256-7
stock-recruitment 85-6, 275,
325, 327, 351, 430, 572

stocking activities 356-61
stocking densities 351-3
stocking of eggs, fry, parr
    and smolts 53-4, 92, 98,
    181, 189, 203-4,207,
    345-72, 406-7, 419, 422
stop boat fishing 74
Store Hellefiske Bank 516
Strait of Belle Isle 496,
    504
straying of fish 311, 314,
    376, 382
stream remedial measures
    55, 63, 354, 403-4
Sukkertoppen 441
sunspot periodicity 241
survival rates 294-5, 297,
    308-10, 316-19, 347-8,
    350, 363, 423, 452, 489-90
Svabo, Jens Christian 461
Systeme d'Evaluation du
    Saumon Atlantique et
    Autres Migrateurs
    D'Estuaire 63

TACS (see catch quotas)
tag loss 264
tag reprinting efficiency 151
tag rewards 462
tagging (of caught) fish 23,
    61, 88, 139-40, 182, 540-1,
    571-2
tagging (marking) fish 36, 149,
    157, 167, 172-3, 176, 263,
    277, 284, 287-9, 462, 472
tagging return rate 462, 472
tattooing 36, 263-4
tetracycline 222
Third United Nations Law of
    the Sea Conference 2
Three-spined stickleback
    (Gasterosteus aculeatus)
    149, 162
time-share ownership 95, 574
tracking 31, 55, 473-80
transfer of adult fish (see
    adult transfers)
Tours 389

trimethopin sulphametoxazol
    222
Truite, Ombre, Saumon 32
Turner's Falls 415
Turner's Falls Dam 417-18,
    421

UDN 186, 188, 206, 220, 222,
    570
undeclared catches 114, 260-1
US Fish and Wildlife Service
    417-18, 420
Utsira 558-9

variation in sea survival
    308-10
Vernon Dam 417, 421
Vezezoux 391
Vibert box 203, 349
Vic-le-Comte 391
Vichy 391
visitor centres 411

Welsh Water Authority 428
West Greenland fishery 9-10,
    15, 18-21, 147, 166, 438-55
Western Massachusetts
    Electric Company 418
Whitefish (Coregonus
    lavaretus) 148
Wilder Dam 417-18
Wildlife, Conservation and
    Development Act, 1983
    547
Wolf trap 149, 151, 155
Wye salmon fishery 77, 79-80

zonas libres 182-3, 196